Aging of the Autonomic Nervous System

Edited by

Francesco Amenta, M.D.
Professor of Human Anatomy
School of Pharmacy
University of Camerino
Camerino, Italy

CRC Press
Boca Raton Ann Arbor London Tokyo

Library of Congress Cataloging-in-Publication Data
Aging of the autonomic nervous system / edited by Francesco Amenta.
p. cm.
Includes bibliographical references and index.
ISBN 0-8493-6981-9
1. Autonomic nervous system--Aging. I. Amenta, Francesco, 1953–.
[DNLM: 1. Aging--physiology. 2. Autonomic Nervous System—physiology. 3. Autonomic Nervous System Diseases—physiopathology. WL 600 A267]
QP368.A45 1993
612.8′9--dc20
DNLM/DLC
for Library of Congress 92-48899
CIP

Direct all inquiries to CRC Press, Inc., 2000 Corporate Blvd., N.W., Boca Raton, Florida 33431.

International Standard Book Number 0-8493-6981-9

Library of Congress Card Number 92-48899

Printed in the United States of America 1 2 3 4 5 6 7 8 9 0

Printed on acid-free paper

DEDICATION

As I finish my work on this book, I want to dedicate it to my parents Giuseppe Amenta and Giovanna Ganga Amenta as a sign of my deep gratitude for all they did for me. I also want to dedicate it to them as a gift from their son on the occasion of their fortieth wedding anniversary.

Francesco Amenta

PREFACE

Because of the progressive growth in the aging populations in Western countries and because of neurological and psychological disturbances which are ever recurring in the elderly, it has become extremely important to increase research into the underlying causes of the molecular, biochemical, physiological, and anatomical bases of age-related changes in cerebral functions. Moreover, increasing efforts and funding have been centered on the identification of treatments of disturbances linked to brain aging. A large body of evidence has been accumulated in the last few years concerning the morphological, biochemical, functional, and pharmacological changes occurring in the central nervous system of several mammals, including man. The results of these investigations are currently being published in gerontological and neurogerontology journals, the number of which is remarkably increasing. Moreover, several books and series of books, proceedings of meetings, and other editorial initiatives are issued each year.

In contrast, in spite of the sensivity of the autonomic nervous system to the aging process, just a few review articles and no books or proceedings have been dedicated to the key subject of the neurobiology of aging of the autonomic nervous system.

The autonomic, or more correctly "vegetative", nervous system represents a conspicuous part of the peripheral nervous system. The autonomic nervous system includes all the neurons which innervate the viscera, the cardiovascular system, smooth muscles, salivary and lacrimal glands, intrinsic muscles of the eye, and muscles of the hair. Of course, the definition of "autonomic" (e.g., free of influence from the central nervous system) nervous system is rather arbitrary, although it is now well embedded in the current scientific terminology. In fact, the autonomic nervous system has its own nuclei and pathways in the central nervous system.

In view of the role of the autonomic nervous system in the regulation of visceral functions (blood pressure, respiration, digestion, urogenital, etc.) and of the age-related changes in a variety of these functions, we have collected into a book the available information concerning the aging of the autonomic nervous system. We hope it will be of practical interest for investigators in basic research, for clinicians with gerontological interests, and for pharmacologists.

The book is composed of four sections: (1) general aspects; (2) main neurotransmitter systems; (3) age-dependent changes of neuroeffector mechanisms in target organs of the autonomic nervous system; and (4) clinical and therapeutic perspectives. Each section represents the efforts of diverse specialists to review the most recent findings on various topics and to analyze the possible consequences of these findings. I heartily thank the authors for their efforts, their competence, and their patience. They have worked to insure, even in chapters discussing similar subjects, that there would be no overlap.

As editor, it is my duty and my pleasure to thank those who made special contributions to this work including Professor Zs.-Nagy for his introduction and CRC Press for their invaluable support and editorial assistance. I am also very grateful to Drs. Palma Shaw and Wade L. Collier, who helped with stylistic revisions of the contributions, and to Professor José A. Vega and Dr. Alberto Ricci, who were of invaluable help in collecting and reorganizing material for publication.

Francesco Amenta

INTRODUCTION

One of the most important central problems of human aging is the loss of brain function, which causes a general loss of performance at all levels of physiology. Since the function of the central nervous system is largely dependent on an extensive regulation realized by the autonomic nervous system, it seems obvious that any information regarding the aging process of this part of the body deserves all our attention. Everybody can quote examples of the consequences of some pathological changes of the autonomic nervous system, like, for example, Parkinson's disease. However, it is still unclear how deeply the autonomic nervous system is involved in the loss of brain cortical functions such as short- and long-term memory functions, Alzheimer's disease, etc. One can predict that synthesizing information on the age-dependent changes of the autonomic nervous system will have an increasing significance for all those scientists who wish to understand some basic interrelationships between the loss of elementary nerve cell functions and the loss of organ or organic performances. Therefore, we have to appreciate very much the efforts of Dr. Francesco Amenta, who has undertaken the difficult task of editing a book like this.

It is obviously an inherent difficulty in such an editorial work that the various authors come from different fields of interest and have different cultural backgrounds in biology and medicine; therefore, such a multiauthor volume can never have the uniformity in style and manner of expression as a book written by a single author. Yet, this type of multiauthor book has a great advantage for the reader; namely, it can offer an enormously wide overview, such as in this case going from the basic cell physiology or transmitter biochemistry through various problems of the target organ and to the clinical applicability and perspectives. All the readers who have not specialized in the particular fields of the human autonomic nervous system can find a well-established summary of the relevant knowledge in various chapters of the book, written by excellent, representative specialists of the given areas. We should be grateful to both the authors and the editor of the book, since their effort facilitates the diffusion of a wide, detailed knowledge about age-dependent changes of the autonomic nervous system.

Finally I would like to also mention the fact that this book is being published by CRC Press, which has a long tradition of preparing scientific books at a very high professional standard.

I wish success to the authors and the editor of this book, and I am sure the readers will find an intellectual pleasure in it.

Professor Imre Zs.-Nagy
VILEG, Hungarian Section
University Medical School
Debrecen, Hungary

THE EDITOR

Francesco Amenta, M.D. is Professor of Human Anatomy in the School of Pharmacy at the University of Camerino, Camerino, Italy.

Professor Amenta attended the University of Rome, where he graduated as Doctor of Medicine and Surgery with the highest honors in 1977. He received his specialist degree in neurology in 1981, again with the highest honors.

Since 1978 Professor Amenta has been teaching and doing research. From 1984 to 1992 he was a member of the Department of Public Health and Cell Biology at the University of "Tor Vergata", Rome. In November 1992 he assumed his present position at the University of Camerino. He also has professional associations with universities in the U.S., U.K., Germany, and Hungary. He has been a consultant to several pharmaceutical companies and to the European Economic Community.

Professor Amenta has many scientific interests. Recently, he has been concentrating on techniques of neurochemistry and histochemistry, neuroeffector mechanisms primarily in the vascular system, autoradiography of the sites of binding of neurotransmitters and of drugs affecting the central and peripheral nervous system, and analysis of pharmacological therapy of aging.

Professor Amenta has written over 180 scientific articles. He has written a book in Italian, edited two books in English, and translated one book from English into Italian. He has also made frequent contributions to the Italian Medical Encyclopedia.

CONTRIBUTORS

Kurt D. Ackerman
Department of Psychiatry
University of Pittsburgh
Pittsburgh, Pennsylvania

Francesco Amenta, M.D.
Facoltà di Farmacia
Università di Camerino
Camerino, Italy

Daryll M. Baker, Ph.D., F.R.C.S.
Department of Surgery
Queen's Medical Centre
Nottingham, United Kingdom

Denise L. Bellinger, Ph.D.
Department of Neurobiology and Anatomy
University of Rochester School of Medicine
Rochester, New York

Giorgio Bernardi, M.D., Ph.D.
Dipartimenta di Sanita Pubblica e Biologia Cellulare
Università Tor Vergata
Rome, Italy

Charles Brink, Ph.D.
Hopital Marie Lannelongue
Le Plessis Robinson, France

Paolo Calabresi, M.D.
Dipartimento di Sanità Pubblica e Biologia Cellulare
Università Tor Vergata
Rome, Italy

Begoña Calzada, M.D.
Departamento de Morfología y Biología Celular
Facultad de Medicina
Universidad de Oviedo
Oviedo, Spain

Wade L. Collier, M.D.
Department of Natural Sciences
Manatee Community College
South Campus
Venice, Florida

Letizia M. Cupini, M.D.
Clinica Neurologica
Ospedale S. Eugenio
Rome, Italy

Miguel E. Del Valle, M.D.
Departamento de Morfología y Biología Celular
Facultad de Medicina
Universidad de Oviedo
Oviedo, Spain

James R. Docherty, Ph.D.
Department of Physiology
Royal College of Surgeons in Ireland
Dublin, Ireland

Sue P. Duckles, Ph.D.
Department of Pharmacology
College of Medicine
University of California, Irvine
Irvine, California

David L. Felten, M.D., Ph.D.
Department of Neurobiology and Anatomy
University of Rochester School of Medicine
Rochester, New York

Suzanne Y. Felten, Ph.D.
Department of Neurobiology and Anatomy
University of Rochester School of Medicine
Rochester, New York

Fabio Ferrante, M.D.
Dipartimento di Scienze Cardiovascolari e Respiratorie
Università La Sapienza
Rome, Italy

Gary A. Ford, M.R.C.P.
Wolfson Unit of Clinical Pharmacology
Claremont Place
The University
Newcastle-Upon-Tyne
United Kingdom

Pierangelo Geppetti, M.D.
Cardiovascular Research Institute
School of Medicine
University of California, San Francisco
San Francisco, California

Brian B. Hoffman, M.D.
GRECC
Veterans Administration Medical Center
Palo Alto, California

Joel Horwitz, Ph.D.
Department of Pharmacology
Medical College of Pennsylvania
Philadelphia, Pennsylvania

Mark D. Johnson, Ph.D.
Department of Pharmacology
Medical College of Pennsylvania
Philadelphia, Pennsylvania

George Kuchel, M.D.
Departments of Neurobiology and Geriatrics
Mount Sinai School of Medicine
New York, New York

Otto Kuchel, M.D., Ph.D.
Clinical Research Institute of Montreal
University of Montreal
Montreal, Quebec, Canada

Dianne Lorton, Ph.D.
Department of Biochemistry
Gliatech, Inc.
Cleveland, Ohio

Kelley S. Madden, Ph.D.
Department of Neurobiology and Anatomy
University of Rochester School of Medicine
Rochester, New York

Nicola B. Mercuri, M.D.
Dipartimento di Sanità Pubblica e Biologia Cellulare
Università Tor Vergata
Rome, Italy

Alberto Ricci, M.D.
Dipartimento di Scienze Cardiovascolari e Respiratorie
Università La Sapienza
Rome, Italy

Jay Roberts, Ph.D.
Department of Pharmacology
Medical College of Pennsylvania
Philadelphia, Pennsylvania

Robert M. Santer, Ph.D.
Department of Anatomy
University of Wales College of Cardiff
Cardiff, United Kingdom

Palma Shaw, M.D.
Department of Anatomy
Hahnemann University
Philadelphia, Pennsylvania

José A. Vega, M.D., Ph.D.
Departamento de Morfología y Biología Celular
Facultad de Medicina
Universidad de Oviedo
Oviedo, Spain

David L. Snyder, Ph.D.
Department of Pharmacology
Medical College of Pennsylvania
Philadelphia, Pennsylvania

Damiano Zaccheo, M.D.
Istituto di Anatomia Umana
Università di Genova
Genova, Italy

TABLE OF CONTENTS

General Aspects

Chapter 1

AUTONOMIC CENTERS AND PATHWAYS: AGE-RELATED CHANGES

Palma Shaw, Fabio Ferrante, and Francesco Amenta

TABLE OF CONTENTS

0-8493-6981-9/93/$0.00 + $.50

I. THE AUTONOMIC NERVOUS SYSTEM

In looking at the autonomic nervous system (ANS), one can make three divisions: sympathetic, parasympathetic, and enteric. The sympathetic or thoracolumbar division opposes the parasympathetic or craniosacral division in maintaining a stable internal environment. In the event that there is some sudden change in the environment to which the body must respond, the hypothalamus will activate the sympathetic system, creating increased outflow to the heart, peripheral vasculature, other viscera, sweat glands, and other piloerector and ocular muscles. The parasympathetic system, on the other hand, acts to restore and maintain basal heart rate, respiration, and metabolism.[1]

The ANS works in parallel with the somatic nervous system to help the body make adjustments to changes in the environment. These systems differ in that there is voluntary control of movements in the somatic nervous system, while those of the autonomic system are automatic. It is for this reason that the ANS is also called the involuntary system. Further differences include the fact that all somatic motor neurons are located within the central nervous system, in either cranial motor nuclei or the ventral horn of the spinal cord. Also, the efferent pathway to skeletal muscle is monosynaptic, so the central motor neurons project directly to the target site. This is unlike autonomic neurons, which are located peripherally within autonomic ganglia. The efferent pathway is disynaptic: a synapse occurs in a peripheral ganglion between the central nervous system and the target organ.[1] The organization of the sympathetic neuroeffector pathways is shown in Figure 1.

A. THE SYMPATHETIC SYSTEM

Preganglionic neurons originate from the cell bodies, situated in the intermediolateral grey matter, of all 12 thoracic and the first two to three lumbar segments of the spinal cord. The axons exit the spinal cord via the corresponding ventral root entering its spinal nerve and then forming the myelinated white communicating rami of the thoracic and lumbar nerves, through which they arrive at the paravertebral chain ganglia (Figure 2). Once they have entered the ganglia, more than one event can occur. They can synapse at that level with other ganglia, go up or down the sympathetic trunk to synapse at a level above or below, or pass through the sympathetic trunk to an intermediary sympathetic ganglion, i.e., celiac ganglion (Figure 3). Furthermore, splanchnic nerves, consisting of fibers derived from the lower seven thoracic ganglia, synapse on the celiac and superior ganglia; these postsynaptic axons go to the celiac plexus and eventually innervate abdominal targets.[1,2] In parallel, the splanchnic nerves stemming from the lumbar area of the sympathetic trunk synapse within the inferior mesenteric ganglion to subsequently innervate lower abdominal and pelvic organs through the hypogastric plexus.[2] Some preganglionic fibers innervate cells in the adrenal medulla through the thoracic splanchnic nerve.[1]

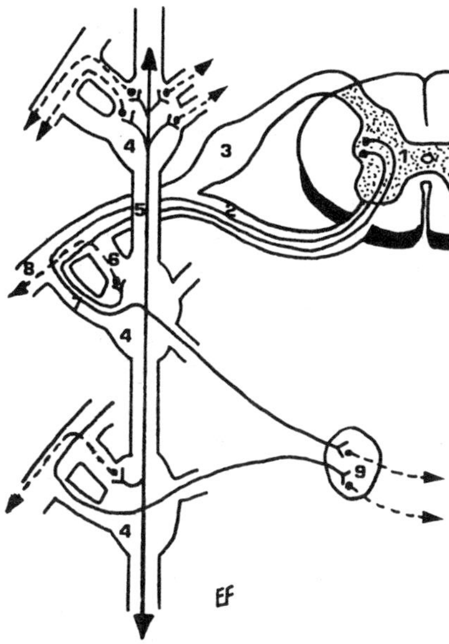

FIGURE 1. Sympathetic neuroeffector pathways. 1, Grey matter of the thoracolumbar spinal cord; 2, ventral nerve root; 3, dorsal nerve root and spinal ganglion; 4, sympathetic ganglion; 5, sympathetic trunk; 6, grey communicating ramus; 7, white communicating ramus; 8, spinal nerve; 9, prevertebral ganglion.

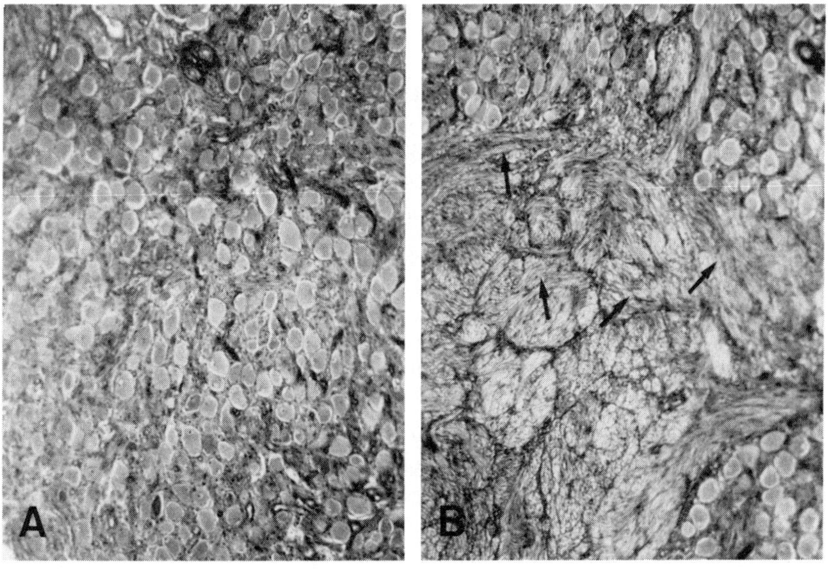

FIGURE 2. Section of a human prevertebral ganglion stained with the technique of Azan-Mallory. (A) Ganglionic area containing nerve cell bodies; (B) ganglionic area rich in nerve fibers (arrows). (Original magnification × 180.)

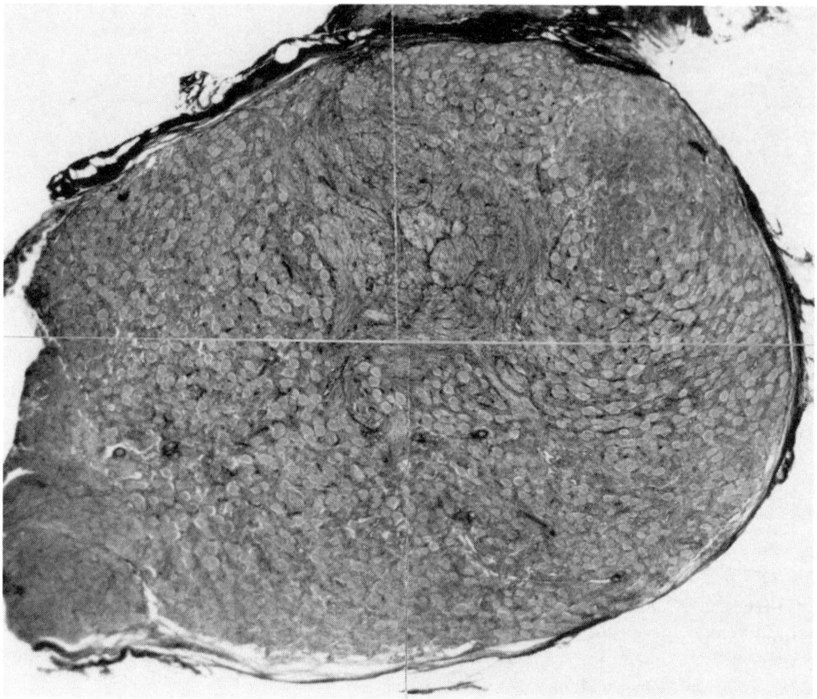

FIGURE 3. Section of human coeliac ganglion stained with the technique of Azan-Mallory. (Original magnification × 25.)

Postganglionic neurons are present in a ratio of ten or greater to every one preganglionic neuron. Their effects are diffuse relative to the parasympathetic system (ratio of 3:1). These postganglionic nonmyelinated fibers exit the paravertebral ganglia to form the grey communicating rami. Those that innervate structures of the head originate from the superior cervical sympathetic ganglion and form sympathetic plexuses which travel with branches of the common carotid artery to their targets. Grey communicating rami run with spinal nerves to innervate end organs, such as vasomotor, pilomotor, and sweat glands.[1,2] The organization of the sympathetic system is shown in Figure 4.

B. THE PARASYMPATHETIC SYSTEM

The preganglionic fibers of the parasympathetic division originate from cell bodies in the grey matter of the brainstem and the spinal cord at sacral levels 2, 3, and 4. These are long axons which tend to synapse close to or within the target organ. Parasympathetic nuclei in the brainstem project through the cranial nerve (CN) to their respective postganglionic neurons, e.g., the Edinger-Westphal nucleus which, through the oculomotor nerve (CN III), synapses at the ciliary ganglion. The superior and inferior salivatory nuclei

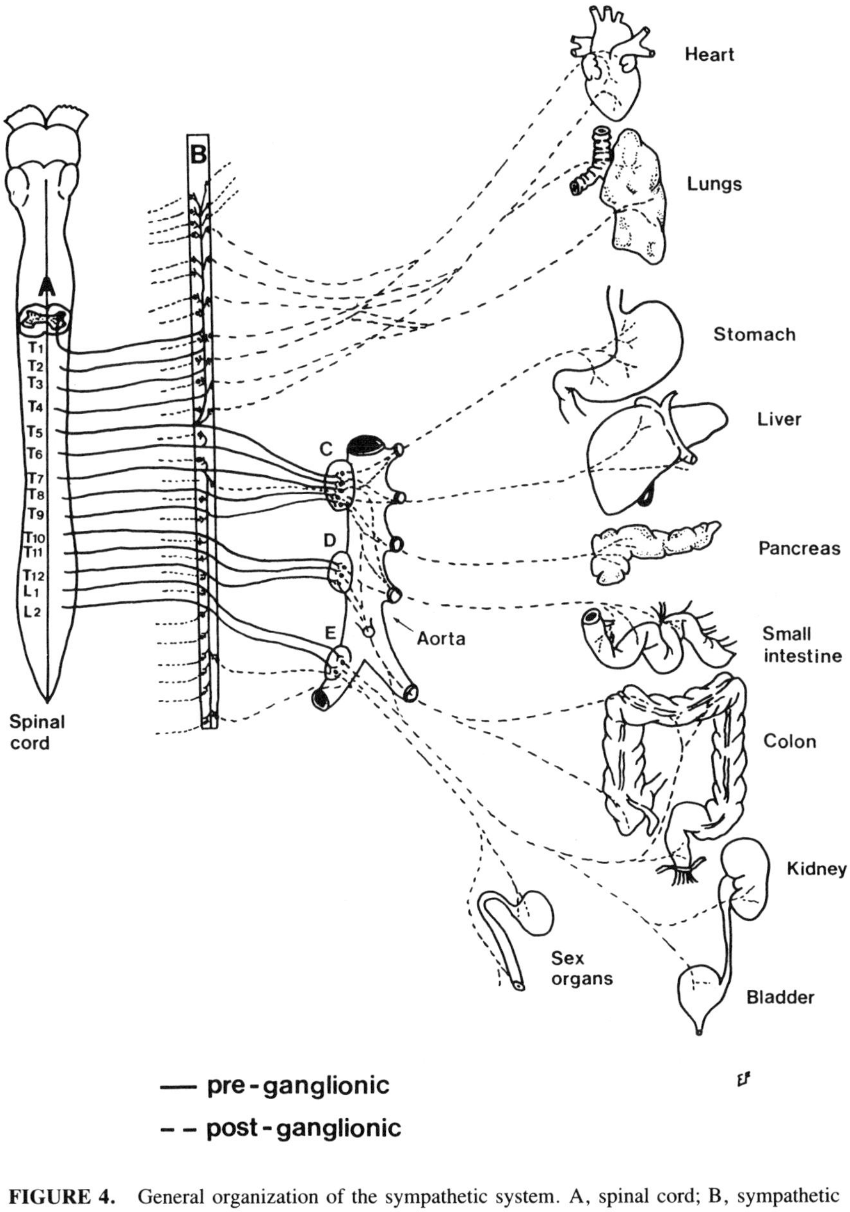

FIGURE 4. General organization of the sympathetic system. A, spinal cord; B, sympathetic chain; C, coeliac ganglion; D, superior mesenteric ganglion; E, inferior mesenteric ganglion.

are associated with the facial nerve (CN VII) and the glossopharyngeal nerve (CN IX), respectively. The vagus nerve (CN X) receives fibers from the dorsal vagal nucleus, which distributes fibers to the thoracic and abdominal organs, and from the nucleus ambiguus, which contributes to the innervation of the cardiac ganglion[1,2] (Figure 5).

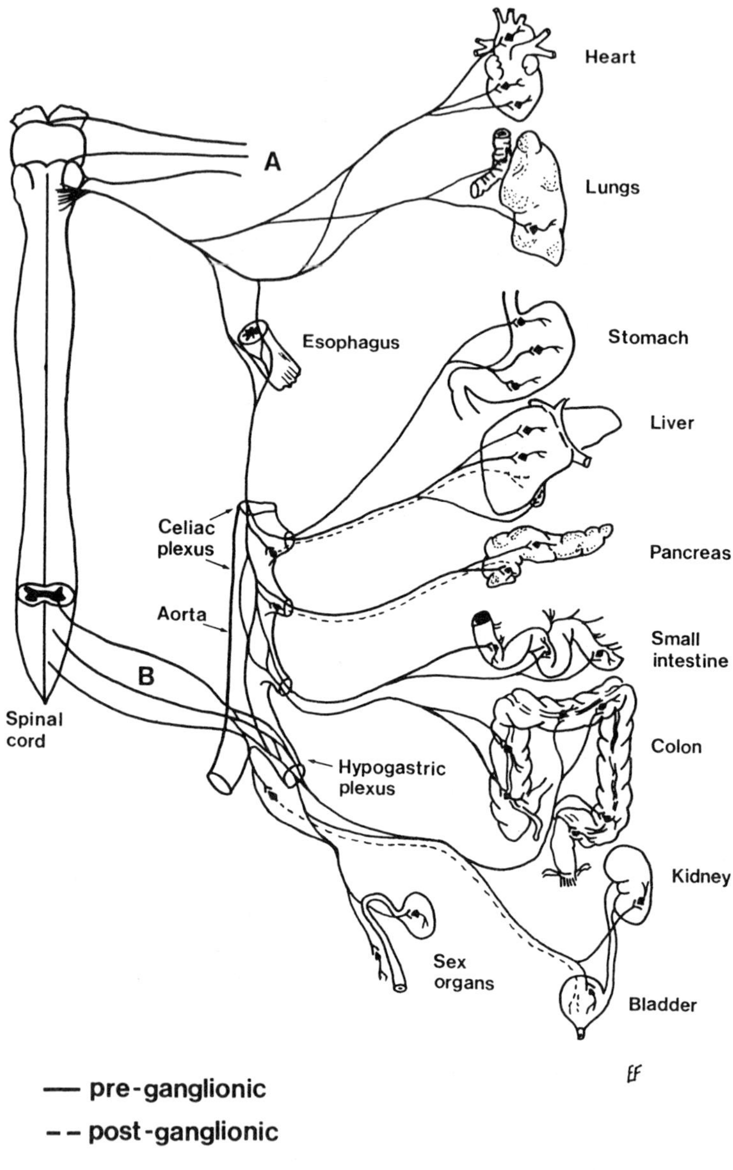

FIGURE 5. General organization of the parasympathetic system. A, cranial portion; B, sacral portion.

The parasympathetic axons of the sacrum contribute to the formation of the pelvic nerve and project to synapse in the hypogastric plexus. This ultimately innervates the descending colon, bladder, and the external genitalia[1] (Figure 5).

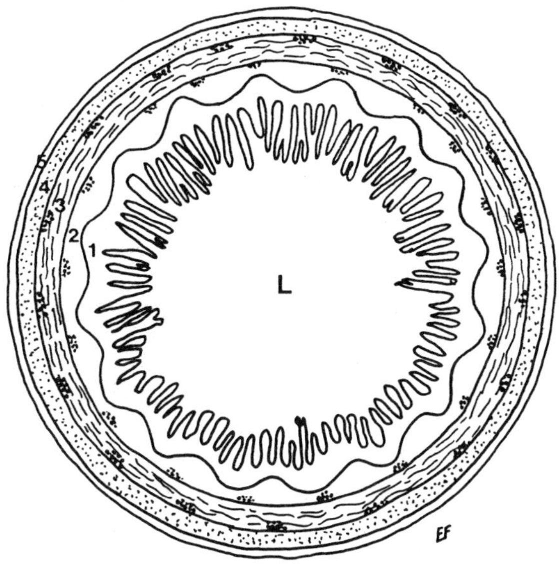

FIGURE 6. Schematic section of a portion of the small intestine showing the localization of the myenteric (of Auerbach) and the submucous (of Meissner) plexus. L, intestinal lumen; 1, mucosa; 2, submucosa; 3, circular muscle coat; 4, longitudinal muscle coat; 5, serosa.

C. THE ENTERIC NERVOUS SYSTEM

The enteric division involves the innervation of the gastrointestinal tract, the pancreas, and the gall bladder. It is made up of local sensory neurons, interneurons, and motor neurons. Sensory neurons are sensitive to changes in gut wall tension and in the chemical environment. Playing a major role in homeostasis through its control of gastrointestinal blood vessel tone, motility, gastric secretion, and fluid transport, this system can function independently, although it usually receives input from the central nervous system. There are two major plexuses, the myenteric (Auerbach's) and the submucous (Meissner's) (Figure 6). These receive extrinsic innervation from the sympathetic and parasympathetic nervous systems. This extrinsic input can override that of the enteric system during situations of stress.[1]

II. NEUROTRANSMITTERS RELATED TO THE AUTONOMIC NERVOUS SYSTEM

A neurotransmitter, by definition, must fulfill certain criteria. First, the enzymes necessary for the synthesis of this substance must be present within the nerve terminal. Second, the release of this substance must be triggered upon stimulation of the nerve and followed by diffusion across the synaptic cleft, with subsequent binding to its appropriate postsynaptic receptor; this, in turn, causes a physiological response characterized by changes in the membrane permeability to certain ions. Finally, there must exist a means to cease the effects of this substance rapidly.[3]

Some of the functional neuroanatomical methods currently being used include autoradiography, histochemistry, and immunocytochemistry. Through these techniques, an increasing number of compounds are being recognized as potential neurotransmitters. The following section describes the recognized neurotransmitters and their properties.

A. ACETYLCHOLINE

Acetylcholine (ACh) is contained within various regions of the nervous system; for example, some well-known areas are the neuromuscular junction, autonomic ganglia, preganglionic neurons, and postganglionic parasympathetic and certain sympathetic nerve terminals.[3] Postganglionic neurons utilize ACh and one or more peptide neurotransmitters to act upon their targets. The enzyme choline acetyltransferase (CAT) synthesizes ACh, and acetylcholinesterase (AChE) hydrolyzes this molecule.[3] AChE is not confined to cholinergic nerves, but it has a major role in the breakdown of nonneuronal ACh as well as some neuropeptides.[4,5]

The release of ACh and its interaction with cholinergic nicotinic receptors evokes a fast excitatory postsynaptic potential (EPSP). This may be sufficient to generate an action potential in the postganglionic neuron, and it is considered to be the main synaptic pathway in sympathetic and parasympathetic ganglia.[1] The rate of excitation differs with respect to the type of receptor affected by ACh. The muscarinic receptors produce a slow EPSP by closing potassium channels, which depolarizes the membrane, and by opening sodium and calcium channels. On the other hand, the activation of muscarinic receptors on some postganglionic neurons opens potassium channels and creates a slow inhibitory postsynaptic potential (IPSP).[1] The effects of ACh are relatively concentrated and are basically at the site of release of the neurotransmitter.[1]

B. CATECHOLAMINES: DOPAMINE, NORADRENALINE, AND ADRENALINE

Catecholamines (dopamine, noradrenaline, and adrenaline) are substances that have a catechol nucleus (3,4-dihydroxylated benzene ring) in their molecule. Although catecholamine-containing neurons are relatively few in number and located in a few, limited cerebral regions, they project diffusely throughout the cerebral cortex and subcortical telencephalic nuclei, the diencephalon, the cerebellar cortex, and the spinal cord. From a functional point of view, catecholaminergic neurons are involved in a variety of cerebral functions.

Figure 7 summarizes the catecholamine biosynthetic pathways. As can be seen, five enzymes are involved in catecholamine biosynthesis. Nerve cells which use adrenaline as a neurotransmitter as well as adrenal medulla cells express all biosynthetic enzymes. Noradrenaline neurons do not express phen-

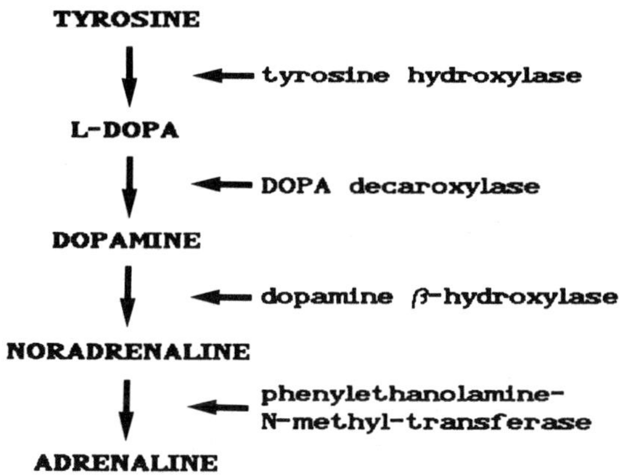

FIGURE 7. Catecholamine biosynthetic pathways.

ylethanolamine-*N*-methyl transferase, whereas dopaminergic neurons do not express phenylethanolamine-*N*-methyl transferase and dopamine β-hydroxylase (Figure 7).

The location of dopamine (DA)-containing neurons within the brain is organized into tracts and has been classified according to length.[3] Long-length DA fibers make up the following tracts: nigrostriatal, mesocortical, and mesolimbic.[6] The nigrostriatal pathway contains the highest DA concentrations (approximately 75% in the human brain) and plays a major role in extrapyramidal motor control.[7] Mesocortical and mesolimbic pathways have target sites which are nuclear and cortical components of the limbic system. Some examples are nucleus accumbens, septal nuclei, olfactory tubercle, amygdala (for the former), and cingulate, medial prefrontal, and entorhinal (for the latter).[3] The impact of the two pathways on the autonomic function needs further investigation. Moreover, the existence of a peripheral dopaminergic system involved in the regulation of cardiovascular and renal functions has been demonstrated. Further studies are necessary to assess whether it is part of the noradrenergic sympathetic neuroeffector system or if it works independently from noradrenergic sympathetic neurons.

In humans, noradrenaline (NA) is a catecholamine that can be isolated from most sympathetic postganglionic endings. It and its methyl derivative, adrenaline, are secreted by the adrenal medulla.[2] NA-containing fibers are widely distributed within the brain and spinal cord, but originate from two major areas, the locus coeruleus and medullary reticular formation.[6] From the locus coeruleus, which extends throughout the grey matter of the pons, arise five noradrenergic tracts. Three of these extend to all regions of the cerebral cortex, limbic system, thalamus, and hypothalamus.[3] Another tract extends to the superior cerebellar peduncle to terminate in the cerebellar cortex.[3] The final tract descends into the medulla and spinal cord.[8] Some authors, using Falck's fluorescence histochemical technique, have demonstrated the extensive division of the postganglionic sympathetic noradrenergic neurons within both Auerbach's and Meissner's plexuses. Arterioles, mucosa, and the circular muscular layer in some intestinal regions, particularly sphincter areas, are innervated by noradrenergic nerves.[9] The noradrenergic nerves at the ganglionic level in the enteric system play an inhibitory role.[10] Moreover, noradrenergic neurons supply the heart, blood vessels, kidney, and lower urinary tract as well as reproductive organs. With the exception of sympathetic postganglionic fibers to the arteries supplying skeletal muscles, the large majority of autonomic fibers innervating peripheral effectors use NA as a neurotransmitter.

Adrenaline is synthesized by using the enzyme phenylethanolamine-*N*-methyl transferase (PNMT). The highest concentrations of adrenaline and PNMT in the brain are found in certain hypothalamic nuclei and the ventrolateral medullary reticular formation.[3]

C. SEROTONIN

Serotonin, 5-hydroxytryptamine (5-HT), appears to be contained within nine clusters of cells which lie in the vicinity of the midline or raphe regions of the brainstem.[3] The more caudal group of cells projects primarily to the medulla[11] and the spinal cord in the intermediolateral cell columns.[12] The more rostral cell groups project to the cerebral cortex, thalamus, striatum, and limbic areas.[3] Investigation supports a role for central 5-HT neurons as being primarily depressor.[3] 5-HT-containing nerve cell bodies and fibers have been identified in the cerebral vessels, gastrointestinal tract, and reproductive system.

D. HISTAMINE

Much data support the role of histamine as a neurotransmitter. Its distribution within the central nervous system is diffuse, with the largest concentrations being in certain hypothalamic nuclei.[13] Specific data determining the precise location of histamine-containing tracts are lacking because presently there is no adequate histochemical technique to demonstrate them.[3]

E. AMINO ACIDS

γ-Aminobutyric acid (GABA) acts as a major inhibitory neurotransmitter. It and its synthetic enzyme, glutamic acid decarboxylase (GAD), have been implicated in the pathophysiology of Huntington's chorea and schizophrenia.[14] The existence of a widely distributed neural and nonneuronal GABAergic system has been recently demonstrated.[15] Its sensitivity to aging is not yet known.

F. NEUROPEPTIDES

1. Substance P

Substance P (SP) was initially discovered in 1931 by von Euler and Gaddum, and noted for its hypotensive and muscle-contracting properties.[3]

It is found in the afferent nerves of the skin and spinal cord and in sensory neurons of the gastrointestinal tract.[10] In mammals, the highest concentrations of SP have been demonstrated in the substantia nigra, trigeminal sensory nucleus, and dorsal horn of the spinal cord.[3] An extensive description of the localization and sensitivity to aging of SP-containing nerve fibers is given in Chapter 5 of this volume.

2. Angiotensin II

In the blood, angiotensin II is a vasoconstrictor factor which stimulates aldosterone secretion from the adrenal cortex. It is formed by an enzyme, angiotensinogen, which exists in high concentrations in the lungs. Angiotensinogen is acted upon by renin to produce angiotensin I, which circulates to the lungs then to be cleaved forming angiotensin II. In addition, there may be another neuropeptide called angiotensin III that is even more potent.[3] Angiotensin II has been shown to mediate the control of drinking behavior and blood pressure. By direct action on the arteriolar smooth muscle, angiotensin II elevates blood pressure.[16]

III. AUTONOMIC CENTERS

A. HYPOTHALAMUS

The hypothalamus is derived from the diencephalon and lies below the anterior thalamus, forming the floor and lower walls of the third ventricle. The right and left sides of the hypothalamus can be further divided into a medial hypothalamic area which is formed by specific nuclei and a lateral hypothalamic area which contains fiber systems and diffuse lateral nuclei (Figure 8). The medial hypothalamic areas are each subdivided into three portions proceeding anterior to posterior, as follows: the supraoptic portion, the tuberal portion, and the mammillary portion. The supraoptic, suprachiasmatic, and paraventricular nuclei are in the supraoptic portion; the ventromedial, dorsomedial, and arcuate nuclei, and median eminence form the tuberal portion; and several mammillary nuclei and the posterior nucleus form the mammillary portion.[2]

The fiber connections consist of both afferent and efferent tracts. The afferent input to the hypothalamus involves part of the medial forebrain bundle; thalamo-hypothalamic fibers form the medial and midline thalamic nuclei, fornix, stria terminalis, pallidohypothalamic fibers, and inferior mammillary peduncle. Efferent output involves the hypothalamo-hypophyseal tract, mammillotegmental tract, mammillothalamic tract, periventricular system including the dorsal fasciculus, tuberohypophyseal tract, and fibers from the septal region via the fornix to the hippocampus.[2]

The hypothalamus is involved in the regulation of eating, body temperature, water balance, anterior pituitary function, and control of the circadian rhythm as well as expression of emotion. Moreover, it plays an important role in the functioning of the ANS.[2] In the brain, the ANS is influenced by

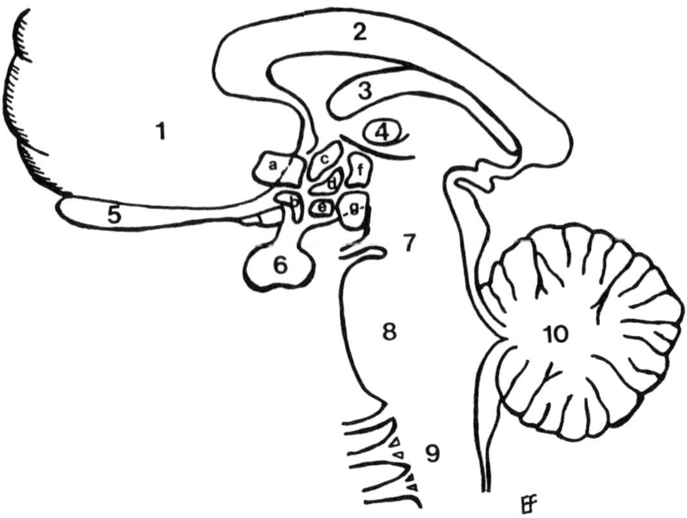

FIGURE 8. General organization of the hypothalamus. 1, frontal cortex; 2, corpus callosum; 3, fornix; 4, adhesio interthalamica; 5, olfactory bulb and tract; 6, pituitary gland; 7, midbrain; 8, pons; 9, medulla oblongata; 10, cerebellum; a, preoptic nucleus; b, supraoptic nucleus; c, paraventricular nucleus; d, dorsomedial nucleus; e, ventromedial nucleus; f, posterior nucleus; g, mammillary nucleus.

many regions which project to the hypothalamus, where the information is integrated. The result is an autonomic response.[1] The sympathetic-activating region seems to be located in the dorsomedial and posterolateral areas. The parasympathetic-activating region is located more anteriorly.[2] The hypothalamus regulates the ANS in two ways. The first is as mentioned above; stimulation in the sympathetic area can lead to piloerection, blood pressure and heart rate elevation, sweating, and dilation of pupils. The second is through regulation of the hormones which affect the ANS.[1] Functions such as respiration and cardiovascular maintenance do not require hypothalamic input to operate. This suggests that there are nuclei in the brainstem which function independently in coordinating the ANS, e.g., the nucleus of the solitary tract.[1,17]

B. NUCLEUS OF THE SOLITARY TRACT

The nucleus of the solitary tract acts in the brainstem to adjust the information received from various major organs in the following ways.[1] First, the nucleus uses a set of reflex arches to control the more simple autonomic function.[1] It receives fibers from the solitary tract which is formed by visceral efferents from cranial nerves IX and X.[2] Second, the more elaborate homeostatic maintenance is managed by transmitting information to higher and lower areas of the brain. In a relay fashion, these regions integrate the necessary information and then send it back to the nucleus of the solitary tract. Within this nucleus is an area called the commissural nucleus that receives information relayed from the visceral afferents. The commissural nucleus then projects to nuclei in the brainstem and the forebrain, including the amygdala, bed nucleus of the stria terminalis, and paraventricular hypothalamic nucleus which, in turn, relay back to the nucleus of the solitary tract and other lower brainstem nuclei.[1]

IV. AGING OF THE AUTONOMIC CENTERS

The impairment of ANS function with advancing age stimulated researchers to identify morphological and functional changes in the central and peripheral portions of the ANS. The hypothalamus received the main attention of investigators. Below, we describe the main changes reported in the aging hypothalamus.

A. HYPOTHALAMUS

The age-dependent changes in the number of neurons in the various nuclei of the hypothalamus have been examined and various results have been demonstrated that differ between as well as within species. Peng and Hsu found no neuronal loss from the hypothalamic nuclei of aged male Sprague-Dawley rats.[18] They do claim a difference with respect to gender in the medial preoptic area (MPOA), anterior hypothalamic area (AHA), and arcuate nucleus (ARN).[19] Furthermore, they measured neuron density, volume, and total neuron number in the supraoptic nucleus (SON), paraventricular nucleus (PVN), ventromedial nucleus (VMN), dorsomedial nucleus (DMN), MPOA, AHA, and ARN, finding no significant statistical differences between young and old male rats.[18] Sartin and Lamperti subsequently used young, middle-aged, and aged Sprague-Dawley male rats to observe changes in the number of neurons per unit area of the MPOA, PVN, VMN, DMN, ARC, subventricular portion (SV) of the ARC, suprachiasmatic nucleus (SCN), and medial mammillary nucleus (MMN).[20] Their results demonstrate a significant reduction in the number of neurons per unit area in the ARC, SV, and VMN of 24-month-old rats. The authors also refer to the progressive decline of endocrine and neuroendocrine function with age, concluding with the idea that the decrease in the number of VMN neurons may correlate with the altered endocrine and neuroendocrine function in aging male Sprague-Dawley rats.[20]

Other studies, such as those of Lamperti and Blaha using young and reproductively senescent female golden hamsters, demonstrate no morphological differences using light-microscope methods in the hypothalamus of the aged hamster.[21]

1. Supraoptic and Paraventricular Nuclei

Goudsmit et al. looked at the volume and total cell number in the supraoptic and paraventricular nuclei in the hypothalamus of control and Alzheimer's disease (AD)-affected male and female subjects. They were able to demonstrate that the total number of cells remained constant during the aging process in the SON and PVN of humans less than and greater than 60 years of age.[22] Some authors have shown that there is an activation of the vasopressin cells of these nuclei during senescence.[23-25] It has been suggested that with the activation of these nuclei and secretion of vasopressin, there is a prevention of neuronal loss.[26] No correlation was observed between the total cell number in the SON and PVN and brain weight.[22] Moreover, the sexual dimorphism which was demonstrated in the cortical structures was not evident in the above nuclei.[27-29] Goudsmit et al. feel that they have demonstrated the absence of SON and PVN degeneration in AD.[22]

Vogels et al.[30] had previously found similar results regarding the stability of cell number, but also noted that there was a remarkable neuronal hypertrophy in PVN cells of AD brains. This hypertrophy was less remarkable in non-AD-affected brains. Furthermore, the SON of both AD and normal brains demonstrated significant neuronal hypertrophy, which was more exaggerated in the former. Inquiry was made into the cause of activation of the vasopressin-producing hypothalamo-neurohypophyseal system (HNS) in aging. The results of their work has refuted the theory of neuronal cell loss as a stimulus for increased vasopressin secretion.[30] An age-related decrease in vasopressin binding sites in the kidney has been reported.[20,31] Moreover, having been noted in aging rat, the reduction of the noradrenergic innervation from the locus coeruleus to SON and PVN neurons has been suggested.[32-36]

2. Suprachiasmatic Nucleus

Changes in the vasopressin cells of the SCN have been studied by Roozendaal et al. using both young and old Brown Norway rats. They noted changes of the circadian rhythm with old age and were interested in establishing a morphological correlate in the SCN. Their results demonstrated a significant decrease (31%) in the number of vasopressin cells.[37]

Goudsmit et al. have described a marked cell loss in the SCN after 80 years of normal aging. Moreover, the cell count of the SCN in AD patients was 60% of control. They felt that the degeneration in the SCN develops abruptly as opposed to a gradual decline with aging.[22,37,38]

3. Hypothalamic-Pituitary-Adrenal System

The increase of hypothalamic-pituitary-adrenal (HPA) activity in the aging rat and its subsequent increase in circulating levels of glucocorticoids have been investigated by Issa et al. using young, aged cognitively impaired (AI), and aged cognitively unimpaired (AU) Long-Evans rats.[39] These elevated levels of corticosteroids have been implicated in hippocampal pathology. Histological analysis demonstrated neuronal loss in pyramidal cell fields in both AU and AI rats, with a significantly greater loss in the latter. Moreover, clear evidence of increased HPA activity was obtained from the AI rats. The AI group also possessed significantly higher basal levels of adrenocorticotropic hormone (ACTH) than the AU or the young groups. Reduced hippocampal type I and II corticosteroid binding sites, present in AI and AU groups, was much more exaggerated in the former. They concluded that HPA dysfunction in the aged animals was only selectively associated with hippocampal neuron loss and spatial memory impairments, thus not occurring simply as a function of age.[40]

Finally, Sahu et al. have demonstrated a widespread age-related decline in the levels of neuropeptide Y and its release in the hypothalamus, further suggesting that this occurrence may play a role in the reduction of testosterone secretion and contribute to the decline in reproductive function in aged male rats.[40]

V. CONCLUDING REMARKS

The hypothalamus, which represents the main center of the ANS, is sensitive to aging which causes several morphological and functional changes at the level of this diencephalic center. Hypothalamic changes are noticeable in the so-called "physiological aging", e.g., the gradual and natural impairment of several functions occurring with advanced age, as well as in AD or senile dementia of the Alzheimer's type (SDAT). Both AD and SDAT, which represent about 70% of all cases of senile dementia, are characterized by memory loss associated with neuronal degeneration and death, primarily in the hippocampus, generation of amyloid plaques in the cerebral gray matter, abnormal cytoskeletal structure formation, and severe cholinergic neurotransmission deficiency.[41] ANS function also is impaired in AD and SDAT. Since the changes described in the hypothalamus of the AD patients are more severe than those observed in nondemented aged subjects, it is possible that the impairment in ANS function in senile dementia represents expression of the structural disarrangement of the hypothalamus noticeable in demented subjects.

Further studies aimed at a better characterization of hypothalamic changes in senile dementia and at the identification of relationships between these modifications and tests of autonomic function would be useful for a possible early diagnosis of senile dementia and related disorders.

REFERENCES

1. **Dodd, J. and Role, L. W.,** The autonomic nervous system, in *Principles of Neural Science,* 3rd ed., Kandel, E. C., Schwartz, J. H., and Jessel, T. M., Eds., Elsevier, New York, 1991, 761.
2. **deGroot, J. and Chusid, J. G.,** *Correlative Neuroanatomy,* Appleton & Lange, East Norwalk, 1988.
3. **Spokes, E. G. S.,** Neurochemistry of autonomic failure, in *Autonomic Failure,* Bannister, R., Ed., Academic Press, London, 1981, 284.
4. **Jacobowitz, D. and Koelle, G. B.,** Histochemical correlations of acetylcholinesterase and catecholamines in postganglionic autonomic nerves of the cat, rabbit and guinea pig, *J. Pharmacol. Exp. Ther.,* 148, 225, 1965.

5. **Lehmann, J. and Fibiger, J. C.,** Acetylcholinesterase and the cholinergic neuron, *Life Sci.,* 25, 1939, 1979.
6. **Moore, R. and Bloom, F.,** Central catecholamine neuron systems: anatomy and physiology of the dopamine systems, *Annu. Rev. Neurosci.,* 1, 129, 1978.
7. **Spokes, E.,** An analysis of factors influencing measurements of dopamine, noradrenaline, glutamate decarboxylase and choline acetylase in human post-mortem brain tissue, *Brain,* 102, 333, 1979.
8. **Commissiong, J.,** Evidence that the noradrenergic coerulospinal projection decussates at the spinal level, *Brain Res.,* 212, 145, 1981.
9. **Llewellyn-Smith, I., Wilson, A., Furness, J., Costa, M., and Rush, R.,** Ultrastructural identification of noradrenergic axons in the enteric plexuses of the guinea pig ileum, *Neurosci. Abstr.,* 6, 247, 1980.
10. **Amenta, F. and Cavallotti, C.,** The musculature and innervation of the alimentary canal, in *Ultrastructure of the Digestive Tract,* Motta, P. and Fujita, H., Eds., Martinus Nijhoff, Boston, 1988, 167.
11. **Basbaum, A., Clanton, C., and Fields, H.,** Three bulbospinal pathways from the rostral medulla of the cat: an autoradiographic study of pain modulating systems, *J. Comp. Neurol.,* 178, 209, 1978.
12. **Loewy, A. and McKellar, S.,** Serotoninergic projections from the ventral medulla to the intermediolateral cell column in the rat, *Brain Res.,* 211, 146, 1981.
13. **Schwatz, J.-C.,** Histaminergic mechanisms in brain, *Annu. Rev. Pharmacol. Toxicol.,* 17, 325, 1977.
14. **Spokes, E. G. S.,** GABA, in *Metabolic Disorders of the Nervous System,* Rose, F. C., Ed., Pitman, London, 1981.
15. **Erdo, S. L.,** *GABA Outside the CNS,* Springer-Verlag, Berlin, 1992.
16. **Barnes, K. and Ferrario, C.,** Angiotensin and central nervous system regulation of blood pressure, *Clin. Exp. Hypertens.,* 2, 465, 1980.
17. **Loewy, A. D. and Spyer, K. M.,** *Central Regulation of Autonomic Functions,* Oxford University Press, New York, 1990.
18. **Pent, M. T. and Hsu, H. K.,** No neuron loss from hypothalamic nuclei of old male rats, *Gerontology,* 28, 19, 1982.
19. **Hsu, H. K. and Peng, M. T.,** Hypothalamic neuron number in old female rats, *Gerontology,* 24, 434, 1978.
20. **Sartin, J. L. and Lamperti, A. A.,** Neuron numbers in hypothalamic nuclei of young, middle-age and aged male rats, *Experientia,* 41, 109, 1985.
21. **Lamperti, A. and Blaha, G.,** The numbers of neurons in the hypothalamic nuclei of young and reproductively senescent female golden hamsters, *J. Gerontol.,* 35, 335, 1980.
22. **Goudsmit, E., Hoffman, M. A., Fliers, E., and Swaab, D.,** The supraoptic and paraventricular nuclei of the human hypothalamus in relation to sex, age and Alzheimer's disease, *Neurobiol. Aging,* 11, 529, 1990.
23. **Fliers, E. and Swaab, D. F.,** Activation of vasopressinergic and oxytocinergic neurons during aging in the Wistar male rat, *Peptides,* 4, 165, 1983.
24. **Hoogendijk, J. A., Fliers, E., Swaab, D. F., and Verwer, R. W. H.,** Activation of vasopressin neurons in the human supraoptic and paraventricular nucleus in senescence and senile dementia, *J. Neurol. Sci.,* 69, 291, 1985.
25. **Fliers, E., Swaab, D. F., Pool, Chr. W., and Verwer, R. W. H.,** The vasopressin and oxytocin neurons in the human supraoptic and paraventricular nucleus; changes with aging and in senile dementia, *Brain Res.,* 342, 45, 1985.
26. **Swaab, D. F.,** Brain aging in Alzheimer's disease: "wear and tear" versus "use it or lose it", *Neurobiol. Aging,* 12, 123, 1991.
27. **Paul, F.,** Biometrische Analyse der Frischvolumina der Grosshirnrinde und des Prosencephalon von 31 menschlichen, adulten Gehirnen, *Z. Anat. Entwicklungsgesch.,* 133, 325, 1970.
28. **Swaab, D. F., Fliers, E., and Partiman, T.,** The suprachiasmatic nucleus of the human brain in relation to sex, age and senile dementia, *Brain Res.,* 342, 37, 1985.

29. **Wessely, W.,** Biometrische Analyse der Firschvolumina das Rombencephalon, des Cerebellum und der Ventrikel von 31 adulten menschlichen Gehirnen, *J. Hirnforsch.,* 12, 11, 1970.
30. **Vogels, O. J. M., Broere, C. A. J., and Nieuwenhuys, R.,** Neuronal hypertrophy in the human supraoptic and paraventricular nucleus in aging and Alzheimer's disease, *Neurosci. Lett.,* 109, 62, 1990.
31. **Ravid, R., Swaab, D. F., and Pool, C. W.,** Immunocytochemical localization of vasopressin binding sites in the rat kidney, *J. Endocrinol.,* 105, 113, 1985.
32. **Bondareff, W., Mountjoy, C. Q., and Roth, M.,** Selective loss of neurons of origin of noradrenergic projections to cerebral cortex (nucleus locus coeruleus) in senile dementia, *Lancet,* 1, 783, 1981.
33. **Marcyniuk, B., Mann, D. M. A., and Yates, P. O.,** The topography of cell loss from locus coeruleus in Alzheimer's disease, *J. Neurol. Sci.,* 76, 335, 1986.
34. **McGreer, E. G.,** Aging and neurotransmitter metabolism in the human brain, in *Aging,* Vol. 7, *Alzheimer's Disease, Senile Dementia, and Related Disorders,* Katzman, R., Terry, R. D., and Bick, K. L., Eds., Raven Press, New York, 1978, 427.
35. **Sladek, J. R., Khachaturian, H., Hoffman, G. E., and Scholer, J.,** Aging of central endocrine neurons and their aminergic afferents, *Peptides,* 1 (Suppl. 1), 141, 1980.
36. **Sladek, J. R., Scholer, J., and Armstrong, W. E.,** Norepinephrine-vasopressin interactions during aging, in *Structure and Function of Peptidergic and Aminergic Neurons,* Sano, Y., Ibata, Y., and Zimmerman, E. A., Eds., Japan Scientific Societies Press, Tokyo, 1983, 289.
37. **Roozendaal, B., van Gool, W. A., Swaab, D. F., Hoogendijk, J. E., and Mirmiran, M.,** Changes in vasopressin cells of rat suprachiasmatic nucleus with aging, *Brain Res.,* 409, 259, 1987.
38. **Swaab, D., Roozendaal, B., Ravid, R., Velis, D. N., Gooren, L., and Williams, R. S.,** Suprachiasmatic nucleus in aging, Alzheimer's disease, transsexuality and Prader-Willi syndrome, in *Progress in Brain Research,* Vol. 72, *Neuropeptides and Brain Function,* de Kloet, E. R., Wiegants, V. M., and de Wied, D., Eds., Elsevier, Amsterdam, 1987, 301.
39. **Issa, A. M., Rowe, W., Gauthier, S., and Meaney, M. J.,** Hypothalamic-pituitary-adrenal activity in aged, cognitively impaired and cognitively unimpaired rats, *J. Neurosci.,* 10, 3247, 1990.
40. **Sahu, A., Kalra, P. S., Crowley, W. R., and Kalra, S. P.,** Evidence that hypothalamic neuropeptide Y secretion decreases in aged male rats: implications for reproductive aging, *Endocrinology,* 122, 2199, 1988.
41. **Goldman, J. and Cot, L.,** Aging of the brain: dementia of the Alzheimer's type, in *Principles of Neural Science,* Kandel, E. R., Schwatz, J. H., and Jessel, T. M., Eds., Elsevier, New York, 1991, 974.

Chapter 2

THE PERIPHERAL NERVE

Damiano Zaccheo

TABLE OF CONTENTS

0-8493-6981-9/93/$0.00 + $.50

I. AGE-DEPENDENT ALTERATIONS IN NEURONAL MORPHOLOGY AND IN CELL POPULATIONS

A. INTRODUCTION

It has been known for some time that the weight and volume of the human nervous system reaches its maximum at puberty, remains at this level for several decades, and then begins a progressive decrease.[1] This decrease in volume is accompanied by a loss of nerve cells.[2] Recently, Anderson[3] has demonstrated that in the elderly, both grey matter and white matter in the frontal and temporal lobes undergo a decrease in volume. The area of the individual neuron is reduced in both lobes, with larger neurons exhibiting a greater reduction. Similar changes have been reported in the hypothalamus, substantia nigra, reticular formation, and amygdaloid nucleus.[4]

There is general agreement that there is a decrease in the volume and cell number in the aging nervous system, and there is general agreement that these changes vary in different areas of the nervous system; in fact, these changes may actually not occur in some areas such as the parietal lobes and the brainstem nuclei.[5]

Various authors have reported morphological changes in the aging neuron as well. These include changes in normal neuronal organelles as well as the genesis of structures not normally found in neurons or the accumulation of products as the individual ages.

For example, Spoerri and co-workers[6] reported a decrease in nonmembrane-associated ribosomes and deterioration of the rough endoplasmic reticulum. Buell and Coleman[7] demonstrated that the arborization of human pyramidal cells of layer 2 were larger in brains obtained from elderly patients. An earlier investigation[8] described a decrease in the number of axosomatic synapses in the dentate gyrus of aged rats.

The progressive increase in the amount of lipofuscin in the nervous system is considered to be the most consistent and reliable sign of aging,[9,10] although lipofuscin is present in some areas of the brain during childhood in the inferior olive and dentate nucleus. Moreover, not all neurons accumulate lipofuscin at the same rate, so it is demonstrable in some cells only in old age.[11] With the progressive accumulation of lipofuscin, there is a corresponding decrease in the amount of RNA,[12] leading to a progressive, decreased protein synthesis.

Another age-dependent alteration is the appearance of neurofibrillary tangles, especially in the hippocampus. This accumulation is more frequently seen in the brains of patients suffering from Alzheimer's disease.

The aging brain may also include areas of granulovacuolar degeneration in the hippocampus. These inclusions are rarely seen before the age of 65 years, but in older individuals their frequency increases.[13] Another cell abnormality are the inclusions first described by Hirano et al.[14] in patients who had a Parkinson's dementia syndrome. These Hirano bodies are found in young individuals[15] and in older individuals not suffering from dementia, their numbers increasing as the individual ages.

TABLE 1
The Percentage of Human Posterior Root Ganglion (PRG) and Anterior Horn (AH) Cells Which Contained Lipofuscin at Various Ages

Age	Sex	% PRG	% AH
7 months	F	0	0
13–27 years	M & F	26	23
52–69 years	F	70	73
80–91 years	F	64	56

Data from Ref. 64.

There have been numerous studies of the age-dependent morphological changes in the peripheral nervous system. These alterations include increased numbers of lysosomes in Schwann cells,[16] concurrent demyelination and remyelination,[17] and concurrent degeneration of myelinated fibers undergoing peripheral nerve regeneration.[18]

B. MODIFICATION OF THE NEURONAL STRUCTURE

The fact that lipofuscin, the "wear and tear" pigment, progressively accumulates in neurons as the individual ages has been discussed previously. Pick and co-workers[19] feel that the progressive increase in the amount of the pigment is the most obvious change observed in the neurons of the peripheral is the most obvious change observed in the neurons of the peripheral nervous system, including the autonomic ganglia. Not only does the quantity of lipofuscin vary with time, but its staining properties and physical characteristics also change with age. Nandy[20] reported lipofuscin as being Sudan black and PAS (periodic acid Schiff) positive and having a greenish-yellow fluorescence when excited at a wavelength of 365 Å. In older individuals, the pigment was more readily colored by Nile blue and possesses an orange-yellow fluorescence. The inclusions housing lipofuscin also undergo an evolution; electron microscopic examination showed that the first inclusion, a lysosome-like body, eventually became the mature pigment granules observable under the light microscope.[21]

Nandy,[20] using the mouse as his experimental animal, also observed that the percentage of motor cells which had lipofuscin in their cytoplasm varied from none at birth to more than 90% in mice 30 months of age. Table 1 shows data from humans. As may be seen, there is virtually no lipofuscin in cells of the posterior root ganglia and the anterior horns at 7 months of age, while by the end of the third decade, the amount of the pigment had reached the levels reported in the elderly.

Scaravilli,[64] reported morphological changes in the appearance of the lipofuscin granules as the cells aged, with the lipofuscin becoming coarser and more irregular in older individuals. He likewise described granular material whose morphological and histochemical characteristics were similar to lipofuscin in the cytoplasm of extraneuronal cells.

Melanin is another pigment which progressively accumulates in aging neurons. Its appearance precedes that of lipofuscin and it tends to be concentrated in the trigeminal and spinal ganglia[22] as well as in other areas of the nervous system. Under the electron microscope, melanin resembles lipofuscin, but the two pigments can be differentiated by common histochemical techniques. There is some possibility that in aged sensory ganglia, some of the lipofuscin granules undergo melaninization.[23]

The cytostructure of the nervous system is also altered by the development of neurofibrillary tangles.[24] These structures, which are argyrophilic and congophilic intracytoplasmic inclusions, under electron microscopic examination resemble paired filaments wrapped around each other in a double helix.[25] The filaments have a periodicity of 80 nm and have the same immunohistochemical properties as similar structures in the central nervous system.

C. DECREASE IN CELL NUMBERS

It is well known that as humans age, the number of cells in the nervous system progressively decreases. However, it is difficult to estimate the amount of decrease or its effects on the individual.[26] The aging of nerve cells is a slow, progressive process of degeneration. In the motor cells of the spinal cord, the gliomesodermal reaction is of small consequence compared to the deposits of lipofuscin and growth of other cytoplasmic vacuoles. In the posterior root ganglia of the same individuals, however, nerve cells were replaced by compact groups of residual satellite cells called Nageotte nodules.[27] These nodules were not observed in young normal human subjects.

Aldskogius and Risling[28] suggest that it is crucial to establish whether the number of cells in normal conditions is consistently the same at the same level of the nervous system and on the two sides of the nervous system. These investigators reported striking changes of about 20 to 25% in the L^7 spinal ganglia of kittens; in the rat, differences were reported to be as high as 45%. In man, there is not yet sufficient data to have confidence in any conclusion, and the studies reported were performed well over half a century ago.[29]

When the motor neurons of the spinal cord are considered, there is evidence of a small decrease in number as the age of the individual increases. In their study involving the lumbar area of spinal cords of 47 patients whose age ranged from 13 to 95 years, Tomlinson and Irving[30] reported no real loss of motor neurons until the age of 60. Over 60, there was a noteworthy decrease in the number of motor cells. This study basically confirmed a previous report[31] where a 15 to 20% loss was found in the number of motor cells in the murine lumbar spinal cord of old animals. Neither study reported any histological alterations, with the exception of infrequent "chromolytic" or vacuolated cells (Figure 1).

Data are scant as far as nerve cell loss in the autonomic nervous system is concerned. In 1938, Amprino[32] reported a large decrease in aged human cervical sympathetic ganglia. Dyck et al.[33] described an age-dependent cell loss in the intermediolateral column of the spinal cord.

D. ALTERATIONS IN CELL SIZE

It is difficult to study the age-dependent alterations in cell size due to the fact that autopsies are performed at various times after the patient's death. Thus, observed changes may be due to postmortem effects rather than normal aging. Notwithstanding, it is generally agreed that aging causes retraction of the cell body from the surrounding glial tissue. This may be due to the fact that the nerve cell grows slower as it ages and appears to be receding.

II. AGE-DEPENDENT ALTERATIONS IN PERIPHERAL NERVES

A. INTRODUCTION

As the individual ages, peripheral nerve function decreases. This deterioration is progressive and multifactorial. For the most part, the alterations in peripheral nerves occur predominantly in the nerve fibers themselves, with little involvement in the endoneurium, perineurium, or epineurium. Basically, there are three fundamental age-induced changes in the peripheral nerve: (1) a decrease in both myelinated and unmyelinated nerve fibers, (2) a mild but progressive degeneration of the axon with concurrent evidence of regeneration, and (3) a progressive demyelination, again accompanied by evidence of remyelination (Figure 2).

B. AGE-DEPENDENT ALTERATIONS IN NERVE FIBERS

1. Myelinated Fibers

Peripheral neuropathy is often due to the cumulative effects of physical trauma and chronic ischemia. In older peripheral nerves, there has been a well-documented history of myelinated fiber loss, changes in the myelin itself, and an increase in the associated connective tissue.[34-36] These changes are more prevalent in the spinal roots than in the nerve trunks; the ventral roots are more affected than are the dorsal. The most prominent alteration is the demyelination-remyelination cycle. Because the demyelination phase is rather short, the remyelination segment is the one most often observed when a histological examination is performed.[36]

Using electron microscopy, the neurologist has been able to clarify what has been occurring as the individual ages. In aged individuals, there are cellular elements which have been identified as denervated Schwann cell bands.

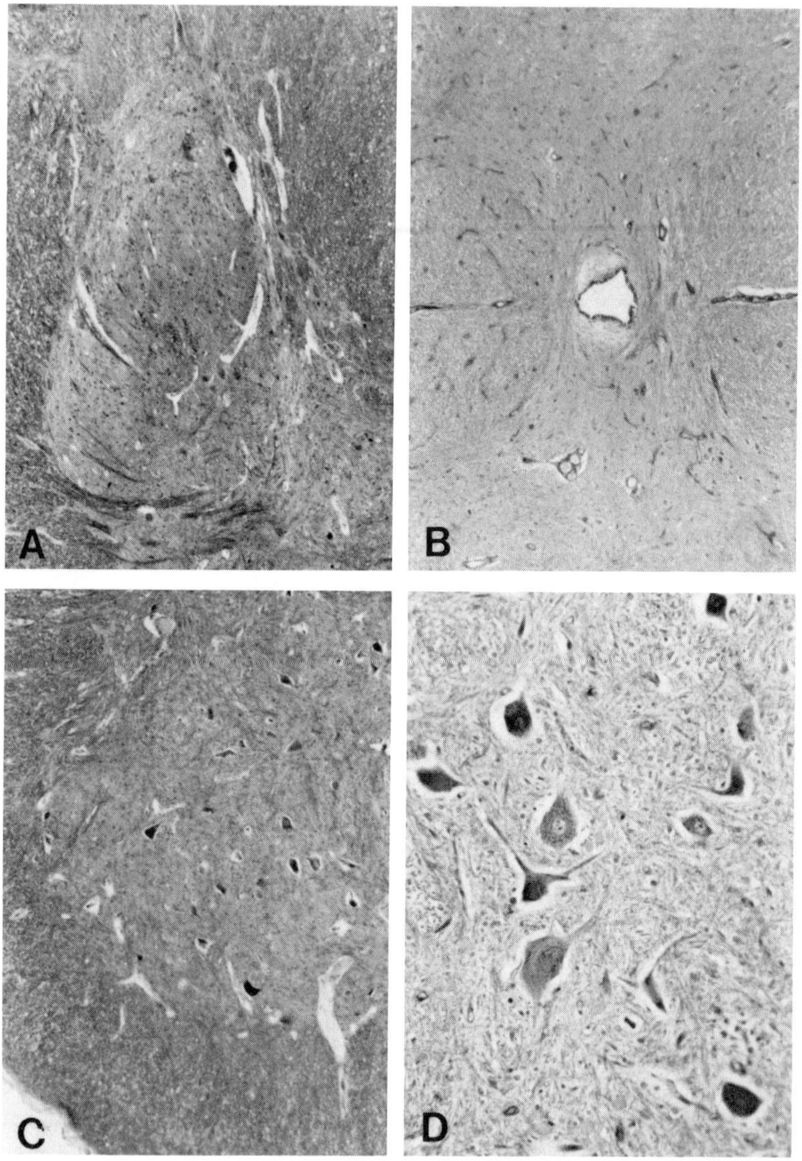

FIGURE 1. Light-microscope pictures showing the different portions of the spinal cord, from where peripheral nerves originate (human material stained with thionin). (A) Posterior (dorsal) grey matter horn. (Original magnification × 40.) (B) Central canal, which represents the central portion of the spinal cord. (Original magnification × 60.) (C) Anterior (ventral) grey matter horn. (Original magnification × 40.) (D) Higher magnification of C, showing the main morphological features of spinal motor neurons. (Original magnification × 160.)

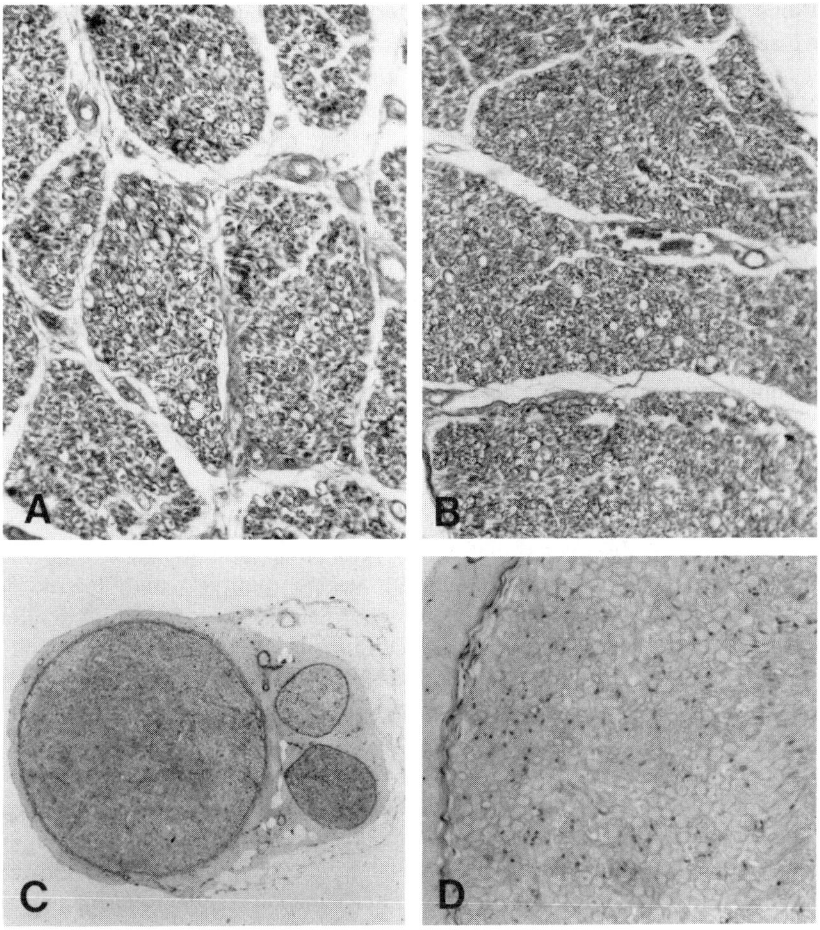

FIGURE 2. Light-microscope pictures showing the microanatomical organization of the peripheral nerve and its components. (Human material stained with Azan-Mallory [A, B] or hematoxylin and eosin [C and D]). (A) Posterior (dorsal) sensory root. (Original magnification × 100.) (B) Anterior (ventral) motor root. (Original magnification × 100.) (C) Trunk of a peripheral nerve. (Original magnification × 25.) (D) Higher magnification of C, showing primarily myelinated nerve fibers constituting the peripheral nerve. (Original magnification × 100.)

These bands are nothing more than the hollow remains of myelinated nerve fibers which have lost their axons and myelin due to degeneration. Ochoa and Mair[37] have used these bands as markers; their studies show that the bands increase with aging. Also described in the literature are "clusters" which increase with aging, and are nothing more than areas of axonal repair.

2. Unmyelinated Fibers

It was not possible to study age-dependent morphological alterations prior to the now widespread use of electron microscopy, which occurred in the late

1960s.[18-38] These pioneer studies utilized biopsied human skin nerves; now, we are able to study nerves taken at autopsy.[36]

It is virtually impossible to collect quantitative data concerning the aging of nonmyelinated fibers, for several reasons. First, it is difficult to count all unmyelinated axons in one nerve trunk because of technical problems. Also, comparative investigations on the unit cross-sectional area of a nerve shows so much individual variation that any comparison becomes valueless.[37]

Quantitative alterations in aging nonmyelinated fibers begin with proliferation of Schwann cell processes.[37] As the individual ages, empty denervated Schwann cell bands which are similar to, but in reality different from, those seen in the axon of degenerating myelinated nerve fibers. These alterations become prominent in old age, Likewise, investigators report an increase in the amount of collagen which accumulates in pockets.[38,39] Regenerating axon buds also occur with aging,[36] providing a subtle but reliable sign of unmyelinated pathology.

C. THE LONG-TERM EFFECTS OF TRAPPED NERVES

When it was discovered that pigs develop a carpal tunnel syndrome with aging, investigators were presented with an opportunity to study the effects of chronically trapped peripheral nerves.[40,41] These workers noticed an alteration of the myelin sheath proximally to the site of compression, causing a tapering of the myelin internodes. Consistently, there was a polarity with the bulb pointing proximally, away from the wrist. Additional studies involving electron microscopy[42] showed that unlike what is seen in acutely compressed nerves, the distortion observed was not a result of acute dislocation of myelin synthesis. Thus, chronic accumulation of myelin is not the long-term result of repeated acute compressions. Researchers also discovered that distal to the lesion site, the myelin was pathologic due to cycles of demyelination and remyelination leading to frank Wallerian degeneration. Subsequent examination of the ultrastructure of the altered internodes showed that there is a progressive displacement of the deep myelin lamellae away from the point of entanglement and a progressive accumulation of excess myelin, producing a bulb-like structure at the proximal end of the internode.

Therefore, initially there is a simple physical rearrangement of myelin, causing partial exposure of the nerve's axon. As time goes by, the rearranged myelin making up the bulb degenerates, exposing more of the axon. Originally, investigators thought that the observed demyelination was the result of compression-induced ischemia. However, the above observations show that this cannot be the case because the degeneration affects different sides of the entrapped nerve.

Neary and co-workers[43] reported on a series of naturally entrapped human median and ulnar nerves which had been removed from autopsies of individuals of various ages. Their method of study was similar to that employed by Ochoa and Marotte[42] in their study on guinea pigs. Neary et al.[43] were able to confirm the myelin distortion, demyelination, remyelination, and axonal degeneration and regeneration. They concluded that the pathological changes can be used as evidence of subclinical entrapment in human subjects who were apparently unaffected. Although this study did not conclusively demonstrate that this alteration increases with aging, studies by Fullerton and Gilliatt[40] and Anderson et al.[41] were able to do so.

D. THE VASA NERVORUM IN AGED NERVES

It is difficult to distinguish between primary arteriosclerotic changes and changes caused by aging of the vasa nervorum. Progressive scarring of small blood vessels supplying the nerves is common in aging. Electron microscopic studies reveal that there is also an age-dependent alteration of the basal lamina. Thus, there is an impaired delivery of blood to the peripheral nerve which aids in the morphologic changes seen in the aging peripheral nerve.[44]

III. AGE-DEPENDENT NEUROPATHIC CHANGES IN LABORATORY ANIMALS

A. INTRODUCTION

It is not surprising that the peripheral nervous system in man and in animals has a similar response to inflammation, trauma, and aging because, except for minor exceptions, the structural and biochemical composition of myelin and the axon are quite similar.[45,46] Neurologists have been studying the effects of aging on the peripheral nervous system (PNS) for some time using laboratory animals as experimental models; the following is a review of what has been learned about the most important laboratory species.

B. RAT

Because the laboratory rat is the most commonly used animal in experimental neurology, more has been written about it than about any other animal. Investigations centered on age-dependent neuropathy have focused on two important areas of investigation: (1) old laboratory rats have been reported to develop spontaneous hindlimb paralysis without other accompanying pathology and (2) various long-term studies required the gathering of baseline data which could be compared with data acquired in subsequent experiments.

The first report about spontaneous hindlimb paralysis was made in 1956 by Berg,[47] who published a second investigation 6 years later.[48] These studies reported anomalies in the nerve roots and peripheral nerve trunks of old rats which were more frequent in the lumbosacral plexus than in the brachial plexus. Gilmore[49] was able to demonstrate differences in severity between the dorsal and ventral roots, the ventral root being affected earlier and more intensely.

Not only were radicular changes noted, but changes within the nerve trunks were also seen. At first, observers concentrated on lesion location, investigating lesions which were distal or proximal. In time it became apparent that this distinction was misleading, as radicular changes are now thought to represent different spatial changes occurring on the same axon. In order to clarify the age-dependent alterations which occur in the nerve roots and peripheral nerve trunks, each phenomenon is discussed separately.

1. Radicular Changes

The initial reports of polyradiculopathy described the appearance of vacuoles in the myelin and simultaneous degeneration of the axons in both the dorsal and ventral nerve roots in paraffin-embedded nerves. The rats used in the study were 16 to 18 months of age. As histological techniques improved, the more modern methods of glutaraldehyde fixation, plastic embedding, single teased fibers, and morphometric quantitation were used.[50-53] Krinke and co-workers[53] used not only light microscopy, but also electron microscopy to investigate polyradiculopathy in the spinal cord, spinal nerve roots, and peripheral nerves of 2-year-old rats. They reported myelin sheaths which were split and distended. The enclosed axons were thinner than normal. The authors noted that the described structural changes had also been reported in the PNS in rats which had been intoxicated with various agents, such as hexachlorophene and acetyl ethyl tetramethyl tetralin, as well as in animals with metabolic disorders and autoimmune diseases.[54]

In a subsequent investigation, Krinke[55] studied the timing of the radiculopathy and saw that the earliest alterations in the ventral nerve root were seen at 13 or 14 months of age. At that time, there was no indication of myelin swelling. By month 16, however, the myelin sheath had presented distinct bubbling. Teased individual fibers exhibited various alterations consistent with demyelination and remyelination, but there was no degeneration to the axon distal to the myelin changes. In older rats, myelin delamination of the sciatic nerve was observed to have spread distally to the level of the knee, but the plantar nerves were not involved. Krinke also reported other peripheral lesions such as axonal sequestration of organelles by the Schwann cell.

Krinke performed morphometric examinations of the L^4 ventral root and reported that the root underwent a reduction in the size of the large myelinated fibers after the third year of life, with an increase in the number of small axons. Thus, Krinke was able to conclude that there was a difference in both the spatial distribution and spread of axonal degeneration and in myelin changes — specifically, that the degeneration of the axon was more apparent distally with a proximally oriented progression, whereas bubbling of the myelin occurred in the opposite manner. Since the demyelination reported in the nerve roots commenced at the second year of age *after* the distal alterations were noted, he was able to conclude that the changes seen in the myelin sheath were secondary to the segmental demyelination seen in the axon.[56]

Similar age-dependent changes in rats of approximately the same age were reported in the lumbar, dorsal, and ventral roots.[50] These investigators also discussed widespread tearing of the myelin sheaths, often in close association with atrophy of the axons. Mitsumori's group also reported the presence of macrophages within myelin bubbles and an increased density of neurofilaments and microtubules. This group also concluded that the myelin abnormalities seen in the myelin sheath were secondary to axonal pathology.

In 1986, Mitsumori et al.[51] came to different conclusions. In this later study, they discussed the alterations observed in the sciatic nerves of rats which had previously undergone radicular changes. Examination of these nerves showed numerous areas of axonal degeneration and the loss of myelinated fibers as well as the occasional myelin bubble. They also noticed demyelinated and remyelinated fibers as well as small onion-like bulbs. There were also clusters of regenerating fibers and atrophic axons. Because of these new findings, they concluded that the atrophic axons seen with the myelin bubbles had caused distal fiber degeneration and that the demyelination seen in the nerve roots were actually secondary to the axonal pathology.

2. Changes in the Peripheral and Distal Nerves

Teased fiber studies by Sharma et al.[57] demonstrated the presence of demyelination and degeneration in the tibial nerve of laboratory rats. This process was seen in animals from 18 to 24 months of age, and was progressive. In the plantar nerve, demyelination was first observed as early as 6 months of age; by month 15, axonal degeneration was noticed. At 2 years of age, there were more degenerating fibers in the plantar nerve than in the tibial nerve. Sharma and co-workers reported that the fiber diameters decreased in both the tibial and plantar nerves, but there was no decrease in the number of fibers.

Ultrastructural investigations were undertaken by Thomas et al.[52] They, too, reported prominent degeneration/regeneration and demyelination/remyelination in the plantar nerves. They were also able to see numerous axonal inclusions such as collections of glycogen and Schwann cell protrusions into the axons. A similar situation was observed in the tibial nerve, except that there were fewer inclusions and axonal degeneration was rare. They suggested that the demyelination seen in the plantar nerve as early as 6 months of age was the result of pressure and that inclusions observed within the axons were of a primary axonal abnormality. A second study showing the probability of a distal alteration in the axons of aged rats investigated the tail nerves, structures which are unlikely to be chronically compressed. Thus, axonopathy in the PNS in the rat does not necessarily depend on trauma.

Similar results were published by Grover-Johnson and Spencer in 1981.[58] These authors studied the sciatic tibial/plantar nerve complex in laboratory rats 2 to 30 months of age. They were able to show that the most serious alterations occurred in the plantar nerves, and that the lesions worsened and spread proximally with aging. They also noticed axonal inclusions similar to those reported by Thomas' group and swollen demyelinated fibers which had accumulations of organelles. These alterations were most often observed in the sciatic nerves and became more prominent as the animal aged. Grover-

Johnson and Spencer concluded that these changes were due to chronic pressure which led to proximal degeneration. To strengthen their hypothesis, they noted that the tibial nerve fibers to the gastrocnemius muscle were unaffected by aging.[59]

At this time, there is no unequivocal age-dependent link between a proximal polyradiculoneuropathy and distal progressive alterations in the tibial and plantar nerves. Most investigators believe that the myelin bubbles occur with the initial shrinking of the axon; less clear is the actual sequence of events.

C. DOG

As might be expected, several investigators have reported age-dependent changes in the peripheral nervous system of the dog. Griffiths and Duncan[60] studied the L^6 and L^7 dorsal and ventral nerve roots using light microscopy, electron microscopy, and individual teased nerve fibers. These investigators reported that myelin bubbles were first noticed at 4 months of age and that they dramatically increased as the animal aged. In some cases, macrophages were also noticed inside the myelin blisters. When teased fibers were studied, an increasing number of short remyelinated internodes were noticed from the age of 18 months onward. Ultrastructural examination revealed that many of the fibers had undergone remyelination. These authors did not report on any changes within the axon. They concluded that age-dependent alterations in the canine nerve roots begin as early as 4 months of age.

In 1982, Braund et al.[61,62] investigated the age-dependent changes of the common peroneal and ulnar nerves in normal dogs 3 months to 15 years of age; 10% of the teased fibers in dogs 10 years old demonstrated degeneration, intercalated internodes of specific localizations which had undergone myelin swelling. These changes occurred more frequently in the peroneal than in the ulnar nerves. Transverse sections of the two canine nerves showed no additional pathology as the animals aged. Also, there was no significant loss of myelinated fibers or reduction in nerve fiber density in the nerves studied.

D. HORSE

Studies of the lateral nerve in horses 6 months to 15 years of age were done by Wheeler in 1987[63] using electrophysiological as well as structural analyses. Wheeler was able to report noticeable age-dependent alterations at the fifth year. As might be expected, these alterations included a slight decrease in the number of large-diameter myelinated fibers along with evidence of degeneration and regeneration and cycles of demyelination and remyelination. Examination of single, teased fibers using light microscopy confirmed what had been reported at the ultrastructural level. Equine idiopathic laryngeal hemiplegia is now a commonly recognized equine peripheral neuropathy, and these observations by Wheeler are evidence of its pathogenesis.

IV. CONCLUSION

In-depth and updating studies concerning age-dependent changes in the PNS need to be undertaken. Neurobiologists are fortunate in having access to additional techniques such as autoradiography and immunohistochemistry to aid in their research. It is well known that there are age-dependent changes in the PNS. It is also known that in many countries, population age ratios are increasing. Consequently, age-dependent neuropathies will become more of a clinical challenge.

REFERENCES

1. **Ellis, R. S.,** A preliminary quantitative study of the Purkinje cell in normal, subnormal and senescent human cerebella with some notes on functional localization, *J. Comp. Neurol.,* 30, 229, 1919.
2. **Hall, T. C., Miller, A. K. H., and Corsellis, J. A. N.,** Variations in the human Purkinje cell population according to the age and sex, *Neuropathol. Appl. Neurobiol.,* 1, 267, 1975.
3. **Anderson, J. M., Hubbard, B. M., Coghill, G. R., and Slidders, W.,** The effect of advanced old age on the neurone content of the cerebral cortex. Observations with an automatic image analyser point counting method, *J. Neurol. Sci.,* 58, 233, 1983.
4. **McNeil, T.,** Neural structure and aging, in *Review of Biological Research in Aging,* Rothstein, M., Ed., Alan R. Liss, New York, 1983, 163.
5. **Konigsmark, B. W. and Murphy, E. A.,** Neuronal populations in the human brain, *Nature (London),* 228, 1335, 1970.
6. **Spoerri, P. E., Gless, P., and Spoerri, O.,** Neuronal regression during ageing: an ultrastructural study of the human cortex, *J. Hirnforsch.,* 22, 441, 1981.
7. **Buell, S. J. and Coleman, P. D.,** Quantitative evidence for selective dendritic growth in normal human aging but not in senile dementia, *Brain Res.,* 214, 23, 1981.
8. **Geinisman, Y.,** Loss of axosomatic synapses in the dentate gyrus of aged rats, *Brain Res.,* 168, 485, 1979.
9. **Sulkin, N. M.,** The properties and distribution of PAS positive substances in the nervous system of the senile dog, *J. Gerontol.,* 10, 135, 1955.
10. **Barden, H.,** Relationship of Golgi thiaminepyrophosphatase and lysosomal acid phosphatase to neuromelanin and lipofuscin in cerebral neurons of the aging rhesus monkey, *J. Neuropathol. Exp. Neurol.,* 29, 225, 1970.
11. **Mann, D. M. A. and Sinclair, K. G. A.,** The quantitative assessment of lipofuscin pigment, cytoplasmic RNA and nucleolar volume in senile dementia, *Neuropathol. Appl. Neurobiol.,* 4, 129, 1978.
12. **Mann, D. M. A. and Yates, P. O.,** Lipoprotein pigments; their relationship to aging in the human nervous system. I. The lipofuscin content of nerve cells, *Brain,* 97, 481, 1974.
13. **Tomlinson, B. E. and Kitchener, D.,** Granulovacuolar degeneration of hippocampal pyramidal cells, *J. Pathol.,* 106, 165, 1972.
14. **Hirano, A., Malamud, N., Elizan, T. S., and Kurland, L. T.,** Amyotrophic lateral sclerosis and Parkinsonism-dementia complex of Guam, *Arch. Neurol.,* 15, 35, 1966.
15. **Gibson, P. and Tomlinson, B. E.,** The numbers of Hirano bodies in the hippocampus of normal and demented subjects with Alzheimer's disease, *J. Neurol. Sci.,* 33, 199, 1977.

16. **Robson, J. T.,** Protagon granules in the normal sciatic nerve with some observations on the greater splanchnic nerve, *J. Neuropathol. Exp. Neurol.,* 10, 77, 1951.
17. **Lascelles, R. G. and Thomas, P. K.,** Changes due to age in internodal length in the sural nerve in man, *J. Neurol. Neurosurg. Psychiatry,* 29, 40, 1966.
18. **Ochoa, J. and Mair, M. G. P.,** The normal sural nerve in man. I. Ultra-structure and numbers of fibres and cells, *Acta Neuropathol.,* 13, 197, 1969.
19. **Pick, J., de Lemos, C., and Gerdin, C.,** The fine structure of sympathetic neurons in man, *J. Comp. Neurol.,* 122, 19, 1964.
20. **Nandy, K.,** Properties of neuronal lipofuscin pigment in mice, *Acta Neuropathol.,* 19, 25, 1971.
21. **Sekhon, S. S. and Maxwell, D. S.,** Ultrastructural changes in neurons of the spinal anterior horn of ageing mice with particular reference to the accumulation of lipofuscin pigment, *J. Neurocytol.,* 3, 59, 1974.
22. **De Castro, F.,** Sensory ganglia of the cranial and spinal nerves. Normal and pathological, in *Cytology and Cellular Pathology of the Nervous System,* Penfield, W., Ed., Hoeber, New York, 1932, 93.
23. **Lieberman, A. R.,** Sensory ganglia, in *The Peripheral Nerve,* Landon, D. N., Ed., Chapman and Hall, London, 1976, 188.
24. **Kawasaki, H., Murayama, S., Tomonaga, M., Izumiyama, N., and Shimda, H.,** Neurofibrillary tangles in human upper cervical ganglia. Morphological study with immunohistochemistry and electron microscopy, *Acta Neuropathol.,* 75, 156, 1987.
25. **Wisniewski, H., Narang, H. K., and Terry, R. D.,** Neurofibrillary tangles of paired helical filaments, *J. Neurol. Sci.,* 27, 173, 1976.
26. **Tomlinson, B. E. and Corsellis, J. A. N.,** Ageing and the dementias, in *Greenfield's Neuropathology,* 4th ed., Adams, J. H., Corsellis, J. A. N., and Duchen, K. W., Eds., Edward Arnold, London, 1984, 951.
27. **Thomas, P. K., Landon, D. N., and King, R. H. M.,** Diseases of the peripheral nerves in *Greenfield's Neuropathology,* 4th ed., Adams, J. H., Corsellis, J. A. N., and Duchen, L. W., Eds., Edward Arnold, London, 1984, 807.
28. **Aldskogius, H. and Risling, M.,** Effects of sciatic neurectomy on neuronal number and size distribution, *Exp. Neurol.,* 74, 597, 1981.
29. **McKinniss, M. E.,** The number of ganglion cells in the dorsal root ganglia of the second and third cranial nerves in human fetuses of various ages, *Anat. Rec.,* 65, 255, 1936.
30. **Tomlinson, B. E. and Irving, D.,** The numbers of limb motor neurons in the human lumbosacral cord throughout life, *J. Neurol. Sci.,* 34, 213, 1977.
31. **Wright, E. A. and Spink, J. M.,** A study of the loss of nerve cells in the central nervous system in relation to age, *Gerontologia,* 3, 277, 1959.
32. **Amprino, R.,** Modifications de la structure des neurones sympathétiques pendant l'accroissement et la sénescence. Recherches sur le ganglion cervical supérieur, *C. R. Assoc. Anat.,* 33, 3, 1983.
33. **Dyck, P. J., Jedrzejowska, H., Karnes, J., Jawamura, Y., Low, P., O'Brien, P. C., Offord, K., Ohnishi, L., Ohta, M., Poolock, J., and Stevens, J.,** Reconstruction of motor, sensory and autonomic neurons based on morphometric study of sampled levels, *Muscle Nerve,* 2, 399, 1979.
34. **Corbin, K. B. and Gardner, E. D.,** Decrease in number of myelinated fibers in human spinal roots with age, *Anat. Rec.,* 68, 63, 1937.
35. **Samosiuk, I. Z.,** Aging characteristics of the peripheral nervous system, *Vrach, Delo,* 3, 111, 1972.
36. **Jacobs, J. M. and Love, S.,** Qualitative and quantitative morphometry of human sural nerve at different ages, *Brain,* 108, 897, 1985.
37. **Ochoa, J. and Mair, M. G. P.,** The normal sural nerve in man. II. Changes in the axons and Schwann cells due to aging, *Acta Neuropathol.,* 13, 217, 1969.

38. **Ochoa, J.,** The Structure of Developing and Adult Sural Nerve in Man and the Changes That Occur in Some Diseases. A Light and Electron Microscopic Study, Ph.D. thesis, University of London, 1970.
39. **Schröder, J. M. and Gibbels, M.,** Marklose Nervenfasern in Senium und im Spätstadium der Thalidomid-Polyneuropathie: Quantitativelektronen-mikroskopische Untersuchungen, *Acta Neuropathol.,* 39, 271, 1977.
40. **Fullerton, P. M. and Gilliatt, R. W.,** Median and ulnar neuropathy in the guinea-pig, *J. Neurol. Neurosurg. Psychiatry,* 30, 393, 1967.
41. **Anderson, M. N., Fullerton, P. M., Gilliatt, R. W., and Hern, J. E. C.,** Changes in the forearm associated with median nerve compression at the wrist in the guinea-pig, *J. Neurol. Neurosurg. Psychiatry,* 33, 70, 1970.
42. **Ochoa, J. and Marotte, L.,** The nature of the nerve lesion caused by chronic entrapment in the guinea-pig, *J. Neurol.,* 19, 491, 1973.
43. **Neary, D., Ochoa, J., and Gilliatt, R. W.,** Sub-clinical entrapment neuropathy in man, *J. Neurol. Sci.,* 24, 283, 1975.
44. **Dick, P. J., Hansen, S., Karnes, J., O'Brien, P., Yasuda, H., Windebank, A., and Zimmerman, B.,** Capillary number and percentage closed in human diabetic nerve, *Proc. Natl. Acad. Sci. U.S.A.,* 52, 2513, 1985.
45. **Duncan, I. D. and Griffiths, I. R.,** Peripheral neuropathies of domestic animals, in *Peripheral Neuropathy,* 2nd ed., Dyck, P. J., Thomas, P. K., Lambert, E. H., and Bunge, R., Eds., W. B. Saunders, Philadelphia, 1984, 707.
46. **Dyck, P. J., Thomas, P. K., Lambert, E. H., and Bunge, R.,** *Peripheral Neuropathy,* 2nd ed., W. B. Saunders, Philadelphia, 1984.
47. **Berg, B. N.,** Muscular dystrophy in ageing rats, *J. Gerontol.,* 11, 134, 1956.
48. **Berg, B. N., Wolf, A., and Simms, H. S.,** Degenerative lesions of spinal roots and peripheral nerves in aging rats, *Gerontologia,* 6, 72, 1962.
49. **Gilmore, S. A.,** Spinal nerve root degeneration in aging laboratory rats: a light microscopic study, *Anat. Rec.,* 174, 251, 1972.
50. **Mitsumori, K., Maita, K., and Shirasu, Y.,** An ultrastructural study of spinal nerve root and dorsal root ganglia in aging rats with spontaneous radiculoneuropathy, *Vet. Pathol.,* 18, 714, 1981.
51. **Mitsumori, K., Maita, K., Nakashima, N., and Shirasu, Y.,** Electron microscopy of sciatic nerves in aging rats with spontaneous radiculoneuropathy, *Jpn. J. Vet. Sci.,* 48, 219, 1986.
52. **Thomas, P. K., King, R. H. M., and Sharma, A. K.,** Changes with ageing in the peripheral nerves of the rat; an ultrastructural study, *Acta Neuropathol.,* 52, 1, 1980.
53. **Krinke, G., Suter, J., and Hess, R.,** Radicular myelinopathy in aging rats, *Vet. Pathol.,* 18, 335, 1981.
54. **Spencer, P. S. and Thomas, P. K.,** The examination of isolated nerve fibres by light microscopy with observations on demyelination proximal to neuromas, *Acta Neuropathol.,* 16, 177, 1970.
55. **Krinke, G.,** Spinal radiculoneuropathy in aging rats: demyelination secondary to neuronal dwindling?, *Acta Neuropathol.,* 59, 63, 1983.
56. **Dyck, P. J., Oviatt, K. F., and Lambert, E. H.,** Intensive evaluation of referred unclassified neuropathies yields improved diagnosis, *Ann. Neurol.,* 10, 222, 1981.
57. **Sharma, A. K., Bajada, S., and Thomas, P. K.,** Age changes in the tibial and plantar nerves of the rat, *J. Anat.,* 130, 417, 1980.
58. **Grover-Johnson, N. and Spencer, P. S.,** Peripheral nerve abnormalities in aging rats, *J. Neuropathol. Exp. Neurol.,* 40, 155, 1981.
59. **Spencer, P. S. and Schaumburg, H. H.,** Central-peripheral distal axonopathy. The pathogenesis of dying-back neuropathies, in *Progress in Neuropathology,* Vol. 3, Zimmerman, H., Ed., Grune & Stratton, New York, 1976, 253.
60. **Griffiths, I. R. and Duncan, I. D.,** Age changes in the dorsal and ventral lumbar nerve roots of dogs, *Acta Neuropathol.,* 32, 65, 1975.

61. **Braund, K. G., McGuire, J. A., and Lincoln, C. E.,** Age-related changes in peripheral nerves of the dog. I. A morphologic and morphometric study of single-teased fibers, *Vet. Pathol.,* 19, 365, 1982.
62. **Braund, K. G., McGuire, J. A., and Lincoln, C. E.,** Age-related changes in peripheral nerves of the dog. I. A morphologic and morphometric study of cross-sectional nerve, *Vet. Pathol.,* 19, 379, 1982.
63. **Wheeler, S. J.,** Studies on the Equine Lateral Palmar Nerve, Ph.D. thesis, University of London, 1987; **Whiteley, S. J., Townsend, J., Tomlinson, D. R., and Brown, A. M.,** Fast orthograde axonal transport in sciatic motoneurones and nerve temperature in streptozotocin-diabetic rats, *Diabetologia,* 28, 847, 1985.
64. **Scaravilli, F.,** Modificazioni della struttura neuronale e delle popolazioni cellulari in rapporto all' invecchiamento, in *Alterazioni Del Nervo Periferico Nell' Eta' Avanzata,* Thomas, P. K., Ed., Abano Terme, 1988, 21.

Chapter 3

AGE-INDUCED CHANGES IN THE MAMMALIAN AUTONOMIC AND SENSORY GANGLIA

José A. Vega, Begoña Calzada, and Miguel E. Del Valle

TABLE OF CONTENTS

0-8493-6981-9/93/$0.00 + $.50

I. INTRODUCTION

In the peripheral nervous system, the neurons are clustered, forming the autonomic ganglia and the sensory ganglia. Two major divisions are currently considered in the autonomic ganglia: the sympathetic (prevertebral and paravertebral) and parasympathetic (cranial and sacral portions) ganglia. In addition, a third division must be considered: the "terminal" ganglia, which include the neurons which are found nearby or within their target organs, the most important being a complex set of neurons within the gastrointestinal tract, the so-called enteric nervous system. All these structures contain the ganglionic or postganglionic neurons of the autonomic nervous system. On the other hand, the sensory ganglia are formed of primary sensory neurons, and are localized in the dorsal root of the spinal nerves (spinal ganglia) or in some cranial nerves.

Mammalian aging is attended by a progressive decline of neuronal function which involves both the central and peripheral nervous system. But whereas the events which occur in the central nervous system with aging have been largely studied and is now well-enough understood, the aging process of the peripheral ganglia is poorly known, and most research has been focused on sympathetic ganglia. Conversely, studies on the aging-induced changes of the parasympathetic ganglia are almost nonexistent. With regard to the sensory ganglia, some classical studies are available, but only within the last decade have the aged primary sensory neurons been extensively studied. On the other hand, most of the studies devoted to the aging of the peripheral ganglia emphasize the neuronal changes, and the modifications which occur in the nonneuronal cells, i.e., the satellite glial cells, have been rather neglected.

Using structural, histochemical, immunohistochemical, and ultrastructural techniques, it was reported that aging of the peripheral ganglia is accompanied by changes in the number and packing of the neurons and in their normal appearance and ultrastructure, modifications of the neuronal biochemical and metabolic profiles, and variations in the levels of classical neurotransmitters, putative neurotransmitter molecules, or their synthesizing and degradating enzymes. Only recently have molecular biology methods been introduced to study the age-related changes of the peripheral ganglia cells. However, whether peripheral ganglia aging represents a universal pattern is not clear. In fact, most authors have shown that there are different and noncomparable age-related changes in the peripheral ganglia neurons, particularly among the sympathetic ganglia neurons of different animal species, and among different ganglia (prevertebral, paravertebral) of the same species; in this way, the morphological age-related changes in the autonomic ganglia seem to be more obvious in man than in other animals, and in the superior cervical ganglion than in other autonomic ganglia.[1] Moreover, it has not been established whether the neuronal variations induced by aging are related to the functional state, requirement differences, biochemical composition, neurotrophic molecules, or genetic alterations.

The autonomic and sensory ganglia of the peripheral nervous system are considered in this chapter with regard to age-induced changes.

II. SYMPATHETIC GANGLIA

A. NEURONS

1. Number and Size

Nerve cell loss is a known feature of the aging peripheral nervous system, although in the majority of cases it is difficult to appreciate and quantify[2] since the population of sympathetic neurons cannot be regarded as a stable entity.[3]

In the rat, the number of neurons in sympathetic ganglia is established in the first month postpartum,[4] and during postpartum maturation there is an increase in neuron diameter and in that of their nuclei,[4,5] while little further changes occur to adulthood.[6] However, in progressing from the adult state to old age, there is a significant reduction in neuron density.[7,8]

The density of neurons significantly decreases with aging in the coeliac-mesenteric ganglion complex of F344 rats, but not in the superior cervical ganglia.[8] A reduction in the number of postganglionic neurons was also found in the hypogastric ganglia of aged Wistar rats.[8] Moreover, it has been reported[3] that neuronal packing density declines dramatically between 12 and 18 months of age in the rat coeliac-superior mesenteric ganglia, whereas in the cervical superior ganglion, the packing density decreases between 18 and 24 months of age. During these periods, a significant increase occurs in the neuron somata area in the coeliac-superior mesenteric ganglia which is much less marked in the superior cervical ganglion, accompanied by a decline in the nuclear area in the coeliac-superior mesenteric ganglia between 18 and 24 months, and 12 and 18 months of age, in the cervical superior ganglion.

In man, only little and discordant data on the neuron number of the autonomic nervous system in relation to aging have been provided. These include severe cell depletion in the cervical sympathetic ganglia[10] and decreased numbers of neurons in the intermediolateral column of the spinal cord, but not in those of the ganglia.[11] Based on indirect features, it was also suggested that the number of neurons in the superior cervical ganglia remain unchanged with aging.[12] However, in human lumbar sympathetic ganglia, we have observed a readily apparent decrease in both the number and packing of neurons, and some of the remaining neurons showed signs of cytolysis (Figure 1A). Moreover, in these ganglia, accumulation of nuclei forming Nageotte's-like bodies were frequently found. These structures are formed by residual satellite glial cells, react positively with antibodies against S-100 protein (Figure 2C, D), and represent an indirect evidence of neuron loss (Figure 1D, E).

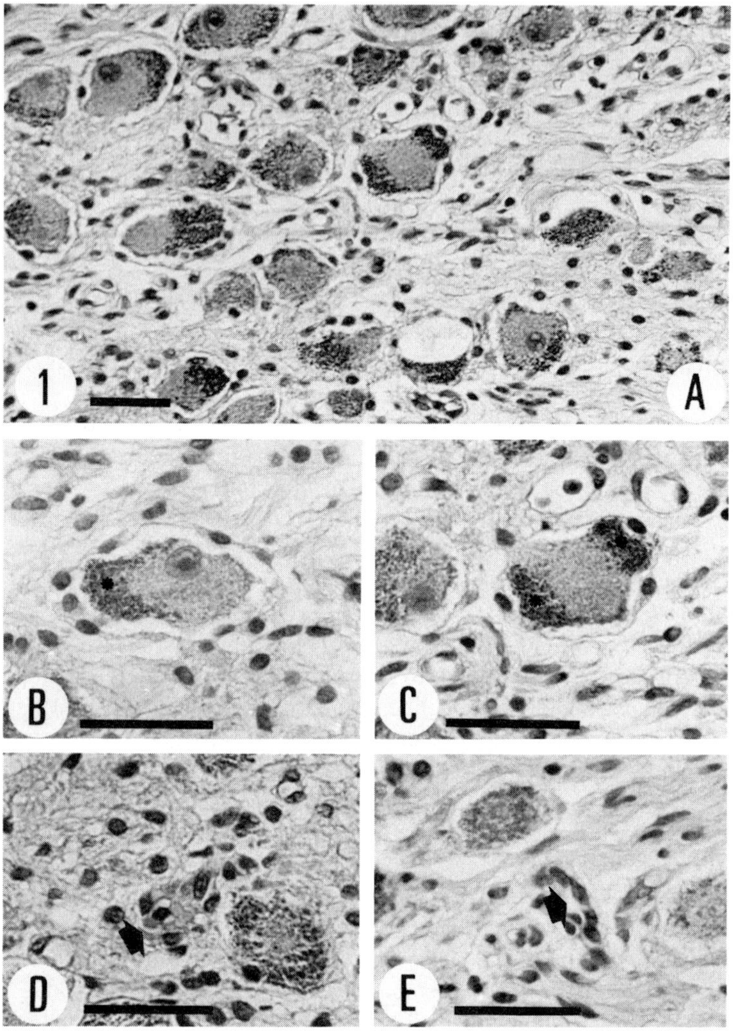

FIGURE 1. Lumbar sympathetic ganglia from a 78-year-old man. All neurons contain large amounts of lipofuscin, and some of them show signs of chromatolysis (A). Lipofuscin granules primarily are concentrated polarly within the perikarya (B and C, asterisks). Moreover, rather abundant clustered nuclei of satellite glial cells forming Nageotte's bodies are present (D and E, arrows). (Hematoxylin-eosin; bar = 50 μm.)

2. Ultrastructure

The ultrastructural changes reported in aged neurons of the sympathetic ganglia include alterations in normal components of the neuron body and the appearance of new structures not typical of younger neurons.

Although most neuronal organelles changed with aging, the more evident changes occurred in the mitochondria, rough endoplasmic reticulum, and

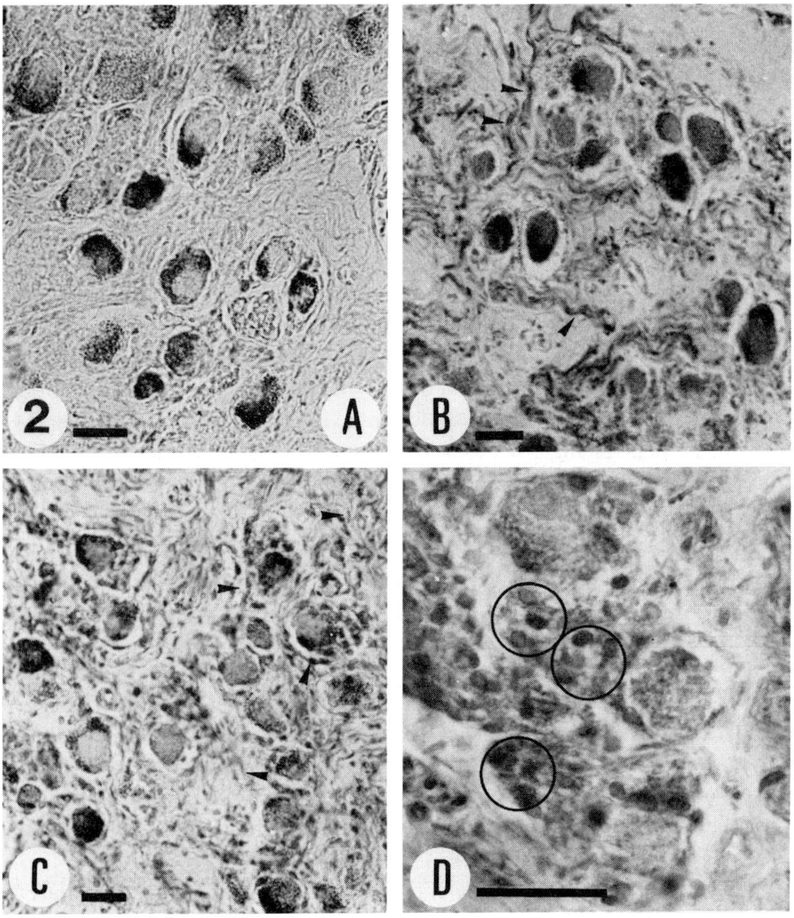

FIGURE 2. Lumbar sympathetic ganglia from an 85-year-old man. The lipofuscin granules of all sympathetic neurons show the histochemical characteristics of melanin (A, Masson-Fontana staining), and some of them display a light immunoreactivity for neurofilament protein triplet, with positive results in nerve fibers (B, arrowheads). An antibody against S-100 protein selectively labels satellite glial cells and Schwann cells (C, arrowheads), and accumulations of S-100 immunoreactive cells, resembling Nageotte's bodies, are frequently found (D, circles). (Bar = 50 μm.)

Golgi apparatus. Pick and co-workers[13,14] showed cytoplasmic residual bodies and altered mitochondria in the sympathetic ganglia of old men as well as the occurrence of large numbers of lamellar bodies. Moreover, various inclusion bodies included a cylinder-shaped type that had a varying pattern of rod-like structures inside it. Myelin figures (laminar bodies) were sometimes found to fill neurite profiles, occasionally with random mitochondrial accumulations.[15]

Similarly, in the coeliac-mesenteric ganglion of the rat, a large accumulation of residual bodies, distorted or swollen mitochondria, and disarrangement and dilatation of the rough as well as the Golgi apparatus cisternae[16] were observed.

On the other hand, Schmidt et al.[17] found significant specific, age-related neuropathologic lesions in the human superior mesenteric ganglia which were much more frequent in the paravertebral superior cervical ganglion. The more evident feature was the presence of markedly swollen dystrophic preterminal axons which compressed or displaced the perikarya of principal neurons. Ultrastructurally, these axons contain abundant disoriented neurofilaments surrounded by peripherally margined dense core vesicles.

We have also studied the ultrastructural aspect of the paravertebral (cervical superior) and prevertebral (coeliac-mesenteric) sympathetic neurons of Wistar male rats 28 to 30 months of age. In the superior cervical ganglia, the more evident changes were accumulation of dense bodies with different osmiophilic properties, presence of altered mitochondria, and apparent decrease and dilation of the rough endoplasmic reticulum as well as of the Golgi apparatus. Occasionally "myelin figures" or laminar bodies were encountered within the perikarya. However, the ganglionic fibers showed a normal aspect (Figure 3). Similar, but much more apparent alterations were observed in the coeliac-mesenteric ganglion neurons, although the accumulation of lipopigment-dense bodies was higher than in the cervical superior ganglion, and showed a heterogeneous composition. In these ganglia, the preganglionic nerve fibers have a normal appearance, containing a large number of synaptic vesicles, whereas the axons of the ganglionic neurons contain disarranged neurofilaments and sometimes dense bodies (Figure 4). In addition, in all the sympathetic ganglia studied, there was a remarkable increase of connective tissue.

An interesting structural feature of the aged sympathetic ganglia is the presence of neurofibrillary tangles, as in the central nervous system neurons. They consist of argyrophilic and congophilic intracytoplasmic structures which, ultrastructurally, appear as paired helical filaments wound in a double helix.[18] The neurofibrillary tangles react positively with antibodies against τ-protein, neurofilament epitopes, microtubule-associated proteins II and V, ubiquitin, and casein kinase II (for references, see Reference 19), but the exact composition of these structures remains unclear. The neurofibrillary tangles seem to be exclusively present in aged sensory and/or sympathetic neurons, and not in motoneurons.[20] Paired helical filaments have been found in the superior cervical ganglion of an aged man having similar ultrastructure and the same immunohistochemical properties as similar structures in the central nervous system.[21] However, the occurrence of neurofibrillary tangles in sympathetic ganglia of other mammalian species has not been reported. To demonstrate the accumulation of neurofilament proteins in the human sympathetic ganglia, we studied immunohistochemically the lumbar sympathetic ganglia of an 85-

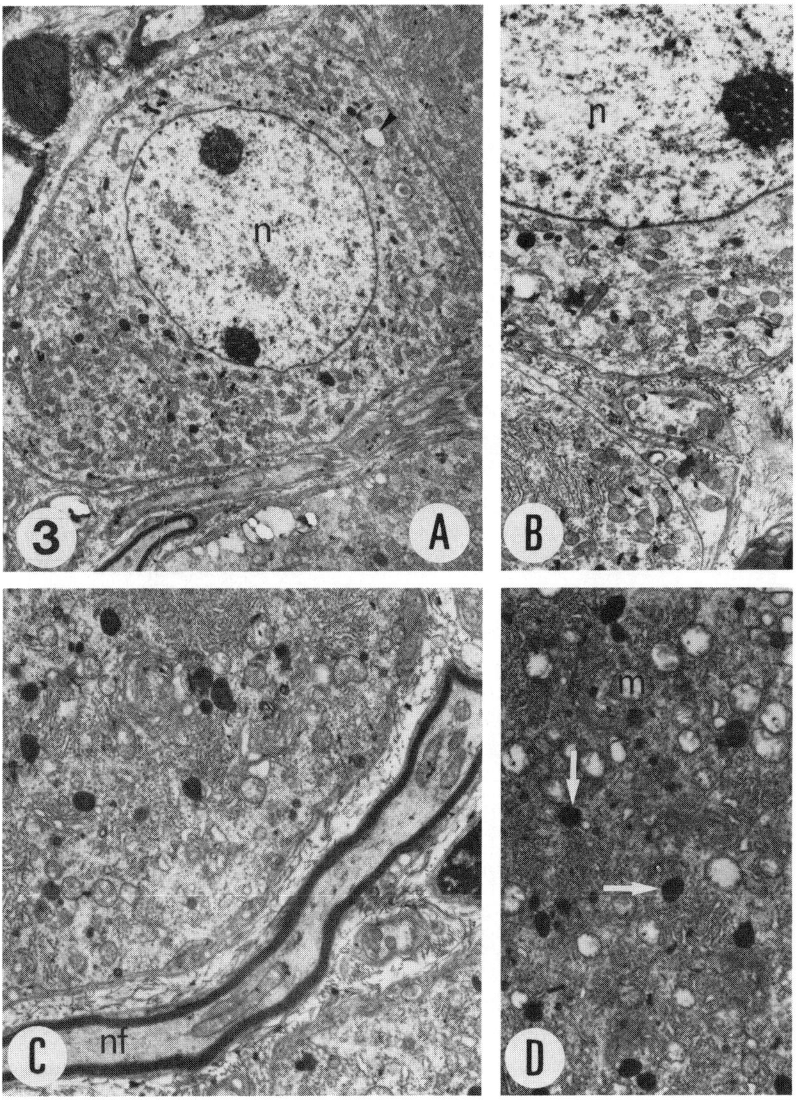

FIGURE 3. Cervical superior sympathetic ganglion from a 30-month-old Wistar male rat. Neurons contain scarce dense bodies with different osmiophilic characteristics (C and D, arrows), the mitochondria are severely altered, and the rough endoplasmic reticulum and Golgi apparatus dilated. The nerve fibers have a normal appearance. m, mitochondria; n, nucleus; nf, nerve fiber. (A, original magnification × 3600; B–D, original magnification × 8400.)

year-old patient showing a strong positive immunoreactivity in nerve fibers and weak positive immunoreactivity in some perikarya (Figure 2B), which was masked by the great amounts of lipopigment.

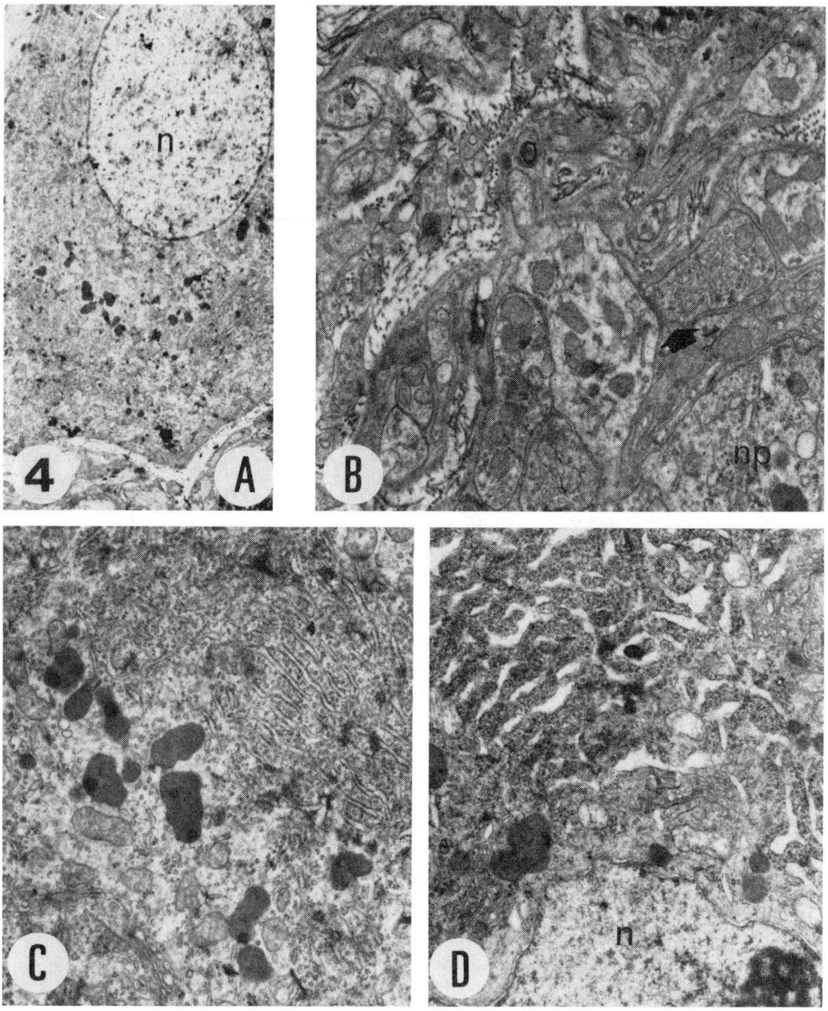

FIGURE 4. Coeliac-mesenteric ganglion from a 30-month-old Wistar male rat. The neuronal cytoplasm contains abundant dense bodies with different patterns of density (C, D). The rough endoplasmic reticulum and the cisternae of the Golgi apparatus are enlarged and dilated, and the mitochondria are altered. Preganglionic nerve fibers show a normal appearance (B, arrow), whereas disarrangement of neurofilaments is evident in the postganglionic ones. n, nucleus; np, neuron perikarya. (A, original magnification × 3600; B–D, original magnification × 14,400.)

3. Lipopigments

Accumulation of lipopigments, primarily lipofuscin, in the neurons is one of the more consistent cytologic changes correlated with mammalian aging (Figures 1 and 2A). Lipopigments are complex condensation products of proteins, phospholipids, and the products of lipid peroxidation.[22]

The increase of lipopigment with age in peripheral neurons has been known for a long time, being considered as a sign of oxidative damage, and it has also been reported that the pigment formation might be considerably increased in rodents fed vitamins or reared under hypoxic conditions.[23]

Lipofuscin formation appears to be a continuous process, increasing progressively with age.[24-27] The age-related increase of lipopigment in human sympathetic ganglia has been documented morphologically both *in vivo*[14,15] and *in vitro*,[27] where it accumulates in a linear fashion.[7,28-30]

In addition to lipofuscin, melanin-like pigment is also present in some nerve cells and also accumulates with age. However, the neuromelanin-like component in the pigment bodies was found exclusively in the adrenergic neurons (not in the noradrenergic ones), indicating that age-related lipopigment accumulation has a differential and selective pattern, although in the rat the rate of lipopigment accumulation was greater in the noradrenergic than in the adrenergic sympathetic neurons.[31]It has been suggested that the gradual melanization of the lipopigments may be due to an auto-oxidation of catecholamines and that the accumulation of the melanized forms of lipopigments may be a sign of advanced cellular trauma.[7,8]

Studies on neuron lipopigments have demonstrated that chemical, staining, fluorescence characteristics, and ultrastructural differences exist between lipopigments from young-adult and old mammalian neurons.

Lipofuscin pigment varies in staining properties in young-adult and old animals. In the former, they are Sudan black and PAS positive, whereas pigment in older animals is more easily stained with Nile blue and ferric-ferricyanide methods. The chemical properties of the two types of neuronal lipofuscin also differ: the early form of the pigment could be more easily digested by lipase and trypsin and markedly dissolved in lipid solvents, whereas the late form was more resistant to such treatments.[32] In addition, the lipopigments of human aged sympathetic neurons are intensely argyrophilic (Figure 5).

The spectral characteristics of the lipopigment autofluorescence also change with aging, suggesting the accumulation of additional components in the residual bodies.[7,8,30] The mammalian sympathetic ganglia contain two types of cells exhibiting formaldehyde-induced fluorescence: the adrenergic neurons and the SIF (small, intensely fluorescent) cells. The adrenergic neurons accumulate pigment granules which have a specific specter of fluorescence and increase with age, while the formaldehyde-induced fluorescence of the neurons decreased, and in the sympathetic ganglia of an older man, most neurons showed pigment fluorescence but not formaldehyde-induced fluorescence.[33] Similar changes are obvious in the adrenergic neurons of the hypogastric ganglion of male rats, where the fluorescence intensity of the neurons decreases with aging and the number of completely nonfluorescent adrenergic neurons increases, these changes being concomitant with elevated nonspecific fluorescence due to lipofuscin pigment.[34]

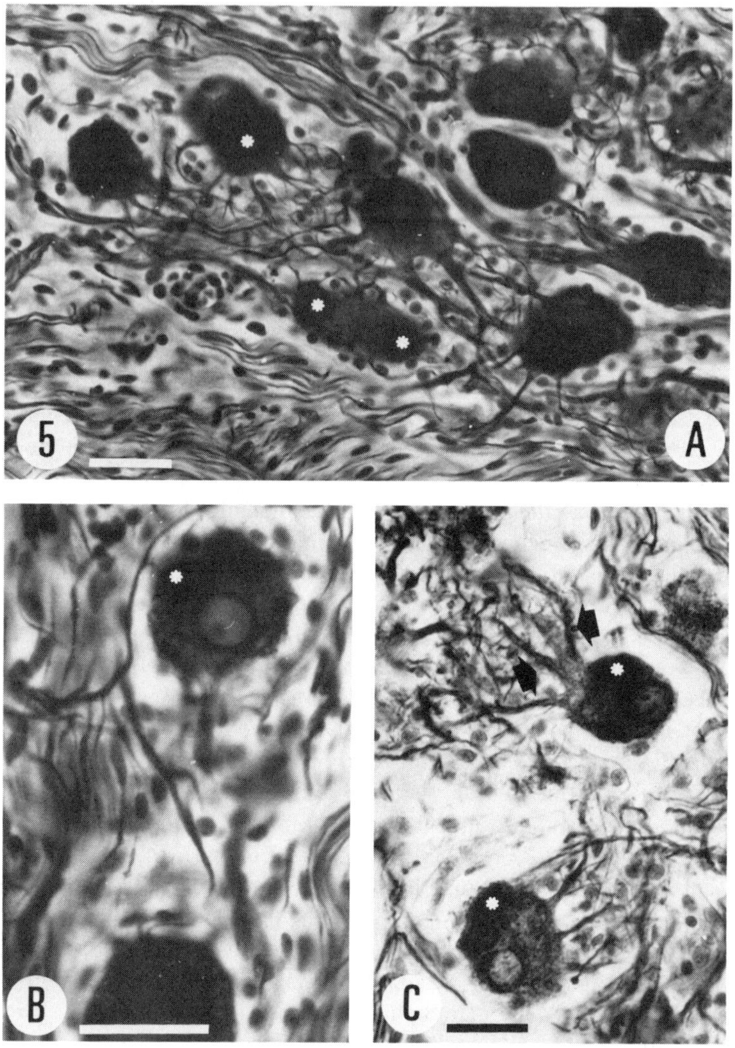

FIGURE 5. Mesenteric ganglia from an 85-year-old man. The lipopigment shows a very strong argyrophilia, and is placed polarily within the perikarya (asterisks). Furthermore disarrangement of nerve cell processes are rather frequent (C, arrows). (Silver impregnation, Tolivia's method; bar = 50 μm.)

Using the glyoxilic acid fluorescence method, Santer et al.[16] studied the coeliac superior mesenteric ganglia of aged rats, showing that the noradrenaline fluorescence of the majority of the neuronal perikarya was very low or absent, and few intraganglionic fluorescent varicosities were seen. By contrast, lipofuscin pigment was very prominent at the nuclear pole regions of neurons and also in dendrites and axonal processes.

Two types of lipopigment have been identified in the human sympathetic ganglia using fluorescence techniques: yellow (lipofuscin-like pigment) and orange or dark (melanin pigment), the yellow pigment being typical of younger individuals, and the orange of the aged ones.[7,8,28-30,35,36] With aging, there is an increase in lipopigment autofluorescence in the human stellate ganglion and superior mesenteric ganglion, changing from yellow to orange, probably due to an accumulation of neuromelanin in adrenergic neurons,[35,36] although the neuromelanins are habitually regarded to be nonfluorescent.[37]

A detailed ultrastructural study conducted by Helen[15] on human sympathetic ganglia revealed that the equivalents of the two classical types of lipopigments show clear differences in their fine structure. The yellow pigment corresponded to a neuronal area containing residual bodies and vacuoles of varying sizes, while the orange pigment was associated with a more complex and large pigment inclusion showing an osmiophilic substructure. At the ultrastructural level, pigment bodies were seen to be composed of three different kinds of osmiophilic properties, consisting of a gray component that had dark patches dispersed into it, and pale, oval, incorporated droplets. In principal ganglion cells, the first two formed the major part, the pale one taking over in the small granule-containing cells.

In aged Wistar rats, the age pigments in the sympathetic neurons showed a tendency to form unipolar or bipolar caps. Ultrastructurally, lipid-like vacuoles and a rather homogeneous matrix were the components shared by pigment bodies; however, frequently there was a third osmiophilic and argyrophilic component which likely represents neuromelanin. The appearance of clusters of highly osmiophilic material throughout the pigment bodies correlated with the orange autofluorescence, and has been considered to be an accumulation of products from different degradation processes and of melanized forms of lipopigments in the original lipofuscin-like granules.[7,8,30] In confirming this opinion, Koistinaho et al.[36] observed in cultured sympathetic neurons that classical age pigment maintains its characteristics, and there is, in addition, a rapid accumulation of ceroid-like pigments which may be caused by the inability of the cultured neurons to cope with increased peroxidative damage.

Since sympathetic neurons are adrenergic, it has been proposed that autooxidation of catecholamines may produce neuromelanin, and since the precursors of neuromelanin are toxic, the orange pigment could represent a sign of advanced cellular damage.[38] Thus, the shift in the pigment type which occurs with aging may be related to changes in cellular metabolism which ultimately may lead to cell death.

4. Catecholamines and Catecholamine-Related Enzymes

Most data on the catecholamine content of aged sympathetic neurons have been obtained from histochemical studies using the formaldehyde-induced fluorescence technique, since catecholamine fluorescence of sympathetic neuron perikarya is an indicator of the cytoplasmic catecholamine concentration.

It is generally assumed that the neuron catecholamine content declines with aging, but the evidence is rather controversial and large differences exist among species or in different ganglia of the same animal species.

Harkonen and Pentilla[39] and Hervonen et al.[33] first described low levels or loss of histochemically demonstrable catecholamines with aging in human sympathetic ganglia, and the rate of synthesis of catecholamines is decreased in the human superior cervical and hypogastric ganglia.[40] These results were later confirmed[8] in human sympathetic ganglia obtained from sympathectomies and postmortems, where it was found that a decline in the cytoplasmic content of noradrenaline, and an increase in the intensity of formaldehyde-induced fluorescence between individual neurons, coincides with large variations in tyrosine hydroxylase activity. However, human aging is generally associated with a progressive increase of plasma catecholamines, mainly noradrenaline. The mechanism underlying this increase is not apparent and appears to be in conflict with evidence based on the tissue noradrenaline levels, which were found to be lower, unchanged, or slightly elevated (for references, see Reference 41).

In rat sympathetic ganglia, there is a decline of the formaldehyde-induced fluorescence for catecholamines in the hypogastric ganglion[8,9,34] and the coeliac-superior mesenteric ganglia,[8,16,42] but the catecholamines remain unmodified in the cervical superior ganglion.[8,9,34] This is in agreement with results obtained from human tissues.

Contrary to the aforementioned studies, Buu and Debinski[43] found in the rat superior cervical ganglia that dopamine and noradrenaline progressively increase with aging. Moreover, after L-DOPA administration, older rats accumulated more dopamine than younger ones, but in both age groups the dopamine increase was not accompanied by any change in noradrenaline, and the 3,4-dyhydroxyphenylacetic acid/dopamine ratio was slightly smaller in older rats. The significance of a dopamine and noradrenaline increase in the sympathetic ganglia of older rats is not known.

It remains unclear whether the decrease and/or very low levels of noradrenaline fluorescence in many aged sympathetic neurons is due to a cessation of catecholamine synthesis or to a reduced level of synthesis, perhaps accompanied by a faster turnover of any catecholamines produced. Santer[42] has observed that monoamine oxidase (MAO) inhibition increased the reduced noradrenaline levels in aged sympathetic neurons of the coeliac-mesenteric ganglion complex, whereas these were not restored after reserpine administration. These results suggest that the low neuron noradrenaline content is due to a decreased synthesis of noradrenaline,[42] possibly accompanied by a rapid transport of the newly synthesized noradrenaline to peripheral nerve terminals.[34] Moreover, indirect data suggest an impairment of the uptake mechanisms and/or storage of noradrenaline in aged sympathetic neurons and their axonal and dendritic processes.[16]

Results obtained from measurement or histochemical detection of catecholamine-synthesizing enzymes, such as tyrosine hydroxylase and dopamine-β-hydroxylase, in aged sympathetic neurons do not fully explain the variations found in neuron catecholamine levels because differences in enzyme activities often do not correspond to changes in catecholamine content.

Reis et al.[44] studied several enzymatic activities related to the metabolism of catecholamines in the sympathetic ganglia of mice and rats. In F344 aged male rats, there was an elevation of tyrosine hydroxylase and aromatic L-amino acid decarboxylase activities, whereas the dopamine-β-hydroxylase activity does not change. Conversely, in the ganglia of CB6F1 aged male mice, no changes in any of the investigated enzymatic activities were detected. Increased tyrosine hydroxylase activity in rat sympathetic neurons has been also reported by McLean et al.[45]

Partanen et al.[40] analyzed the activities of tyrosine hydroxylase and dopamine-β-hydroxylase in sympathetic ganglia of aged F344 rats showing no age differences in dopamine-β-hydroxylase activity of any examined ganglia, or in the activities of any enzymes in the coeliac-mesenteric ganglion complex. However, tyrosine hydroxylase activity was significantly higher in the superior cervical and hypogastric ganglia of aged rats. On the other hand, these authors observed that recoveries of reserpine-depleted catecholamine stores in the superior cervical and hypogastric ganglia were slower in aged than in young rats, and that catecholamine loss following inhibition of synthesis and intraneuronal degradation was faster in the superior cervical ganglia, but not in the hypogastric ganglia, compared with young rats. These findings suggest that neurotransmitter synthesis and release are enhanced with age in the superior cervical ganglion. The lack of age-related changes in the hypogastric ganglion might reflect the different cellular composition as well as physiological role of this ganglion.

5. Other Neurotransmitters

In addition to catecholamines, neurons of the mammalian sympathetic ganglia contain enzymes related to the metabolism of acetylcholine as well as a number of biologically active neuropeptides.[46,47]

With regard to acetylcholine metabolism, Reis et al.[44] showed an increase in the activity of choline acetyltransferase in the cervical superior ganglia of F344 aged male rats, whereas Hervonen et al.[33] and Partanen et al.[40] found unchanged patterns of choline acetyltransferase and acetylcholinesterase activities in the aged neurons of the human and rat coeliac-mesenteric ganglion complex and hypogastric ganglion, while choline acetyltransferase activity was increased in the human superior cervical ganglion.

Enzymatic activities related to other putative neurotransmitters such as glutamic acid have also been analyzed in the rat superior cervical ganglia. Results showed that both aspartate aminotransferase and glutamate dehydrogenase, the major glutamate-metabolizing enzymes, did not change during the life span.[48]

Data regarding age-related changes in the neuropeptide content of sympathetic ganglia are very scarce. The enkephalin-like peptide content is low in the superior cervical and coeliac ganglia of aged rats.[40] In human sympathetic ganglia, there is an age-related decrease in the proportion of neurons innervated by enkephalin-containing fibers, which do not contain osmiophilic pigment, whereas neurons containing neuromelanin were not innervated by enkephalinergic fibers, which leads to speculation that enkephalinergic preganglionic fibers play a role in maintaining the neuronal integrity of the sympathetic neurons.[50] An age-related decrease also occurs in the neuron content of neuropeptide Y in the superior cervical ganglia of rats, whereas the number of immunoreactive neurons remains unchanged.[51] Finally, immunohistochemical studies demonstrate that dystrophic axons of human sympathetic ganglia contain neuropeptide Y-like, but not other putative neuropeptides.[17]

6. Metabolic and Molecular Features

Few studies regarding the metabolic age-related changes of sympathetic neurons are available, and results largely differ among ganglia and animal species. It has been observed that glucose utilization increased significantly between 12 and 30 to 33 months in the superior cervical ganglion of male F344 rats, and between 3 and 24 months in the coeliac-mesenteric ganglion complex. However, no significant age-related changes were observed in the hypogastric ganglion, suggesting that functional activity in sympathetic ganglia increases in older rats, although end-organ effects may be reduced.[52] Moreover, a reduced metabolic activity was found in the coeliac-superior mesenteric ganglia, whereas the paravertebral ganglia have high metabolic activities,[52-54] and in the bovine superior cervical ganglion, the activity of dihydrofolate reductase gradually decreased with aging and the folic acid concentration increased.[55,56] Furthermore, the neurons of the superior cervical sympathetic ganglia of old rats are able to synthesize normal amounts of protein.[57]

Age-related changes also occur in the expression of some genes (e.g., C-*fos*), which encodes a nuclear phosphoprotein and presumably is related to neuronal plasticity.[58,59] The expression of C-*fos* increases with age in rat superior cervical ganglia neurons, which suggests that it is involved in the regulation of genetic events associated with adaptive changes in the neuronal activity of the sympathetic ganglion during aging.[60] Thus, it may be assumed that aging causes an impairment in neuronal plasticity, and indirect data indicate that although remarkable plasticity seems to persist in mature neurons, it may be deficient in aged sympathetic neurons.[61]

Wu et al.[62] have observed a reduced age-dependent decline in the capacity of sympathetic ganglia to generate long-term changes in synaptic efficacy; thus, there exists an impairment of the peripheral synaptic plasticity, but the ganglia have the ability to grow in culture.[63] Recently, it has been reported that in the superior cervical ganglion of aged F344 rats, there is a delay and an enhancement in the expression of a synaptic vesicle protein (p65), but not a drop to values observed in adult animals, which leads one to think that while aged sympathetic neurons retain the ability to respond to alterations in sympathetic activity, they are unable to regulate the activity once a response is initiated.[64]

On the other hand, it seems that neurotrophic molecules play an important role in the events of neuronal aging. It is known that sympathetic neurons are dependent upon nerve growth factor (and probably other neurotrophic molecules[68]) for survival and maintenance,[69] although only a subpopulation of human sympathetic ganglia express nerve growth factor (NGF) receptors.[70] In the superior cervical ganglia of aged rats, there is a loss of both high- and low-affinity components of NGF receptors, and this decrease may lead to the diminished responsiveness to NGF in aged sympathetic ganglion neurons,[71] including changes in ganglionic transmission, since NGF seems to act as an inhibitory neuromodulator of catecholaminergic transmission.[72]

B. SIF CELLS

SIF cells are localized habitually within the connective tissue of the ganglionic capsule, or even clusters partially within the ganglia. Two main types of SIF cells are considered: type I cells are interneurons, whereas type II may be purely endocrine cells.[73] (For a detailed study of mammalian SIF cells, see Reference 74.) The SIF cells contain biologically active substances which can act as neurotransmitters. These include catecholamines, mainly noradrenaline and dopamine,[75,76] and, in some animal species, serotonin.[77] Neuropeptide Y, substance P, calcitonin gene-related peptide, opioid peptides, and CCK-8 have been observed immunohistochemically in SIF cells (for references, see Reference 78), but only Partanen et al.[34] have studied aged SIF cells. They found in aged rats two types of SIF cells which emitted greenish-yellow and yellow-yellow formaldehyde-induced fluorescence, respectively. The former was present in adult and young animals, whereas the latter was predominant in the older ones. Moreover, according to the size of the densely cored vesicles, two kinds of SIF cells were distinguished in the hypogastric ganglion of newborn and young adult rats, whereas a third type of SIF cell was found in aged rats.

C. SATELLITE GLIAL CELLS

The satellite glial cells surrounding the principal sympathetic neurons are also affected by aging. An increase in nonspecific fluorescence due to the accumulation of lipofuscin pigment,[34] which may be related to the removal of cellular debris from neurons, including lipofuscin pigment, has been reported in these cells.[79] However, our data from old Wistar rats show that satellite glial cells of sympathetic ganglia do not experience ultrastructural changes in their organelles with aging, contain only scarce dense bodies (Figures 3 and 4), and in human ganglia, retain their immunohistochemical characteristics (Figure 2C).

The only data available regarding the functional state indicate that the metabolism of folic acid in the satellite cells of the cervical superior ganglion of aged animals preserved or slightly increased the dihydrofolate reductase activity, but folic acid markedly increased in senescence.[56]

III. PARASYMPATHETIC GANGLIA

A. NEURONS

1. Number and Size

Data on age-induced changes in the number and size of parasympathetic ganglion neurons are almost nonexistent, probably due to the difficulty in isolating these ganglia and the heterogeneity of neurons in mammalian parasympathetic ganglia which have been considered as mixed (sympathetic/parasympathetic) vegetative structures.

The ciliary ganglia of aged rats do not differ significantly from those of young rats with respect to the size of the ganglion cell population, number of inputs per ganglion cell, density of synaptic buttons, and amplitude of the postsynaptic potentials recorded intracellularly.[81] However, in man, the neuronal counts remained stable up to 50 years, after which the number of neurons declined notably.[82]

2. Ultrastructure and Lipopigments

Since no data are available regarding the ultrastructure of the parasympathetic neurons of old mammals, we have analyzed the ciliary ganglia of Wistar rats 28 to 30 months old. In these neurons there is a scarce number of small dense bodies, but the occurrence of ''myelin figures'' was rather frequent. Moreover, some neurons showed alterations of the mitochondria and enlargement of the rough endoplasmic reticulum and Golgi apparatus. On the other hand, nerve fibers surrounding the perikarya showed disarrangement of neurofilaments, inclusion bodies, and ''myelin figures'' as well as large axoplasmic vacuoles (Figure 6). No remarkable changes were encountered in glial cells.

3. Neurotransmitters and Related Enzymes

The cranial parasympathetic ganglion neurons produce and release acetylcholine and a large number of neuropeptides, including several considered typical of those in sympathetic or sensory neurons. Furthermore, tyrosine hydroxylase immunoreactivity has been found in a subpopulation of cranial parasympathetic ganglia.[83,84]

No data are available, to our knowledge, on the age-induced changes in neurotransmitters or their related enzymes of the cranial or sacral parasympathetic ganglia of mammals. However, in the avian ciliary ganglion, the activities of both choline acetyltransferase and acetylcholinesterase are reduced at later ages, and the endogenous levels of choline and acetylcholine progressively decrease from 1 to 7 years of age. The aging period, or ''synaptic regression period'', is characterized by a progressive and marked decrease of both choline and acetylcholine as well as a reduction in the choline/acetyl-

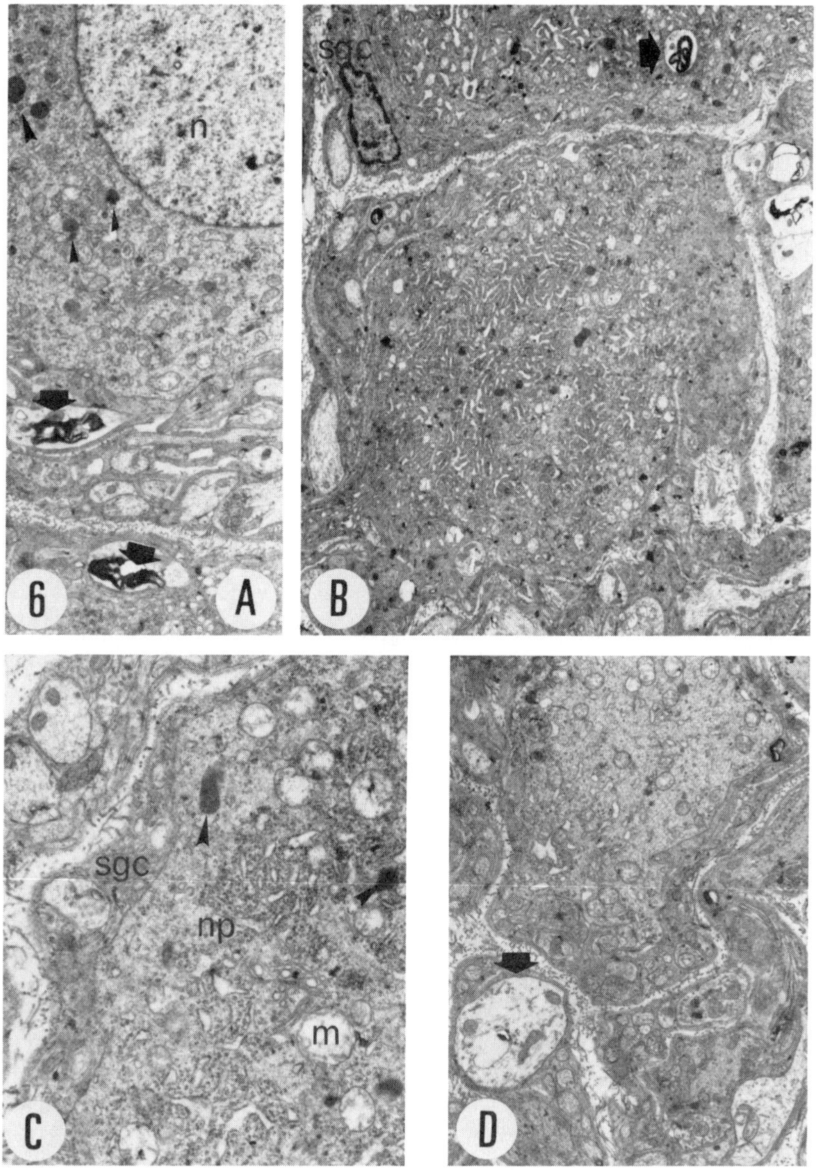

FIGURE 6. Ciliary ganglion from a 28-month-old Wistar male rat. The number of dense bodies present in parasympathetic neurons is rather scarce (A and C, arrowheads), and occasionally ''myelin figures'' can be seen (B, arrow). Mitochondrial alterations, and enlargement of the rough endoplasmic reticulum and Golgi apparatus are the most remarkable features. By contrast, severe alterations occur in the nerve fibers (A and D, arrows). m, mitochondria; n, nucleus; np, neuron perikarya; sgc, satellite glial cells. (A and D, original magnification × 8400; B, original magnification × 3600; C, original magnification × 14,400.)

choline ratio. Moreover, the number of nicotinic cholinergic receptor binding sites remains virtually unchanged with aging.[85-87]

IV. SENSORY GANGLIA

A. NEURONS

1. Number and Size

Evidence of nerve loss with aging has been reported in the sensory ganglia, but remarkable discrepancies exist among authors. The dorsal root ganglia and sensory ganglia of the cranial nerves contain two main, normally distributed subpopulations of primary sensory neurons, the so-called large-light (type A) and small-dark (type B) neurons.[88]

In the spinal ganglia of aged men, signs of neuronal loss could be identified by the presence of compact residual satellite cells, the so-called nodules of Nageotte.[89] However, studies conducted in human sensory ganglia revealed both no changes[90,91] and decreased[92,93] number of sensory ganglion cells with age.

Aldskogius and Risling[94] reported a neuron loss of about 20 to 25% in the L_7, spinal ganglia of kittens, while in rats the difference was as high as 45%.[95]

In a recent study, we analyzed age-related changes in the number of primary sensory neurons of 3- to 27-month-old male Wistar rats. No significant changes were encountered in comparing young and adult rats (12 months old), but there was a marked decrease in the older animals of about 18 to 22%. Moreover, the mean size of the remaining neurons was smaller than in adult and young animals. This may be due to the fact that the loss of neurons mainly affects the large-light neurons, while the decrease of the small-dark neurons was not evident.[131]

2. Ultrastructure

The ultrastructural aspect of the aged sensory neurons has been extensively studied by Van der Bosch de Aguilar and Vanneste.[96-98] They observed that large-light neurons are impaired early and more severely than the small-dark neurons with aging, and distinguished two phases of neuronal changes. During the first period, beginning at about 3 months, lipofuscin pigments gradually accumulated in the neurons. During the second period, beginning at about 24 months, various kinds of inclusions appeared (granulofilamentous bodies, nematosomes) which can originate from altered microtubules and/or filaments. At the same time, lipofuscin pigments became more heterogeneous. Furthermore, they found an "unusual" accumulation of substances in the senescent neuron mitochondria as well as tubular or filamentous inclusions, modifications in matrix density, deposits of glycogen particles, and electron-dense globules. In addition, some neurons contained neurofilament tangles consisting of tangled masses of argyrophilic fibrous elements coursing in large bundles or in irregular strands throughout the neuron perikaryon and extending into the proximal parts of the cell processes.

Recently, we[131] observed in aged sensory ganglia a great increase in the conjunctive tissue, and within the neuronal cytoplasm, numerous and large electron-dense bodies which do not have an homogeneous pattern of density (Figure 7A). Most of them contained highly osmiophilic droplets, and occasionally a laminar organization was observed (Figure 7B). Moreover, the nucleus became irregular and the chromatin lacked their typical organization (Figure 7C), and there was an enlargement and dilation of the Golgi apparatus and rough endoplasmic reticulum which led to the formation of cytoplasmic vacuoles, alterations and abnormal accumulations of mitochondria, abundant laminar bodies and "myelin figures" (Figure 7E), and the neurofibrillar network results were more evident than in adult animals (Figure 7B, D). On the other hand, both free and coupled ribosomes are diminished, and those remaining tend to form densely packed bodies localized perinuclearly.

3. Lipopigments

As reported in both sympathetic and parasympathetic ganglia, age-dependent changes occur in the lipopigment content of sensory neurons. Lipofuscin is present at an early stage in sensory ganglia,[88] but progressive accumulation in large amounts is the most obvious change observed in the sensory neurons of aged rats,[96-99] and in addition to lipofuscin, melanin-like pigment is also present in the trigeminal and spinal ganglia neurons of man[12,100] (Figure 7B).

The age pigments in the vagus and spinal ganglion neurons of Wistar rats are packed in the peripheral area of the perikarya during aging, and ultrastructurally display characteristics identical to those observed in the sympathetic ganglia.[36]

4. Cytoskeletal Proteins

The neuronal cytoskeleton is composed of an extensively cross-linked network of three fibrillar components: microtubules, neurofilaments or intermediate filaments, and microfilaments. It is well established that the cytoskeleton is the major internal structure that defines the morphology of neurons[101] and plays a crucial role in the transport of molecules from their site of synthesis in the perikarya to the end of the axons.[102] Knowledge of their assembly, organization, and metabolism is essential to understanding the molecular basis of neuronal form and function. Although all these aspects are well documented, very little information is available on the age-induced changes in the cytoskeleton of peripheral neurons. However, as mentioned in Sections II.A.2[18,20,21] and IV.A.2,[96-98] it is evident that aging produces important modifications in the normal metabolism and amounts of neuronal cytoskeleton constituents.

The intermediate filaments (neurofilaments) of the neuronal cytoskeleton consist of three subunits with approximate molecular weights of 68, 160, and

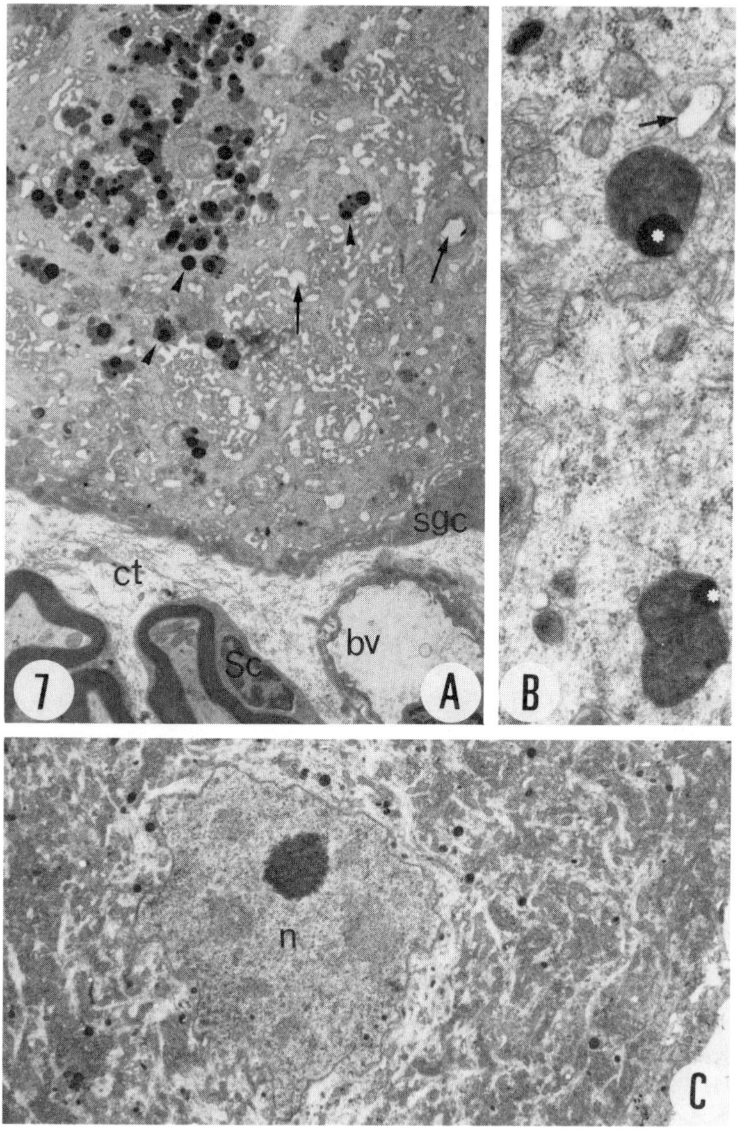

FIGURE 7. Lumbar (L_6) dorsal root ganglion from a 30-month-old Wistar male rat. Large and heterogeneous dense bodies are present in the perikarya of aged sensory neurons (A, arrowheads). The highly osmiophilic parts of the dense granules are localized in the periphery (B, asterisks), and some granules show a laminar ultrastructure (B, inferior granule). Moreover, there is an enlargement and dilation of the Golgi apparatus and endoplasmic reticulum which forms vacuoles (A and B, arrows). The nucleus shows an irregular profile (C), some neurons present abnormal accumulations of mitochondria and laminar bodies or "myelin figures" (E, arrowheads), and the neurofilamentous network is well evident. bv, blood vessel; ct, connective tissue; m, mitochondria; n, nucleus; Nb, Nissl bodies; nf, neurofilaments; Sc, Schwann cell; sgc, satellite glial cell. (A, original magnification × 4500; B, D, and E, original magnification × 30,000; C, original magnification × 6600.)

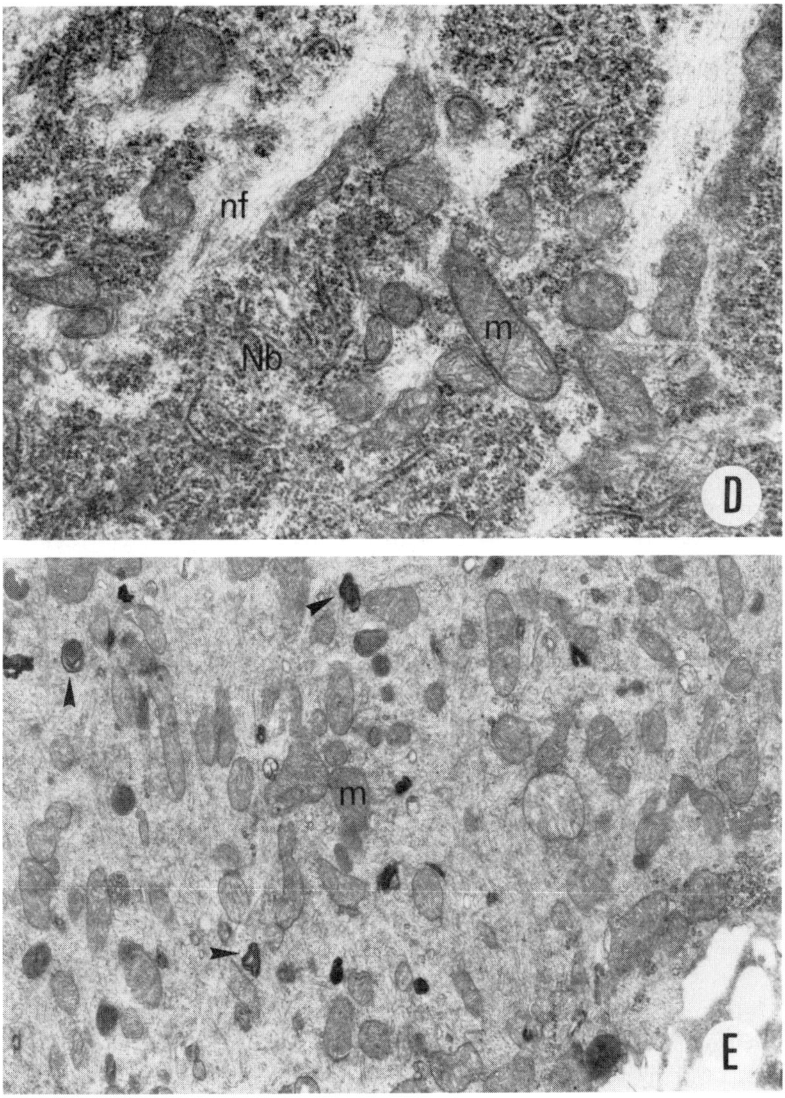

FIGURE 7D and E.

200 kDa, respectively. Neurofilament subunits are synthesized independently, then assembled, forming the so-called neurofilament triplet, and transported to the axon, where they are phosphorylated.[103] Although two main categories of sensory neurons have been considered on the basis of their neurofilament immunoreactivity,[104] all three neurofilament protein subunits can be detected in their perikarya,[105] and the protein triplet is present in a subpopulation of neurons in guinea pig dorsal root ganglia.[106]

In aged sensory neurons of Wistar rats, no changes were encountered with respect to young or adult animals in the distribution of neurofilament triplet immunoreactivity, which was restricted to the axon, and no accumulation of immunoreactive material was found in the perikarya. Moreover, preliminary studies using combined immunohistochemistry and image analysis techniques demonstrate that in aged sensory neurons, there is a decrease in the intensity of immunostaining for the 68- and 160-kDa subunits of the neurofilaments, whereas no apparent changes occur in the 200-kDa subunit. However, all neurons and axons retain their antigenic capacities, and minimal changes occur in the neuronal size/intensity of immunostaining ratio with aging (Figure 8).[132]

5. Neurotransmitters and Related Enzymes

Although a great number of studies have been conducted to identify the chemical molecules responsible for different sensory functions, it is still unknown which, if any, of the classical neurotransmitters or neuropeptides the sensory neurons utilize.[107,108] However, changes in enzymatic activities related to neurotransmitter metabolism have been observed with aging in sensory neurons.

In the sensory neurons of old rats, we have shown a decrease in fluoride-resistant acid phosphatase activity, which was strongly diminished in small-dark neurons and apparently remained unchanged in the large-light ones. Conversely, in young and adult animals, the most evident activity for this enzyme was detected in the small-dark neurons, although some large-light neurons also displayed activity. Thus, since fluoride-resistant acid phosphatase participates in neurotransmitter metabolism,[109,110] it could be speculated that aging modifies the normal metabolism in some putative sensory neurotransmitters. In the same study, the neuronal acetylcholinesterase and non-specific cholinesterase activities, which are present in a subset of primary sensory neurons,[111,112] did not show changes in either the patterns of distribution or the end-product amount with aging.[132]

6. Metabolic and Molecular Features

Neuron-specific enolase, the most acidic glycolytic neuronal enzyme, is present in all primary sensory neurons, although differences exist among neurons in the intensity of immunoreactivity.[113,114] Neuron-specific enolase is closely linked to the functional activity of neurons, and differences in its activity or amounts denote the neuronal metabolism and energetic demands.[115] Recently, we investigated age-related changes in the neuron-specific enolase immunoreactivity of the L_4 to L_6 dorsal root ganglia sensory neurons of Wistar male rats (Figure 8), showing that although a slight decrease occurs in aged neurons, the difference is not significant with respect to that found in young and adult rats.[132] Similar, and heterogeneous, results have been reported in human brain from old people.[116]

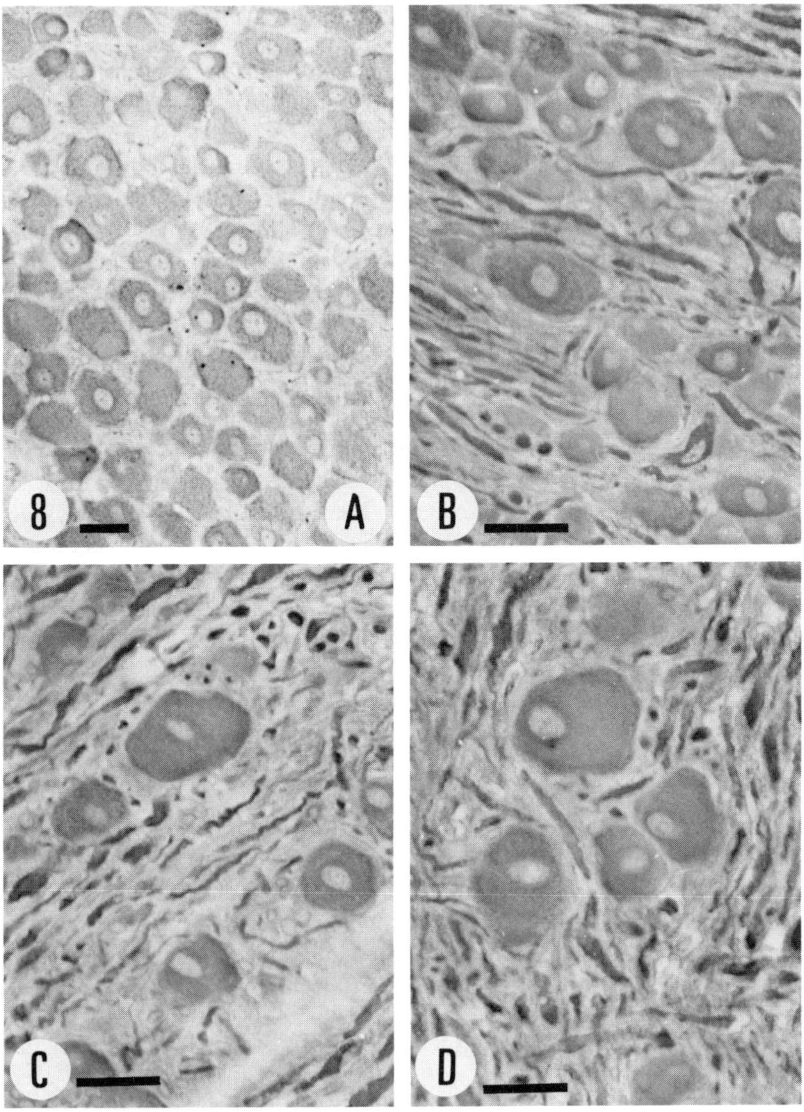

FIGURE 8. Lumbar (L_5 to L_7) dorsal root ganglia from 30-month-old Wistar male rats. All neurons display a weak immunoreactivity for the neuron-specific enolase (A) as well as for the neurofilament protein subunits. (B) 68-kDa neurofilament protein subunit; (C) 160-kDa neurofilament protein subunit; (D) 200-kDa neurofilament protein subunit. (Bar = 50 μm.)

On the other hand, the carbonic anhydrase activity, characteristic of the large-light sensory neurons of mammals[117] and the activity of which is at least partially determined by environmental influences,[118] was weakly increased in sensory neurons of old Wistar rats without significant differences from those typical of young and adult animals.[132]

It is known that a subset of primary sensory neurons in the rat display S-100 protein immunoreactivity.[119,120] Although the role of S-100 protein in sensory neurons remains unclear, it could be related to the intraneuronal homeostasis of calcium,[121] and S-100β plays a neurotrophic role in the peripheral nervous system.[122] The calcium ion plays a role in the activation of some neutral proteases, and changes in S-100 protein could be involved in the neuronal degenerative changes which occur with age. However, we have observed no significant changes in the patterns of neuronal S-100 protein distribution or intensity with aging.[132]

Protein synthesis was found to be diminished in sensory neurons of old animals, although the rate of synthesis varied widely between animals of the same age.[123] In contrast, some authors observed an increase in [^{3}H]-leucine incorporation by sensory neurons with aging.[124]

Results obtained from *in vitro* studies of aged sensory neurons demonstrate that with aging there is a decrease in membrane fluidity parallel to the capacity to extend neurites,[125] and in membrane elasticity;[126] however, the neurons are able to extend their neurites and survive for more than 3 weeks. In addition, although their resting membrane potential was small, they recovered their membrane excitability after continuous hyperpolarization, and both Na and Ca spikes were elicited in these hyperpolarized nerve cells.[127] Moreover, the effects of gangliosides and NGF on neurite elongation of mice sensory neurons were not altered with aging,[128] and the ganglioside content of dorsal root ganglia increased drastically until 12 months after birth and thereafter did not change considerably.[129]

Sensory neurons are dependent and responsive to NGF, and probably other neurotrophic factors, which ensures the survival of mature sensory neurons.[69] Preliminary results from our laboratory indicate that sensory neurons from old rats have decreased immunoreactivity to NGF-epidermal growth factor- and fibroblastic growth-factor receptor. But the functional significance of these findings, if any, must be clarified in future studies.

B. SATELLITE GLIAL CELLS

It is not clear whether satellite glial cells proliferate with aging, and although accumulation of these cells (Nageotte's bodies) are frequent in dorsal root ganglia of aged men,[89] the number of these structures represents a marker of neuron loss, but is not an indicator of satellite cell augmentation. Van der Bosch-Aguilar and Vanneste,[130] using light and electron microscopy with radioactive tracers, have observed that satellite glial cells maintain their functionality and remain relatively intact. Our results in the aged-rat dorsal root ganglia show that satellite glial cells have a normal pattern of distribution and immuno-histochemical properties do not change with aging. Moreover, ultrastructural aspect was almost identical to that of young adult animals.[131,132]

ACKNOWLEDGMENTS

This work was supported in part by grants from the Sandoz Foundation for Gerontological Research and Fundacion para la Investigacion Cientifica Aplicada y la Tecnologia (Asturias, Spain). The photographic assistance of Mr. J. A. García-Sánchez is gratefully acknowledged.

REFERENCES

1. **Gabella, G.,** *Structure of the Autonomic Nervous System,* Chapman and Hall, London, 1976.
2. **Tomlinson, B. E. and Corsellis, J. A. N.,** Ageing and the dementias, in *Greefield's Neuropathology,* 4th ed., Adams, J. H., Corsellis, J. A. N., and Duchen, L. W., Eds., Edward Arnold, London, 1984, 951.
3. **Baker, D. M. and Santer, R. M.,** Morphometric studies on pre- and paravertebral sympathetic neurons in the rat: changes with age, *Mech. Ageing Dev.,* 42, 139, 1988.
4. **Davies, D. C.,** Neuronal numbers in the superior cervical ganglion of the neonatal rat, *J. Anat.,* 127, 43, 1978.
5. **Thurn, C.,** Uber die postnatale Entwicklung des Ganglion stellatum der Katze (Felis domestica), *Z. Anat. Entwicklungsgesch.,* 136, 59, 1972.
6. **Hendry, I. A. and Campbell, J.,** A method to correct adequately for changes in neuronal size when estimating neuronal numbers after nerve growth factor treatment, *J. Neurocytol.,* 5, 337, 1976.
7. **Hervonen, A., Koistinaho, J., Alho, H., Helen, P., Santer, R. M., and Rapoport, S. I.,** Age related heterogeneity of lipopigments in human sympathetic ganglia, *Mech. Ageing Dev.,* 35, 17, 1986.
8. **Hervonen, A., Partanen, M., Helen, P., Koistinaho, J., Alho, H., Baker, D. M., Johnson, J. E., and Santer, R. M.,** The sympathetic neurons, a model of neuronal aging, in *Neurohistochemistry: Modern Methods and Applications,* Panulea, P., Paivarinta, H., and Soinila, S., Eds., Alan R. Liss, New York, 1986, 569.
9. **Partanen, M., Hervonen, A., and Rapoport, S. I.,** Microspectrofluorometric quantitation of histochemically demonstrable catecholamines in peripheral and brain catecholamine-containing neurons in male Fischer-344 rats at different ages, in *The Aging Brain: Cellular and Molecular Mechanisms of Aging in the Nervous System,* Giacobini, E., Filogamo, G., and Vernadakis, A., Eds., Raven Press, New York, 1982, 161.
10. **Amprino, R.,** Modifications de la structure des neurones sympathetiques pendant l'accroissement et la senescence. Recherches sur le ganglion cervical suprieur, *C. R. Assoc. Anat.,* 33, 3, 1938.
11. **Dyck, P. J., Jedrzejowska, H., Karnes, J., Jawamura, Y., Low, P., O'Brien, P. C., Offord, K., Ohnishi, L., Ohta, M., Pollock, J., and Stevens, J.,** Reconstruction of motor, sensory and autonomic neurons based on morphometric study of sampled levels, *Muscle Nerve,* 2, 399, 1979.
12. **Scaravilli, F.,** Changes in neuronal structure and cell populations with ageing, *New Issues Neurosci.,* 1, 95, 1988.
13. **Pick, J., de Lemos, C., and Gerdin, C.,** The fine structure of sympathetic neurons in man, *J. Comp. Neurol.,* 122, 19, 1960.
14. **Pick, J.,** Pigment, abnormal mitochondria and laminar bodies in human sympathetic neurons, *Z. Zellforsch.,* 6, 118, 1967.

15. **Helen, P.,** Fine-structural and degenerative features in adult and aged human sympathetic ganglion cells, *Mech. Ageing Dev.*, 23, 161, 1983.
16. **Santer, R. M., Partanen, M., and Hervonen, A.,** Glyoxylic acid fluorescence and ultrastructural studies of neurones in coeliac-superior mesenteric ganglion of the aged rat, *Cell Tissue Res.*, 211, 475, 1980.
17. **Schmidt, R. E., Chae, H. Y., Parvin, C. A., and Roth, K. A.,** Neuroaxonal dystrophy in aging human sympathetic ganglia, *Am. J. Pathol.*, 136, 1327, 1990.
18. **Wisniewski, H., Narang, H. K., and Terry, R. D.,** Neurofibrillary tangles of paired helicoidal filaments, *J. Neurol. Sci.*, 27, 173, 1976.
19. **Baum, L., Masliah, E., Iimoto, D. S., Hansen, L. A., Hallyday, W. C., and Saitoh, T.,** Casein kinase II is associated with neurofibrillary tangles but is not an intrinsic component of paired helical filaments, *Brain Res.*, 573, 126, 1992.
20. **Anderton, B. A., Breinburg, D., Downes, M. J., Green, R. J., Tomlinson, B. E., Ulrich, J., Wood, J. N., and Kahn, J.,** Monoclonal antibodies show that neurofibrillary tangles and neurofilaments share antigenic determinants, *Nature (London)*, 298, 84, 1982.
21. **Kawasaki, H., Murayama, S., Tomonaga, M., Izumiyama, N., and Simda, H.,** Neurofibrillary-tangles in human upper cervical ganglia. Morphological study with immunohistochemistry and electron microscopy, *Acta Neuropathol.*, 75, 156, 1987.
22. **Donato, H.,** Lipid peroxidation, cross-linking reaction and aging, in *Age Pigments*, Soda, R. S., Ed., Raven Press, New York, 1981, 63.
23. **Nandy, K.,** Aging neurons and pharmacological agents, in *Aging*, Vol. 21, *Brain Aging: Neuropathology and Neuropharmacology*, Cervós-Navarro, J. and Sarkander, H.-I., Eds., Raven Press, New York, 1983, 401.
24. **Botar, I.,** Quantitative und Qualitative Untersuchung des Nervenzellen des Ganglion Coeliacum im Alter. Altersercheinungen der sympathischen Nervenzellen, *Acta Anat.*, 28, 157, 1956.
25. **Samorajsti, T., Ordy, J. M., and Rady-Reimer, P.,** Lipofuscin pigment accumulation in the nervous system of ageing mice, *Anat. Rec.*, 160, 555, 1968.
26. **Hopfner, C., Caulet, T., and Adnett, J. J.,** Etude ultrastructurale du neurone sympathetique de l'homme ag. Activite phosphatasique acide des pigments, anomalies mitochondriales et presence de crops lamellaires, *Virchows Arch. B.*, 5, 326, 1970.
27. **Hervonen, H., Hervonen, A., and Kanerva, L.,** Fine structure of the human fetal sympathetic neurons in vitro, *Histochem. J.*, 10, 271, 1978.
28. **Koistinaho, J., Alho, M., Mäkelä, V., Helen, P., and Hervonen, A.,** Microspectrofluorimetric registration of the accumulation of age pigment in the human sympathetic ganglion and central nervous system, *Neurosci. Lett.*, Suppl. 14, 202, 1983.
29. **Koistinaho, J., Honkaniemi, J., and Hervonen, A.,** The effect of bleaching on the lipopigments in the human sympathetic neurons, *Mech. Ageing Dev.*, 28, 91, 1986.
30. **Hervonen, A., Alho, H., Koistinaho, J., and Helen, P.,** Microspectrofluorometric characterization of the two types of lipofuscin autofluorescences in the human sympathetic ganglia at different ages, *Neurology*, Suppl. 1, 34, 1984.
31. **Koistinaho, J.,** Difference in the age-related accumulation of lipopigments in the adrenergic and noradrenergic peripheral neurons in the male rat, *Gerontology*, 32, 300, 1986.
32. **Nandy, K.,** Properties of neuronal lipofuscin pigment in mice, *Acta Neuropathol.*, 19, 25, 1971.
33. **Hervonen, A., Vaalasti, A., Partanen, M., Kanerva, L., and Hervonen, H.,** Effect of ageing on the histochemically demonstrable catecholamines and acetyl-cholinesterase of human sympathetic ganglia, *J. Neurocytol.*, 7, 11, 1978.
34. **Partanen, M., Santer, R. M., and Hervonen, A.,** The effect of ageing on the histochemically demonstrable catecholamines in the hypogastric (main pelvic) ganglion of the rat, *Histochem. J.*, 12, 527, 1980.
35. **Koistinaho, J., Sorvaniemi, M., Alho, H., and Hervonen, A.,** Microspectrofluorometric quantitation of autofluorescent lipopigment in human sympathetic ganglia, *Mech. Ageing Dev.*, 28, 79, 1986.

36. **Koistinaho, J., Hartikainen, K., Hatanpaa, K., and Hervonen, A.,** Age pigments in different subpopulations of peripheral neurons in vivo and in vitro, *Adv. Exp. Med. Biol.*, 266, 49, 1989.
37. **Barden, H.,** The histochemical relationships and the nature of neuromelanine, in *Aging*, Vol. 1, Borody, H., Harman, D., and Ordy, J. M., Eds., Raven Press, New York, 1975, 79.
38. **Dillard, C. J. and Tappel, A. L.,** Fluorescent products from relation of peroxidizing polyunsaturated fatty acids with phosphatidyl ethanolamine and phenylalanine, *Lipids*, 8, 183, 1973.
39. **Harkonen, M. and Pentilla, A.,** Catecholamines, monoamino oxidase and cholinesterase activity in the human sympathetic ganglion, *Acta Physiol. Scand.*, 82, 310, 1971.
40. **Partanen, M., Waller, S. B., London, E. D., and Hervonen, A.,** Indices of neurotransmitter synthesis and release in aging sympathetic nervous system, *Neurobiol. Aging*, 6, 227, 1985.
41. **Raskind, M. A., Peskind, E. R., Veith, R. C., Beard, J. C., Gumbrecht, G., and Halter, J. B.,** Increased plasma and cerebrospinal fluid norepinephrine in older men: differential suppression by clonidine, *J. Clin. Endocrinol. Metab.*, 66, 438, 1988.
42. **Santer, R. M.,** Fluorescence histochemical evidence for decreased noradrenaline synthesis in sympathetic neurones of aged rats, *Neurosci. Lett.*, 15, 177, 1979.
43. **Buu, N. T. and Debinski, W.,** Vesicular and ganglionic norepinephrine in the rat: progressive increase with age, *J. Neurochem.*, 54, 620, 1990.
44. **Reis, D. J., Ross, R. A., and Joh, T. H.,** Changes in the activity and amounts of enzymes synthesizing catecholamines and acetylcholine in brain, adrenal medulla, and sympathetic ganglia of aged rat and mouse, *Brain Res.*, 136, 465, 1977.
45. **McLean, M. R., Goldberg, P. B., and Roberts, J.,** An ultrastructural study of the effects of age on sympathetic innervation and atrial tissue in the rat, *J. Mol. Cell. Cardiol.*, 15, 75, 1983.
46. **Schultzberg, M. and Lindh, B.,** Transmitters and peptides in autonomic ganglia, in *Handbook of Chemical Neuroanatomy*, Vol. 6, *The Peripheral Nervous System*, Bjrklund, A., Höklelt, T., and Owman, C., Eds., Elsevier, Amsterdam 1988, 297.
47. **Baffi, J., Grcs, T., Slowik, F., Horváth, M., Lekka, N., Pásztor, E., and Palkovits, M.,** Neuropeptides in the human superior cervical ganglion, *Brain Res.*, 570, 272, 1992.
48. **Rothe, F., Schmidt, W., and Wolf, G.,** Postnatal changes in the activity of glutamate dehydrogenase and aspartate aminotransferase in the rat nervous system with special reference to the glutamate transmitter metabolism, *Brain Res.*, 313, 67, 1983.
49. **Govoni, S., Missale, C., Castelletti, L., Spano, P. F., and Trabucchi, M.,** Decreased content of met-enkephalin-like peptides in superior cervical and coeliac ganglia of aged rats, *Neurobiol. Aging*, 4, 147, 1983.
50. **Jarvi, R., Helen, P., Pelto-Huikko, M., Rapoport, S. I., and Hervonen, A.,** Age-related changes of enkephalinergic innervation of human sympathetic neurons, *Mech. Ageing Dev.*, 44, 143, 1988.
51. **Gurusinghe, C. J., Harris, P. J., Abbott, D. F., and Bell, C.,** Neuropeptide Y in rat sympathetic neurons is altered by genetic hypertension and by age, *Hypertension*, 16, 63, 1990.
52. **Partanen, M., London, E. D., and Rapoport, S. I.,** Glucose utilization in sympathetic ganglia of male Fischer-344 rats at different ages, *J. Auton. Nerv. Syst.*, 5, 391, 1982.
53. **Baker, D. M. and Santer, R. M.,** Quantitative succinate dehydrogenase activity of prevertebral sympathetic neurons: variations with age, *J. Anat.*, 146, 252, 1986.
54. **Baker, D. M. and Santer, R. M.,** Development of a quantitative histochemical method for determination of succinate dehydrogenase activity in autonomic neurons and its application to the study of aging in the autonomic nervous system, *J. Histochem. Cytochem.*, 38, 525, 1990.
55. **Diculescu, I. and Onicescu, D.,** Folate metabolism in the superior cervical ganglion histochemical study, *Histochemistry*, 62, 101, 1979.

56. **Onicescu, D.,** Folate metabolism in ageing, *Morphol. Embryol. Bucur.*, 26, 213, 1980.
57. **Willson, D. L., Hall, M. E., and Stone, G. S.,** Test of some aging hypothesis using two-dimensional protein mapping, *Gerontology,* 24, 426, 1978.
58. **Van Bveren, C., Van Straaten, F., Curran, T., Miller, R., and Verma, I. M.,** Analysis of FBJ-MuSv provirus and c-fos (mouse) gene reveals that viral and cellular fos gene products have different carboxyl termini, *Cell,* 32, 1241, 1983.
59. **Morgan, J. and Curran, T.,** Stimulus-transcription coupling in neurons: role of cellular immediate-early genes, *Trends Neurosci.*, 12, 459, 1989.
60. **Yang, G. and Koistinaho, J.,** Age-related increase in *c-fos* expression in the sympathetic neurons of the rat superior cervical ganglion, *Brain Res.*, 533, 338, 1990.
61. **Adler, J. E. and Black, I. B.,** Plasticity of substance P in mature and aged sympathetic neurons in culture, *Science,* 225, 1499, 1984.
62. **Wu, R., McKenna, D. G., and McAffe, D. A.,** Age-related changes in the sympathetic plasticity of rat superior cervical ganglia, *Brain Res.*, 542, 324, 1991.
63. **Fukuda, J., Kasai, H., and Tada, Y.,** Sympathetic ganglion neurons from aged human grown in monolayer culture, *Neurosci. Lett.*, 38, 193, 1983.
64. **Greif, K. F. and Falherty, K. N.,** Changes in expression of a synaptic vesicle antigen in aging sympathetic neurons, *Neurobiol. Aging,* 10, 51, 1989.
65. **Gómez-Pinilla, F., Cotman, C. W., and Nieto-Sampedro, M.,** NGF receptor immunoreactivity in aged rat brain, *Brain Res.*, 479, 255, 1989.
66. **Lärktors, L., Ebendal, T., Whittemore, S. R., Persson, H., Hoffer, B., and Olson, L.,** Decreased level of nerve growth factor (NGF) and its messenger RNA in the aged rat brain, *Mol. Brain Res.*, 3, 55, 1987.
67. **Hellweg, R., Fischer, W., Hock, C., Gage, F. H., Bjrklund, A., and Thoenen, H.,** Nerve growth factor and choline acetyltransferase activity in the brain of aged rats with spatial memory impairments, *Brain Res.*, 537, 123, 1990.
68. **Walicke, P. A.,** Novel neurotrophic factors, receptors, and oncogenes, *Annu. Rev. Neurosci.*, 12, 103, 1989.
69. **Levi-Montalcini, R.,** The nerve growth factor 35 years later, *Science,* 237, 1154, 1987.
70. **Sobue, G., Yasuda, T., Mitsuma, T., and Pleasure, D.,** Nerve growth factor receptor immunoreactivity in the neuronal perikarya of human sensory and sympathetic nerve ganglia, *Neurology,* 39, 937, 1989.
71. **Uchida, Y. and Tomonaga, M.,** Loss of nerve growth factor receptors in sympathetic ganglia from aged mice, *Biochem. Biophys. Res. Commun.*, 146, 797, 1987.
72. **Ueyama, T., Hano, T., Hamada, M., Nishio, I., and Masuyama, Y.,** New role of nerve growth factor — an inhibitory neuromodulator of adrenergic transmission, *Brain Res.*, 559, 293, 1991.
73. **Chiba, T., Black, A. C., Jr., and Williams, T. H.,** Evidence for dopamine-storing interneurons and paraneurons in rhesus monkey sympathetic ganglia, *Neurocytology,* 49, 95, 1977.
74. **Taxi, J., Derer, M., and Domich, A.,** Morphology and histophysiology of SIF cells in the autonomic ganglia, in *Autonomic Ganglia,* Elfvin, L.-G., Ed., John Wiley & Sons, New York, 1983, 67.
75. **Rybarczyk, K. E., Baker, H. A., Burke, J. P., Hartman, B. K., and Van Orden, L. S., III,** Histochemical and immunohistochemical identification of catecholamines, dopamine-β-hydroxylase and phenylethanolamine-N-methyltransferase, in *SIF cells. Structure and Function of the Small Intensely Fluorescent Cells,* Ernko, O., Ed., Fogarty, Int. Cent. Proc. No. 30, DHEW Publ. No. HIN 76-942, U.S. Government Printing Office, Washington, D.C., 1976, 68.
76. **Kanerva, L. and Hervonen, A.,** SIF cells, short adrenergic neurons and vacuolated nerve cells of the paracervical (Frankenhuser) ganglion, in *SIF cells. Structure and Function of the Small Intensely Fluorescent Cells,* Ernko, O., Ed., Fogarty Int. Cent. Proc. No. 30, DHEW Publ. No. HIN 76-942, U.S. Government Printing Office, Washington, D.C., 1976, 36.

77. **Villena, F., Montoya, G. A., Jofr, A., and Goset, C.,** Electron microscopy and histofluorescence studies of toad sympathetic ganglion with reference to SIF cells, *Cell. Mol. Biol.*, 30, 313, 1984.
78. **Papka, R. E., Traurig, H. H., and Klenn, P.,** Paracervical ganglia of the female rat: histochemistry and immunohistochemistry of neurons, SIF cells, and nerve terminals, *Am. J. Anat.*, 179, 243, 1987.
79. **Glees, P. and Gopinath, G.,** Age changes in the central and peripherally located sensory neurons in rat, *Z. Zellforsch.*, 141, 285, 1973.
80. **Stefansson, K., Wollman, R. L., and Moore, B. W.,** Distribution of S-100 protein outside the central nervous system, *Brain Res.*, 234, 309, 1982.
81. **Wigston, D. J.,** Maintenance of cholinergic neurones and synapses in the ciliary ganglion of aged rats, *J. Physiol.*, 344, 223, 1983.
82. **Bigl, V., Arent, T., Fischer, S., Fische, S., Wener, M., and Arendt, A.,** The cholinergic system in aging, *Gerontology,* 33, 172, 1987.
83. **Grimes, P. A., McGlinn, A. M., and Stone, A.,** An immunohistochemically distinct population of cat ciliary ganglion cells, *Brain Res.*, 535, 323, 1990.
84. **Harbedo, J. E., Suzuki, N., Ekblad, E. and Owman, C.,** Vasoactive intestinal polypeptide and acetylcholine coexist with neuropeptide Y, dopamine-β-hydroxylase, tyrosine hydroxylase, substance P and calcitonin gene-related peptide in neuronal subpopulations in cranial parasympathetic ganglia of rat, *Cell Tissue Res.*, 267, 291, 1992.
85. **Marchi, M. and Giacobini, E.,** Development and aging of cholinergic synapses. II. Continuous growth of acetylcholine and choline levels in autonomic ganglia and iris of the chick, *Dev. Neurosci.*, 3, 39, 1980.
86. **Marchi, M., Hoffman, D. W., Giacobini, E., and Fredrickson, T.,** Development and aging of cholinergic synapses. IV. Acetylcholinesterase and choline acetyltransferase activities in autonomic ganglia and iris of the chick, *Dev. Neurosci.*, 3, 235, 1980.
87. **Marchi, M., Yurkewicz, L., Giacobini, E., and Fredrickson, T.,** Development and aging of cholinergic synapses. V. Changes in nicotinic cholinergic receptor binding in ciliary ganglia and irises in the chicken, *Dev. Neurosci.*, 4, 258, 1981.
88. **Lieberman, A. R.,** Sensory ganglia, in *The Peripheal Nerve,* Landon, D. N., Ed., Chapman and Hall, London, 1976, 188.
89. **Thomas, P. K., Landon, D. N., and King, R. H. M.,** Diseases of the peripheral nerves, in *Greefield's Neuropathology,* 4th ed., Adams, J. H., Corsellis, J. A. N., and Duchen, L. W., Eds., Edward Arnold, London, 1984, 807.
90. **Ohta, M., Offord, K., and Dyck, P. J.,** Morphometric evaluation of first sacral ganglia in man, *J. Neurol. Sci.*, 51, 131, 1981.
91. **Emergy, J. L. and Singhal, R.,** Changes associated with growth in the cells of the dorsal root ganglion in children, *Dev. Med. Child. Neurol.*, 15, 460, 1973.
92. **Nagashima, K. and Oota, K.,** A histopathological study of the human spinal ganglia. I. Normal variations in aging, *Acta Pathol. Jpn.*, 24, 333, 1974.
93. **Scharf, J. H. and Blumenthal, H. J.,** Neure aspekte zur altersabhngigen Involution des sensiblen Peripheren nervensystems, *Z. Zellforsch.*, 78, 280, 1967.
94. **Aldskogius, H. and Risling, M.,** Number of dorsal root ganglion neurons and axons in cats of different ages, *Exp. Neurol.*, 106, 70, 1989.
95. **Jacobs, J. M., Scaravilli, F., Duchen, L. W., and Mertin, J.,** A new neurological rat mutant mutilated foot, *J. Anat.*, 132, 525, 1981.
96. **Van der Bosch de Aguilar, P. and Vanneste, J.,** Etude ultrastructurale des neurones ganglionaires spinaux au cours du vieillissement chez le rat, *Acta Anat.*, 110, 59, 1981.
97. **Van der Bosch de Aguilar, P. and Goemare-Vanneste, J.,** Paired helical filaments in spinal ganglion neurons of elderly rats, *Virchows Arch. B*, 47, 217, 1984.
98. **Vanneste, J. and Van der Bosch de Aguilar, P.,** Mitochondrial alterations in the spinal ganglion neurons in ageing rats, *Acta Neuropathol.*, 54, 83, 1981.
99. **Mitsumori, K., Maita, K., and Shirasu, Y.,** An ultrastructural study of spinal nerve roots and dorsal root ganglia in aging rats with spontaneous radiculoneuropathy, *Vet. Pathol.*, 18, 714, 1981.

100. **De Castro, F.,** Sensory ganglia of the cranial and spinal nerves. Normal and pathological, in *Cytology and Cellular Pathology of the Nervous System,* Penfield, W., Ed., Hoeber, New York, 1932, 93.
101. **Brady, D. and Gilbert, D.,** Cytoskeletal elements in neurons, *Annu. Rev. Neurosci.,* 4, 505, 1981.
102. **Schnapp, B. J., Vale, R. D., Sheetz, M. P., and Reese, T. A.,** Single microtubules from squid axoplasm support bidirectional movement of organelles, *Cell,* 40, 455, 1985.
103. **Schlaepfer, W. W.,** Neurofilaments: structure, metabolism, and implications in disease, *J. Neuropathol. Exp. Neurol.,* 46, 117, 1987.
104. **Lawson, S. N. and Waddell, P. J.,** Soma neurofilament immunoreactivity is related to cell size and fiber conduction velocity in rat primary sensory neurons, *J. Physiol.,* 435, 31, 1991.
105. **Trojanowski, J. Q., Walkenstein, N., and Lee, V. M.-Y.,** Expression of neurofilament subunits in neurons of the central and peripheral nervous system: an immunohistochemical study with monoclonal antibodies, *J. Neurosci.,* 6, 650, 1986.
106. **Vickers, J. C. and Costa, M.,** Neurofilament protein triplet immunoreactivity in the dorsal root ganglia of the guinea pig, *Cell Tissue Res.,* 265, 159, 1991.
107. **Dalsgaard, C.-J.,** The sensory systems, in *Handbook of Chemical Neuroanatomy,* Vol. 6, *The Peripheal Nervous System,* Björklund, A., Hökfelt, T., and Owman, C., Eds., Elsevier, Amsterdam, 1988, 599.
108. **Vega, J. A., Amenta, F., Hernández, L. C., and Del Valle, M. E.,** Presence of catecholamine-related enzymes in a subpopulation of primary sensory neurons in dorsal root ganglia of the rat, *Cell. Mol. Biol.,* 37, 519, 1991.
109. **Dalsgaard, C.-J., Ygge, J., Vicent, S. R., Ohrling, M. J., Dokray, G. J., and Elde, R.,** Peripheral projections and neuropeptide coexistence in a subpopulation of fluoride-resistant acid phosphatase reactive spinal primary sensory neurons, *Neurosci. Lett.,* 51, 139, 1984.
110. **Nagay, J. I. and Hunt, S. P.,** Fluoride-resistant acid phosphatase neurons containing substance P or somatostatin, *Neuroscience,* 7, 89, 1982.
111. **Vega, J. A., Rodríguez, C., Medina, M., Martínez-Tellería, A., Bengoechea, M. E., and Perez-Casas, A.,** Acetylcholinesterase and fluoride-resistant acid phosphatase activities in dorsal root ganglia of the rat, *Cell. Mol. Biol.,* 35, 39, 1989.
112. **Dubovy, P., Svizenska, I., and Vega, J. A.,** Non-specific cholinesterase activity in mouse spinal ganglia. The usefulness of histochemical study and image analysis for simple characterization of neuron subclasses, *Cell. Mol. Biol.,* 36, 23, 1990.
113. **Vega, J. A., Rodríguez, C., Medina, M., and Del Valle, M. E.,** Neuron specific enolase (NSE)-like and neurofilament protein (NFP)-like immunoreactivities in the rat dorsal root ganglia and sciatic nerve, *Cell. Mol. Biol.,* 36, 537, 1990.
114. **Bonfanti, L., Bellardi, S., Ghidella, S., Gobetto, A., Polak, J. M., and Merighi, A.,** Distribution of five peptides, three general neuroendocrine markers, and two synaptic-vesicle-associated proteins in the spinal cord and dorsal root ganglia of adult and newborn dog: an immunocytochemical study, *Am. J. Anat.,* 191, 154, 1991.
115. **Marangos, P. J. and Schemechel, D. E.,** Neuron specific enolase, a clinically useful marker for neurons and neuroendocrine cells, *Annu. Rev. Neurosci.,* 10, 269, 1987.
116. **Schemechel, D. E., Marangos, P. J., Martin, B. M., Winfield, D. S., Burkhart, D. S., Roses, A. D., and Ginns, E. I.,** Localization of neuron-specific enolase (NSE) mRNA in human brains, *Neurosci. Lett.,* 76, 233, 1987.
117. **Kazimierczak, J., Sommer, E. W., Philippe, E., and Droz, B.,** Carbonic anhydrase activity in primary sensory neurons, *Cell Tissue Res.,* 245, 487, 1986.
118. **Droz, B., Kazimierczak, J., Barakat, I., and Philippe, E.,** Carbonic anhydrase expression in primary sensory neurons: modulation by environmental interactions, *J. Histochem. Cytochem.,* 36, 960, 1988.
119. **Vega, J. A., Rodríguez, C., Medina, M., Del Valle-Soto, M. E., and Hernández, L. C.,** Expression of cytoskeletal proteins in glial cells of dorsal root ganglia, *Cell. Mol. Biol.,* 35, 635, 1989.

120. **Vega, J. A., Del Valle-Soto, M. E., Calzada, B., and Alvarez-Mendez, J. C.,** Immunohistochemical localization of S-100 protein subunits (α and β) in dorsal root ganglia of the rat, *Cell. Mol. Biol.,* 37, 173, 1991.
121. **Heizmann, C. W. and Hunziker, W.,** Intracellular calcium-binding proteins: more sites and insights, *TIBS,* 16, 98, 1991.
122. **Van Eldik, L. J., Christie-Pope, B., Bolin, L. M., Shooter, E. M., and Whetsell, W. O., Jr.,** Neurotrophic activity of S-100β in cultures of dorsal root ganglia from embryonic chick and fetal rat, *Brain Res.,* 542, 280, 1991.
123. **Van der Bosch de Aguilar, P. and Vanneste, J.,** Ageing of the spinal ganglion neurons in the rat: a radioautographic study following injection of [^{3}H]-lysine, *Neurosci. Lett.,* 18, 225, 1980.
124. **McMartin, D.,** Effects of age on axoplasmic transport in peripheral nerves, in *Aging,* Vol. 21, *Brain Ageing: Neuropathology and Neuropharmacology,* Cervós-Navarro, J. and Sarkander, H.-I., Eds., Raven Press, New York, 1983, 351.
125. **Horie, H., Kawasaki, Y., and Takenaka, T.,** Lateral diffusion membrane lipid changes with ageing in C57BL mouse dorsal root ganglion neurons from a fetal stage to an aged stage, *Brain Res.,* 377, 246, 1986.
126. **Horie, H., Ikuta, S., and Takenaka, T.,** Membrane elasticity of mouse dorsal root ganglion neurons decreases with aging, *FEBS Lett.,* 269, 23, 1990.
127. **Fukuda, J. and Yamaguchi, K.,** Electrophysiological properties of tissue-cultured nerve cells from senescent mouse, *Neurosci. Lett.,* 26, 263, 1981.
128. **Tanaka, M., Kanai, H., and Hirai, S.,** Effects of nerve growth factor and gangliosides on neurite elongation of cultured nerve cells from senescent mouse, *Nippon Ronen Igakkai Zasshi,* 26, 563, 1989.
129. **Ohsawa, T.,** Rat dorsal root ganglion gangliosides during ear development until senescence, *Mech. Ageing Dev.,* 30, 259, 1990.
130. **Van der Bosch de Aguilar, P. and Vanneste, J.,** The microenvironment of the spinal ganglion neuron in the rat during aging, *Exp. Neurol.,* 81, 294, 1983.
131. **Calzada, B., et al.,** unpublished data.
132. **Vega, J. A., et al.,** unpublished data.

Main Neurotransmitter Systems

Chapter 4

CIRCULATING CATECHOLAMINES AND AGING

Otto Kuchel and George Kuchel

TABLE OF CONTENTS

0-8493-6981-9/93/$0.00 + $.50

I. NEURONAL AND ADRENOMEDULLARY CONTRIBUTIONS TO CIRCULATING NORADRENALINE AND ADRENALINE IN AGING

A. BASAL ACTIVITY

It has generally been accepted that circulating noradrenaline (NA) levels reflect the activity of sympathetic neuronal terminals, while adrenaline (A) originates almost exclusively from adrenomedullary secretory activity. Basal plasma NA levels appear to be significantly higher in elderly human subjects. Since the original observation nearly 2 decades ago,[1] a large number of reports have demonstrated increases in these levels of as much as 50% or more in aged compared to young subjects.[2-12] Although no studies have shown a decrease in plasma NA with age, two studies have shown no difference in plasma NA between young and aged subjects[13,14] and one has shown an increase with age in males, but not in females.[15] Similar increases with aging have also been described in urinary and cerebrospinal fluid (CSF) NA levels[12] (Table 1). These age-related increases in plasma NA levels are still present in studies which take into account the observation that elderly subjects are known to require a longer period of time to reach truly basal levels.[16] Elderly subjects tend to exhibit a delayed return to prestimulus levels of sympathetic activity following many stimuli, including orthostasis.[14] In contrast, plasma A concentrations sampled under comparable conditions have not been found to increase with age in humans, with most studies reporting no change[4,8,9,17] or even a decrease.[12] This dissociation between age-related changes in NA and A is apparent not only in plasma, but also in urine[11] and CSF.[12]

Overall, the literature examining the impact of aging on sympathetic physiology has been remarkably consistent. Most evidence to date supports the hypothesis that the specific alterations in sympathetic physiology described in older individuals are representative of the normal aging process. In fact, the suggestion has been made that indices of sympathetic activity such as plasma NA could be used as markers of normal chronological aging.[17] This conclusion stems from studies examining healthy elderly subjects appropriately screened for the presence of diseases and medications which could modify sympathetic activity. Nevertheless, it is not known whether some of the interindividual variability and also discrepancy in plasma NA levels in a small number of the studies described above[13-15] can be accounted for by methodological differences, the spectrum of ages studied, or by selection criteria for normal-aged subjects. The last category of consideration could be particularly important from a gerontological perspective.

It has been observed that many of the changes observed in the course of normal aging or aging without identifiable disease involve heterogeneous patterns of change. It has been suggested that it may be possible and useful to separate normal human aging into two categories, usual and successful. In many cases, extrinsic factors such as diet, exercise, personal habits, and

TABLE 1
Changes in Basal Plasma, Urinary, and CSF NA and A Concentrations in Human Aging

Biological fluid	Change	Ref.
	Noradrenaline	
Plasma	Increase	1–10, 15
Urine	Increase	11
CSF	Increase	12
	Adrenaline	
Plasma	No change	4, 8, 9, 27
Plasma	Lower	12
Urine	No change	11
CSF	No change	12

psychological lifestyle may play a role.[18] This has raised hope that intervention strategies could be devised which would facilitate the transition from the former to the latter category of aging and thus result in aging with fewer functional deficits and disability.[18] In most modern human gerontological studies, subjects are screened for the presence of relevant disease, removing this confounding factor. The finding of elevated basal plasma NA levels in elderly bicycle marathon racers would suggest that physical fitness and muscle strength probably do not significantly contribute to this finding.[19,20] Similarly, these alterations are not dependent on the presence of hypertension,[21] carbohydrate intake,[22] or obesity.[23]

Nevertheless, it is important to consider the possibility that a number of physiological changes described as representing "normal" aging simply on the basis of being present in the great majority of elderly individuals in Western society may contribute to some of these age-related changes. Systolic blood pressures between 120 and 160 mmHg have been felt to represent normal human aging, yet they are not present in all aged individuals and are a relatively uncommon accompaniment of aging in many countries.[24] Most of these countries are nonindustrialized where many lifestyle patterns differ from those in our society, including a lower intake of salt.[24] It has been proposed that atherosclerosis and elevations of blood pressure into the hypertensive range may contribute to age-related alterations in sympathetic function (see below).[16] Thus, it is possible that many of the generally robust and reproducible alterations in the aging human sympathetic nervous system, although extremely common in Western societies, may actually not be inevitable and may be modifiable. Many of these questions will be resolved through the performance of preferably longitudinal studies[25] examining large numbers of healthy elderly subjects in a variety of different settings and societies.

TABLE 2
NA and A Responses to Various Stimuli in Normal Human Aging

Biological fluid	Type of stimulation or inhibition	Effect of aging on response	Ref.
	Noradrenaline		
Plasma	Standing and isometric exercise	Increase	28
	Ergometry	Increase	11
	Posture	Increase	29
	Orthostatism + glucose	Increase	14
	Orthostatism + isometric exercise	Increase	30
	Dynamic exercise	Increase	31
	Mental stress	Increase	27
	Head-up tilt	Increase	32
	Exercise	No change	33
	Clonidine	Normal suppressibility	12
	Insulin-euglycemic clamp	Decrease	34
CSF	Clonidine	Decreased suppressibility	12
	Adrenaline		
Plasma	Dynamic exercise	Increase	31
	Ergometry	Decrease	11
	Mental stress	No change	27
	Oral glucose administration	Normal suppressibility	26
CSF	Clonidine	Normal suppressibility	12

B. STIMULATED ACTIVITY

Similar, but less consistent differences between NA and A were found in the effect of aging on human responses to several stimulatory or inhibitory maneuvers (Table 2). In response to the majority of stimuli, ranging from postural changes to mental stress and oral glucose ingestion,[14,26] plasma NA exhibited a hyperresponsiveness with aging,[11,26,27-32] and only in one study did plasma NA not change.[33] Only following a euglycemic insulin administration was the increase in NA smaller in elderly subjects.[34] This decline may result from decreased tissue sensitivity to insulin observed in the elderly.[26] The activation of NA release by oral glucose administration elevating plasma NA in old, but not in young, subjects[34] has been suggested to operate via splanchnic mechanisms independent of changes in circulating levels of glucose or insulin.[35] Thus, distinct afferent and efferent signals involved in NA stimulation and their different intensities may be responsible for age-related differences in NA release, with splanchnic factors predominating in the elderly. The A responses to stimulation in aging have been inconsistent.[11,27,31] The plasma A response to oral glucose administration[26] as well as of plasma NA and CSF A suppressibility by clonidine[12] were unaffected by age. In contrast, CSF NA showed a decreased suppressibility by clonidine with age.[12]

In contrast to the above results, some animal studies tend to demonstrate a decreased plasma NA and A response to stress with aging.[36] This difference may relate to the ability to use a stronger stimulus in animal as opposed to human studies,[37] which could bring out age-related deficits not readily apparent following milder stimulation. In addition, animal studies have only rarely included consideration for the potential confounding effects of aging, disease, genetics, and environmental effects on sympathetic nervous system (SNS) function. This makes a comparison of animal and human studies very difficult at this time. For this reason, we have limited our review to catecholamines as they relate to human aging.

Studies to date indicate an apparent difference between the effect of aging on the neural and adrenomedullary components of catecholamines released into body fluids and their responsiveness to several stimuli. The elevation of NA with age does not appear to be related to dietary carbohydrate intake[38] nor to adiposity, since supine, standing, and isometric exercise levels are also increased in nonobese elderly subjects.[39] The unique age-related increase of neuronal activity is supported by the CSF data because central nervous system noradrenergic neurons and peripheral sympathetic neurons have morphological, pharmacological, and physiological[40] similarities. Nevertheless, CSF NA is derived from the brain and not plasma, even though a close relationship exists between plasma and CSF NA concentrations.[41] This suggests that the sympathetic and central nervous system share common regulatory mechanisms, and both systems appear to have an increased neuronal activity with aging even though the precise mechanisms involved in the changes in those systems may be different. In the brain, the decreased α_2-adrenergic mechanisms (as tested by clonidine suppressibility of NA) may contribute to increased brain noradrenergic activity with aging.[12] Mechanisms underlying the increased peripheral neural activity will be discussed further. In contrast to increased neuronal mechanisms with aging, the adrenomedullary mechanisms, as reflected by adrenaline, do not appear to react in the same way to advanced age. The only exception may be an A hyperresponsiveness to stressful stimulation exceeding usual, everyday physiological limits such as dynamic exercise in untrained elderly individuals.[31]

II. MECHANISMS UNDERLYING THE AGE-RELATED INCREASED PERIPHERAL NEURONAL ACTIVITY

Plasma NA concentration is determined by rates of NA release into plasma and its subsequent removal. Thus, higher NA levels in the elderly could result from increased NA entry into plasma and/or reduced clearance of NA. Most of the presently available evidence suggests that the predominant mechanism of increased plasma NA is an increased appearance of NA[42-47] and possibly also a decreased neuronal NA uptake by the heart.[48] Decreased neuronal NA uptake could contribute towards increased plasma NA levels in aging by increasing the synaptic spillover of NA and/or decreasing the clearance of plasma NA (Table 3). Hoeldtke and Cilmi[43] have described an increase in the entry of NA into the plasma without any change in NA production in

TABLE 3
NA and A Appearance and Metabolism in Normal Human Aging

		Effect of aging on response	Ref.
Noradrenaline			
	Basal appearance (whole body)	Increase	42–45
	Basal spillover rate (organ specific)	Increase	10
		No change	50
	Appearance after chronic physical activity	Increase	46
	Net renal release	Increase	47
	Clearance	No change	14, 42, 51
		Decrease	50, 52
	Neuronal NA uptake by the heart	Decrease	48
Adrenaline			
	Appearance	No change	52
	Adrenal release	Decrease	53
	Clearance	No change	52
		Increase	8
	Renal uptake	Decrease	47
Changes in NA-Metabolizing Enzyme and Metabolic Products			
Monoamine oxidase (MAO)	In brain, platelets, and plasma	Increase	54
NA sulfate	Plasma concentrations	Increase	55, 101

elderly subjects. They have interpreted this finding as suggesting that human aging is associated with alterations in the local disposition of sympathetic neuronal NA.[44] Human studies do not allow for the direct evaluation of neuronal NA uptake, at least in part because of the relative nonspecificity of drugs such as desipramine.[49] The measurement of the spillover rate of NA showed an increase[10,44] or no change[49] with age. The NA clearance data are also controversial, with some showing no changes[34,42,50,51] and others a decrease.[49] However, a multivariate analysis of the findings of one study has shown that most of the increase in plasma NA with age can be explained by elevations in plasma NA spillover. Studies examining adrenaline point to few consistent age-related changes in A plasma entry or clearance.[8,47,51-53] A appearance did not change with age,[51] and its adrenal release[52] and renal uptake[47] decreased, while A clearance was either reported not to change[51] or increase.[8] Some catecholamine-metabolizing enzyme activities such as monoamine-oxidase (MAO)[54] and circulating metabolites such as NA sulfate[55] were also found to increase with age. The effect of age on catecholamine-synthesizing enzymes, as discussed further, is still controversial.

Most of the evidence suggests that an increased NA appearance is the predominant mechanism resulting in elevated plasma catecholamines, this being a manifestation of increased peripheral neuronal activity with aging. The primary cause of this selective activation of noradrenergic postganglionic terminals is unknown. The reciprocity between the age-related decrease of adrenergic receptor sensitivity and increased NA release has raised the hypothesis that increased plasma NA is due to a compensatory increase in sympathetic nervous activity to the heart due to the reduced responsiveness of adrenergic receptors to stimulation.[6,56] It is not established at this time whether sympathetic hyperactivity of aging results in adrenergic receptor desensitization or whether altered receptor/tissue responsiveness activates reflexes which increase sympathetic activity. However, it appears that not all of the aspects of a decreased β-adrenergic receptor responsiveness in aging can be explained on the basis of homologous desensitization.[56] Such a mechanism could not explain the normal α_2-receptor-mediated suppression of plasma NA and NA appearance in elderly man.[58] Of particular interest is the observation that the capacity to inhibit sympathetic activity after response to a stimulus appears to be blunted in the elderly, resulting in longer peak NA excursions in older individuals than in their younger counterparts.[58]

It appears to be most plausible that the heightened sympathetic nervous activity with age is mediated by increased central nervous stimulation. It has been proposed that decreased baroreceptor sensitivity associated with aging and/or hypertension results in increased sympathetic nerve activity. Peripheral nerve recording studies in healthy elderly subjects suggested an increased nerve activity,[59] which in turn could result in increased NA synthesis. Considerable evidence exists from animal[60,61] and, to a lesser extent, human[62] studies that catecholamine synthesis can be augmented by a variety of stimuli, including increased nerve activity. Tyrosine hydroxylase, the enzyme catalyzing the rate-limiting step in catecholamine, plays a key role in these phenomena. Short-term stimuli tend to increase the activity of this enzyme by its phosphorylation,[61] whereas longer stimulations increase levels of this enzyme and its mRNA.[60] Thus, it is possible that increased sympathetic nerve activity in aged healthy controls results in increased catecholamine synthesis and subsequently increased NA entry into the bloodstream. Animal data suggest that tyrosine hydroxylase activity and mRNA levels are increased with aging in individual rat superior cervical ganglia.[63,114]

In spite of the many manifestations of heightened sympathetic activity in the aging sympathetic nervous system, tissue levels of catecholamines appear to decline. Sympathetic ganglia from aged human subjects exhibit marked depletion of catecholamine histofluorescence.[64] Similar findings were observed in human gingival tissue vessels[65] and for NA concentrations in human posterior tibial arteries.[66] It is not possible to determine at this time whether these findings reflect decreased numbers of nerve fibers or the depletion of catecholamine stores in individual neurons and their terminals.[67] The preganglionic component of the human sympathetic system has revealed only very modest decreases in cell numbers and fibers in aging.[68-70] Nevertheless, these findings are compatible with the hypothesis that increased release of NA is maintained at the cost of augmented NA synthesis and depletion of tissue catecholamine stores. Decreased NA reuptake could tilt the balance even further toward heightened NA spillover and depletion of stores, particularly at the level of the nerve terminals.[71,72] However, in spite of the above findings, healthy older human subjects generally exhibit normal and often heightened sympathetic response to most challenges (Table 2). These observations could be reconciled by the fact that although physiological studies are performed on well-screened healthy controls, tissue samples, particularly those obtained postmortem, are from subjects whose health status is generally less well established. However, these findings could also be reflective of the profound plasticity and perhaps redundancy present in the SNS.[67] The adrenomedullary autonomic mechanisms, reflected by A determination, remain relatively unaffected by aging.

III. DOPAMINE IN PLASMA AND OTHER BODY FLUIDS IN AGING

Dopamine (DA), the third circulating catecholamine, has usually been omitted from pathophysiological studies of the autonomic nervous system, mostly because its circulating plasma concentrations are near the limit of detection.[73] DA concentrations are higher in other body fluids such as urine. However, DA measurements in these fluids are considered to be unreliable markers of events occurring at the level of peripheral dopaminergic terminals.[74] Urinary DA appears to predominantly reflect the extraneuronal renal decarboxylation of plasma dihydroxyphenylalanine (DOPA) entering the kidney.[75] Furthermore, plasma dopamine was considered to be simply a catecholamine precursor rather than a functional catecholamine on its own. Only recently have well-characterized dopaminergic receptors been found to play an important homeostatic role not only in the brain, but also in the peripheral nervous system, including its cardiovascular and renal targets.[76]

Thus, a comprehensive survey of specific alterations in SNS function associated with normal aging[77] is not complete without a discussion of changes in DA and circulating markers of its secretion.[24,78] A large body of knowledge has been accumulated regarding age-related changes in dopaminergic pathways in the central nervous system,[79] yet very little is known regarding the impact of age on the peripheral dopaminergic system. Although plasma DA levels in aging have not yet been evaluated, the urinary excretion of DA appears to decline by nearly 25% in healthy elderly individuals.[80] In addition, elderly hypertensive patients have a lower natriuretic and urinary dopamine response to high salt and high protein intake.[81] Plasma levels of DOPA, the precursor of DA, also do not increase with age and may, in fact, decrease.[82] The divergent changes in levels of NA and its precursors, including DA, in aging cannot be reconciled on the basis of dopamine-β-hydroxylase (DβH) activity alone. Although serum DβH activity has been reported to increase with age, no subjects over the age of 60 were examined in this study.[83] Using the less reliable spectrophotometric determination of DβH, no differences were observed in DβH levels between young and aged normotensive subjects.[84]

Despite limited information on this subject, it appears that, in contrast to the increasing plasma NA with age, there is no change or even a loss in peripheral dopaminergic activity with age. Although much of this work has been limited to animal studies, some of these functional alterations may be due to a decrease in DA-containing nerve fibers in the kidney, as well as a loss of tubular and, to a lesser degree, vascular DA receptor sites.[78] Among other possible explanations for these changes would be degeneration or functional impairment of DA-containing fibers while sparing NA-containing fibers.[85] Further studies will be required to confirm these results and to evaluate the nature of the enhanced vulnerability of the peripheral dopaminergic system, as contrasted with the noradrenergic system, in aging.

IV. CIRCULATING CATECHOLAMINE PRECURSORS AND THEIR AGE-DEPENDENT CHANGES

The increase in plasma NA with progressing age is, as previously discussed, predominantly due to an increased NA release. Measurements of catecholamine precursor availability may have important implications for our understanding of the regulation of SNS activity in aging.[86] The synthesis of catecholamines appears to be controlled, in part, by the availability of the amino acid precursor tyrosine.[87] Wurtman's concept of precursor control of neurotransmitter synthesis[86] raises the possibility that an increased precursor supply could be a primary moving force initiating the increased neuronal turnover and release of catecholamines, particularly NA. However, it is possible that an increased tyrosine precursor supply may be simply a phenomenon secondary to the increased catecholamine turnover necessitating a higher precursor supply.

Tyrosine, the main catecholamine precursor, originates partly in exogenous sources, being fairly abundant in dietary protein. However, most tyrosine originates from the phenylalanine content of ingested protein. At least half of the phenylalanine in protein is converted to tyrosine in each passage through the hepatic circulation by the action of the enzyme phenylalanine hydroxylase.[88] To a limited extent, tyrosine is also formed in the brain from phenylalanine.[89] Among other factors, the degree of catecholamine synthesis depends on the intraneuronal tyrosine available for the action of tyrosine hydroxylase to generate the main catecholamine precursor, DOPA.

Neuronal entry of tyrosine in the brain is dependent on a complicated interaction between tyrosine and other large neutral amino acids competing for the same blood-brain barrier transport system.[86] Tyrosine entry into peripheral neurons appears to be less hindered by such a barrier. The intraneuronal coupling of precursor levels to neurotransmitter synthesis is mediated by the degree of saturation of the rate-limiting tyrosine hydroxylase and several feedback mechanisms acting within the neuron itself (end-product inhibition of tyrosine hydroxylase), an extracellular link (presynaptic autoreceptors), and a multisynaptic reflex arc that diminishes the firing rate of the precursor-dependent neuron when the amount of transmitter release per firing has increased.[86] It has been suggested that feedback regulation exists at the level of the neuron, mostly at the rate-limiting tyrosine hydroxylase step, but not at the level of the circulation.

In recent years, there has been a considerable reevaluation of the physiological role of the extraneuronal circulating DOPA. Evidence suggests that plasma DOPA reflects its neuronal release[90] and is not of alimentary origin,[91] as initially suspected. It has been proposed that plasma concentrations of circulating DOPA in man can be considered to reflect indirectly the level of tyrosine hydroxylase activity.[90] However, this does not appear to be the case in rats, at least under conditions of immobilization stress.[92]

One way to approach the question of whether DOPA exerts a feedback-suppressive action on plasma tyrosine was to study the effect of oral administration of L-DOPA on the plasma tyrosine level. Cuche et al. observed a considerable (up to 50%) suppression of plasma tyrosine 4 h after administration of L-DOPA, but did not discuss this finding.[93] We encountered the age dependency of the L-DOPA action on plasma tyrosine while studying the effect of L-DOPA administration on plasma tyrosine in two categories of essential hypertensive (EH) patients having a difference in mean age of nearly 30 years. The need for age-matched control subjects led us to compare two groups of healthy subjects of two distinct age groups. We found that all subjects had a clear suppressibility of plasma tyrosine by oral L-DOPA administration. We interpreted this finding as indicating a possible inhibitory action

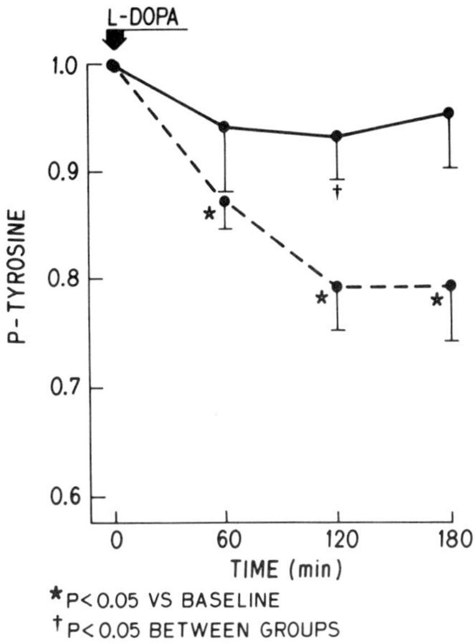

FIGURE 1. Plasma tyrosine suppressibility by a single oral L-DOPA administration (500 mg) in the older age group (full line) and in the younger age group (interrupted line).

(of the orally administered L-DOPA passing through the liver) on phenylalanine hydroxylase. This enzyme is very similar to tyrosine hydroxylase[94] and is responsible for phenylalanine conversion to tyrosine in the liver, a major source of circulating tyrosine. We found that tyrosine suppressibility by L-DOPA was lower in borderline EH patients having high sympathetic activity[95] than in stable EH patients who had a generally lower sympathetic tone.[96]

More pertinent to this review, control subjects of the older age group (middle aged, mean age 52) had a lower suppressibility of plasma tyrosine by L-DOPA than young (mean age 25) control subjects (Figure 1). Since the degree of tyrosine suppressibility by L-DOPA administration appeared to correlate inversely with sympathetic tone, this finding could be interpreted to suggest an increased sympathetic activity in older subjects. It is not clear to what degree physiological conclusions can be derived from the pharmacological plasma tyrosine suppressibility by L-DOPA. Nevertheless, one explanation for a decreased suppressibility of plasma tyrosine by L-DOPA in the older age group is that in the presence of an increased neural firing rate,[86] there is a more rapid precursor repletion of tyrosine following its inhibition by exogenous L-DOPA.

Those differences become of particular interest when considered with other findings of basal and post-L-DOPA-induced changes, apparently by DA generated from DOPA. Older subjects with significantly higher baseline systolic blood pressure (BP) than younger subjects had a greater systolic BP decrease and pulse rate increase, but a smaller diastolic BP decrease, than younger subjects (Table 4). The greater decrease in systolic BP could be due to higher basal systolic BP and a greater intraindividual variability present in older individuals, resulting in "regression to the mean".[97] In addition, this change could also reflect a suppression of the heightened sympathetic outflow associated (as reflected by a higher reflex pulse rate increase in the elderly) with the central inhibitory action of L-DOPA.[98] The lower diastolic BP decrease in the elderly is probably due to a lower vascular responsiveness to the vasodilating effect of DA despite its higher generation from L-DOPA. *Before L-DOPA administration,* at the precursor level, the baseline values of plasma tyrosine tended to be higher, and those of DOPA lower, in the older subjects, although these differences did not reach statistical significance. However, the plasma tyrosine/DOPA ratio was significantly higher in older subjects, compatible with a higher precursor flow of tyrosine and lower DOPA, as the latter is released to a lesser degree and instead utilized for catecholamine synthesis. Plasma 3-*O*-methyl-DOPA, the main metabolite of DOPA, was lower, but DA sulfate, an important metabolite of DA, was higher in older than in younger subjects. At the same time, urinary normetanephrine and NA sulfate excretions were higher in the older age group, while the urinary DOPAC excretion was lower (Table 5). *In response to L-DOPA,* older subjects decarboxylate more DOPA to plasma-free DA (a predominantly extraneuronal process), but have less increase in the main DA metabolites DOPAC and HVA in plasma and less urinary DOPAC increase than younger subjects. The plasma DOPA/DOPAC ratio was higher in elderly subjects. This ratio, which is higher because less intraneuronally generated DA is oxidized and is rather used for NA synthesis, suggests a lower intraneuronal deamination of DA, which is preferentially used for NA synthesis (Figure 2). A similar age-related impairment of DA deamination was recently observed in renal slices of aged rats incubated with L-DOPA.[99] Plasma aldosterone following L-DOPA was lower in older subjects, compatible with an *in vitro*-observed inhibitory action of the elevated plasma DA and its sulfate[100] on aldosterone secretion. In the remaining renal indices, older subjects had a lower creatinine clearance as well as a lower glomerular load of DOPA at the height of the L-DOPA-induced dopamine generation.

These findings suggest that under basal conditions, older subjects tend to have lower plasma levels of 3-*O*-methyl DOPA, main marker of DOPA. The urinary excretion of DOPAC, the main metabolite of DA, was also decreased in the older age group. Plasma DA sulfate levels and urinary excretion of NA sulfate and normetanephrine were increased in the older subjects. In response to L-DOPA, these subjects have more plasma-free DA generation from L-DOPA, and more DA sulfate, but less oxidative deamination and *O*-methylation of dopamine, resulting in less DOPAC and urinary DOPAC and HVA generation, respectively. In contrast to plasma DA, their urinary excretion of free DA excretions is less.

Older healthy subjects thus appear to have higher baseline indices of NA, but, in general, lower indices of dopaminergic activity. *In response to*

TABLE 4
Differences Between the Blood Pressure and Plasma Catecholamines, Their Precursors, and Aldosterone Responses of Young and Older Healthy Subjects to Oral L-DOPA (500 mg) (Mean ± SE)

	Time as related to L-DOPA administration (0) (min)	Young subjects (n = 10)	Older (middle-aged) subjects (n = 5)	Significance level
Mean age (range)		25.5 (19–29)	52.2 (48–57)	
Decrease in BP (Δ mmHg)				
Systolic	+180	3.3 ± 1.2	11.2 ± 3.5	a
Diastolic	+180	6.3 ± 0.4	3.3 ± 0.3	a
Increase in pulse rate (Δ beats/min)	+180	1.1 ± 0.3	3.1 ± 0.4	a
Plasma tyrosine (μg/ml)	−30	5.73 ± 0.5	7.13 ± 0.78	
Plasma DOPA (ng/ml)	−30	2.27 ± 0.13	1.99 ± 0.59	
Plasma tyrosine/DOPA ratio	−30	2682 ± 349	4781 ± 1179	b
Plasma 3-*O*-methyl-DOPA (ng/ml)	0	47.3 ± 5.2	20.5 ± 1.24	a
Plasma DA sulfate (ng/ml)	0	1.12 ± 0.21	2.95 ± 0.61	a
Plasma-free DA (ng/ml)	+120	2.47 ± 0.31	5.29 ± 0.76	a
Plasma DOPAC (ng/ml)	+120	553 ± 102	121 ± 35.7	a
Plasma HVA (ng/ml)	+120	444 ± 64	191 ± 21.4	a
Plasma DOPA/DOPAC ratio	+120	1.41 ± 0.27	6.67 ± 1.49	a
Plasma aldosterone (ng/dl)	+180	7.42 ± 0.64	4.97 ± 0.64	a

[a] $p < 0.05$ by Student's *t* test.
[b] $p < 0.05$ by Wilcoxon test.

TABLE 5
Differences Between Renal Indices and Urinary Catecholamines Baseline and in Response to Oral L-DOPA (500 mg) (Mean ± SE)

	Collection period prior to or after L-DOPA administration (0) (min)	Young subjects (n = 10)	Older (middle-aged) subjects (n = 5)
Creatinine clearance (ml/min)	+0–60	132 ± 5.9	97.7 ± 13.4
	+120–180	117.1 ± 4.5	83.8 ± 9.36
Urinary normetanephrine excretion (μg/h)	−1440	4.74 ± 0.48	13.38 ± 1.55
Urinary NA sulfate excretion (μg/h)	−60–0	1.88 ± 0.49	6.14 ± 1.6
Urinary free DA excretion (μg/h)	+60–120	14895 ± 1904	8599 ± 980
Urinary DOPAC excretion (μg/h)	−60–0	0.066 ± 0.01	0.006 ± 0.003
	+120–180	15.32 ± 2.36	7.44 ± 2.35
Glomerular DOPA load (μg/min)	+0–60	56.91 ± 11.2	27.41 ± 4.6

Note: $p < 0.05$, by Student's *t* test, for all differences between young and middle-aged subjects.

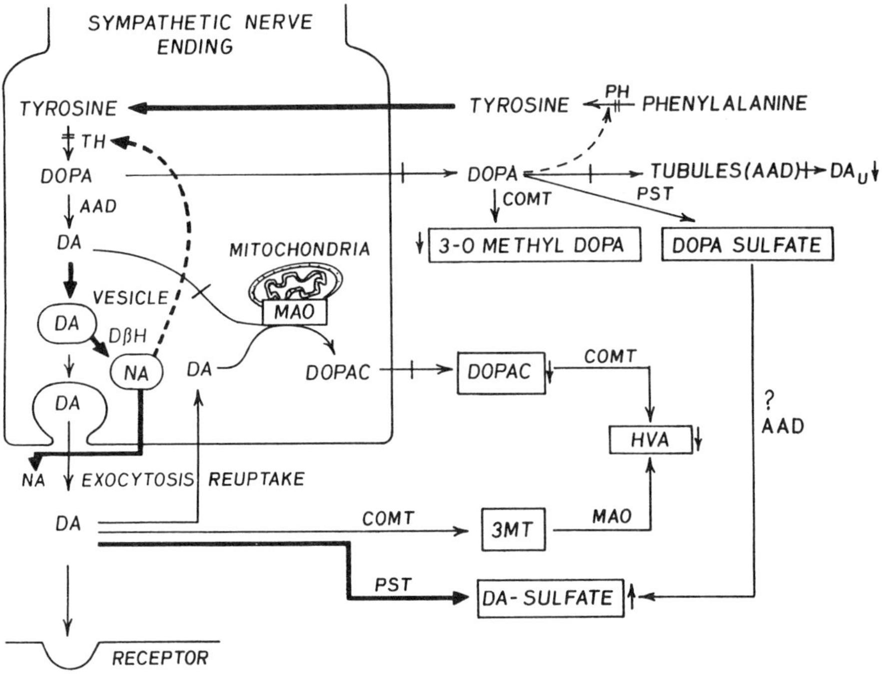

FIGURE 2. Schematic outline of the proposed effect of aging on catecholamines. Full lines represent stimulatory and interrupted lines inhibitory pathways; crossed lines represent a proposed age-related inhibition or strengthening of these pathways. The exogenous L-DOPA-induced tyrosine-suppressive action is extraneuronal, probably due to a decreased tyrosine generation from phenylalanine by the enzyme phenylalanine hydroxylase (PH). The baseline conditions in older subjects are characterized by higher-plasma circulating tyrosine (see strength of the arrows) and lower DOPA, the latter reflected by a decreased concentration of the main DOPA metabolite 3-*O*-methyl DOPA. The excretions of NA metabolites (normetanephrine and NA sulfate) were higher, but that of DOPAC, the main DA metabolite was lower. Following L-DOPA administration, a smaller increase of DOPAC in plasma and urine and HVA in plasma vs. a higher free DA generation from DOPA suggested that less DA generated from DOPA was intraneuronally metabolized in the elderly, probably because of its higher use as a precursor for NA generation. Thus, the observed age-related changes may be secondary to an increased NA appearance causing, by a strengthened intraneuronal feedback, an inhibition of the rate limiting TH activity, which may explain the decreased neuronal spillover of DOPA. The latter may result in a decreased glomerular DOPA load, urinary DA and 3-*O*-methyl-DOPA excretion, as well as attenuated PH inhibition by DOPA resulting in a higher tyrosine generation from phenylalanine. The sulfoconjugation of DA also appears to be accentuated with aging. TH, tyrosine hydroxylase; AAD, aromatic 1-amino acid decarboxylase; MAO, monoamine oxidase; PST, phenolsulfotransferase; DβH, dopamine-β-hydroxylase; COMT, catechol-*O*-methyltransferase; 3-MT, 3-methoxytyramine; DOPAC, dihydroxyphenylacetic acid; HVA, homovanillic acid.

L-DOPA, they exhibit more DOPA decarboxylation to DA. DA appears, however, to be preferentially used as a precursor to maintain elevated NA turnover with age rather than enter the metabolic and renal excretory pathways of DA. The only exception is the more avid sulfoconjugation of dopamine to DA sulfate in older subjects. Heightened NA generation with age is reflected by increased urinary excretion of normetanephrine and in plasma by increased NA sulfate, the latter reminiscent of the observed increase in plasma NA sulfate with age,[55] which was recently confirmed in a wider healthy population.[101] In contrast to plasma NA and DA changes, urinary-free DA excretion decreases with age (this preliminary finding is in agreement with a previous report[80]) and may possibly be related to a lower DOPA load which is partly due to a lower creatinine clearance at the height of the L-DOPA action. The lower DOPAC increase is probably also attributable to a lower increase of plasma DOPAC.

The ages of the two groups studied do not allow us to reach definitive conclusions about changes of these parameters of catecholamine function during the aging process. Nevertheless, a difference between a middle-aged and a young mature group is less likely to represent maturation than early aging changes. Many physiological alterations described with aging can be detected as early as the third decade[102] and may, in fact, be of great interest to gerontological research.[103]

The finding of lower basal levels of 3-*O*-methyl DOPA, an indirect marker of DOPA release in the elderly, is in accordance with Goldstein's proposed trend of decreased plasma DOPA with age[82] and with several other indices of decreased dopaminergic activity.[24] The L-DOPA administration-induced changes replicate, at least partially, the precursor function of endogenous DOPA and underscore distinct responses in the plasma and urinary catecholamines and their metabolites.

These data indirectly suggest that a higher precursor supply may play a role, albeit a secondary one, in the increased release of NA with age. Nevertheless, they do not negate the importance of a later biosynthetic step, including DOPA decarboxylation to DA and its β-hydroxylation in the increased appearance of NA with age.

V. BIOLOGICAL AND CLINICAL IMPLICATIONS OF CATECHOLAMINE CHANGES WITH AGING

The previously described age-related increase in basal and stimulated NA results, with persistent stimulation throughout the day, in a gradual development of a hyperadrenergic state in the elderly.[104] Age-related increases in sympathetic nerve impulse traffic accompanying the increased plasma NA level[105] suggest that this hyperadrenergic state is widespread and affects convincingly at least the two regional sympathetic activities (muscle and skin) which are accessible to microelectrode recording.

The uniformity of this age-dependent change has led to the hypothesis that plasma NA may become an index of biological age.[17] When taking into account all the known vascular, renal, and metabolic actions of NA, including the finding that older subjects had a greater BP response to infused NA when on high salt intake,[106] it would not be surprising to find some harmful cardiovascular repercussions related to a higher incidence of even subclinical cardiovascular pathologies in the apparently healthy aging population.

Against the physiological background of an apparently fragile hyperadrenergic state with aging, normal age-related limitations in cardiovascular homeostasis may become clinically significant under stress, such as orthostatism or moderate diuretic-induced sodium depletion. Response to such stimuli suggests that although the healthy elderly may appear well adjusted to environmental challenges under optimal conditions, decreased cardiovascular reserve renders them susceptible to an excessive postural BP decrease following mild sodium depletion.[107] An additional hypotensive and hypovolemic stress may become operative when postural change follows eating-induced splanchnic vasodilation. Postprandial hypotension had been reported in elderly institutionalized subjects even without syncope,[108] while BP remains unchanged in young healthy subjects after a meal. The causes of this vulnerability of the elderly to orthostatism are certainly more complex, such as greater risk of sodium and volume depletion due to age-related decreases in thirst, renin response, and urinary-concentrating ability.[109] The most recent data suggest a predominant role of a dysfunction in the afferent limb of the baroreflex arc, possibly related to the effect of noncompliant arterial walls on arterial baroreceptors.[110] Nevertheless, it is the NA response after the meal (an elevated and, to a meal responsive plasma NA) in nonsyncopal elderly controls which discriminates them best from syncopal elderly subjects who were unable to maintain an elevated plasma NA level while BP levels were falling after a meal.[108]

Thus, the hyperadrenergic state of the elderly appears to be a very unsettled condition which can easily slip back into a functionally hypoadrenergic pathological state. This probably applies not only to postural adaptation, but also to other responses such as hypovolemia induced by dehydration. A limitation in the cardiovascular response to posture may also serve as a nonosmotic stimulus to vasopressin release predisposing to hyponatremia. Excessive BP decreases and hyponatremia are seen with increasing frequency in elderly subjects under the stress of illness or drugs such as diuretics. These are probably the conditions in which an even, sustained hyperadrenergic state of the elderly is unable to cope with a sudden need for an adaptive adrenergic response, despite the previously outlined evidence of hyperresponsiveness in the elderly to usually milder physiological stimuli.

Recognition of the physiological hyperadrenergic state in the elderly is also of importance for resolving the controversy about sympathetic nervous activity in essential hypertension. With better age matching, no differences between normotensive and hypertensive subjects were found above the age of 40. This suggests that elevated plasma NA alone would not account for hypertension in the elderly.[111]

The finding that elevated NA alone is not accountable for hypertension in the elderly justifies a search for other catecholamine- and noncatecholamine-related abnormalities which may contribute to the higher incidence of hypertension with aging. As far as catecholamines are concerned, the inadequacy of dopamine may be of importance since, as previously outlined, dopaminergic mechanisms appear to weaken with advanced age. The impact of this decrease may be accentuated by the increasing activity of NA, whose action opposes DA at the level of vascular and partly renal and adrenal targets.[24] The predominance of NA over DA with aging may explain the observation of an age dependence of the salt sensitivity of BP; the BP was increased in the older age group only in countries with high salt intake.[112] As previously demonstrated there is also an increasing sensitivity of BP to dietary sodium with increasing age.[113] The dopaminergic deficiency with progressing age, whether absolute or relative to NA, may also account for an increased vulnerability of elderly subjects to civilization factors other than excess salt, such as high protein intake[81] and sedentarism.[24]

REFERENCES

1. **Christensen, N. J.,** Plasma noradrenaline and adrenaline in patients with thyrotoxicosis and myxoedema, *Clin. Sci. Mol. Med.*, 45, 163, 1973.
2. **Pedersen, E. B. and Christensen, N. J.,** Catecholamines in plasma and urine in patients with essential hypertension determined by double-isotope derivative techniques, *Acta Med. Scand.*, 198, 373, 1975.
3. **Ziegler, M. G., Lake, C. R., and Kopin, I. J.,** Plasma noradrenaline increases with age, *Nature (London)*, 261, 333, 1976.
4. **Prinz, P. N., Halter, J., Benedetti, C., and Raskind, M.,** Circadian variation in plasma catecholamines in young and old men: relation to eye movement and slow wave sleep, *J. Clin. Endocrinol. Metab.*, 49, 300, 1979.
5. **Bertel, O., Buhler, F. R., Kiowski, W., and Lutold, B. E.,** Decreased beta-adrenoceptor responsiveness as related to age, blood pressure, and plasma catecholamines in patients with essential hypertension, *Hypertension*, 2, 130, 1980.
6. **Krall, J. F., Connelly, M., Weisbart, R., and Tuck, M. L.,** Age-related elevation of plasma catecholamine concentration and reduced responsiveness of lymphocyte adenylate cyclase, *J. Clin. Endocrinol. Metab.*, 52, 863, 1981.
7. **Brodde, O.-E., Anlauf, M., Graben, N., and Bock, K. D.,** Age dependent decrease of α_2-adrenergic receptor number in human platelets, *Eur. J. Pharmacol.*, 81, 345, 1982.
8. **Wilkie, F. L., Halter, J. B., Prinz, P. N., Benedetti, C., Eisdorfer, C., Atwood, B., and Yamasaki, D.,** Age-related changes in venous catecholamines basally and during epinephrine infusion in man, *J. Gerontol.*, 40, 133, 1985.
9. **Vargas, E., Lye, M., Faragher, E. B., Goddard, C., Moser, B., and Davies, I.,** Cardiovascular haemodynamics and the response of vasopressin, aldosterone, plasma renin activity and plasma catecholamines to head-up tilt in young and old healthy subjects, *Age Ageing*, 15, 17, 1986.
10. **MacGilchrist, A. J., Hawksby, C., Howes, L. G., and Reid, J. L.,** Rise in plasma noradrenaline with age results from an increase in spillover rate, *Gerontology*, 35, 7, 1989.
11. **Lehmann, M., Keul, J., Huber, J. K. G., Bachl, N., and Simon, G.,** Alters- und belastungsbedingtes verhalten der plasmacatecholamine, *Klin. Wochenschr.*, 59, 19, 1981.

12. **Raskind, M. A., Peskind, E. R., Veith, R. C., Beard, J. C., Gumbrecht, G., and Halter, J. B.,** Increased plasma and cerebrospinal fluid norepinephrine in older men: differential suppression by clonidine, *J. Clin. Endocrinol. Metab.*, 66, 438, 1988.
13. **Cryer, P. E., Silverberg, A. B., Santiago, J. V., and Shah, S. D.,** Plasma catecholamines in diabetes, *Am. J. Med.*, 65, 407, 1978.
14. **Young, J. B., Rowe, J. W., Pallotta, J. A., Sparrow, D., and Landsberg, L.,** Enhanced plasma norepinephrine response to upright posture and oral glucose administration in elderly human subjects, *Metabolism*, 29, 532, 1980.
15. **Jones, D. H., Hamilton, C. A., and Reid, J. L.,** Plasma noradrenaline, age and blood pressure: a population study, *Clin. Sci. Mol. Med.*, 55, 73s, 1978.
16. **Rowe, J. W. and Troen, B. R.,** Sympathetic nervous system and aging in man, *Endocrinol. Rev.*, 1, 167, 1980.
17. **Christensen, N. J.,** Is plasma noradrenaline an index of biological age?, in *The Sympathoadrenal System*, Christensen, N. J., Henriksen, O., and Lassen, N. A., Eds., Munksgaard, Copenhagen, 1986, 267.
18. **Rowe, J. W. and Kahn, R. L.,** Human aging: usual and successful, *Science*, 237, 143, 1987.
19. **Lehmann, M. and Keul, J.,** The age- and training-dependent effects of physical work on plasma catecholamines and lactate acidosis, *Dtsch. Z. Sportmed.*, 12, 320, 1981.
20. **Lehmann, M. and Keul, J.,** Age-associated changes of exercise-induced plasma catecholamine responses, *Eur. J. Appl. Physiol.*, 55, 302, 1986.
21. **Lake, C. R., Ziegler, M. G., Coleman, M. D., and Kopin, I. J.,** Age-adjusted plasma norepinephrine levels are similar in normotensive and hypertensive subjects, *N. Engl. J. Med.*, 296, 208, 1977.
22. **Chen, M., Halter, J. B., and Porte, D., Jr.,** Plasma catecholamines, dietary carbohydrate, and glucose intolerance: a comparison between young and old men, *J. Clin. Endocrinol. Metab.*, 62, 1193, 1988.
23. **Sowers, J. R., Rubenstein, L. Z., and Stern, N.,** Plasma norepinephrine responses to posture and isometric exercise increase with age in the absence of obesity, *J. Gerontol.*, 38, 315, 1983.
24. **Kuchel, O. G. and Kuchel, G. A.,** Peripheral dopamine in pathophysiology of hypertension: interaction with aging and lifestyle, *Hypertension*, 18, 709, 1991.
25. **Rowe, J. W.,** Clinical research on aging: strategies and directions, *N. Engl. J. Med.*, 297, 1332, 1977.
26. **Defronzo, R. A.,** Glucose intolerance and aging: evidence for tissue insensitivity to insulin, *Diabetes*, 28, 1095, 1979.
27. **Barnes, R. M., Raskind, M., Halter, J., and Gumbrecht, G.,** The effects of age on the plasma catecholamine response to mental stress in man, *J. Clin. Endocrinol. Metab.*, 54, 64, 1982.
28. **Palmer, G. J., Ziegler, M. G., and Lake, C. R.,** Response of norepinephrine and blood pressure to stress increases with age, *J. Gerontol.*, 33, 482, 1978.
29. **Saar, N. and Gordon, R. D.,** Variability of plasma catecholamine levels: age, duration of posture and time of day, *Br. J. Clin. Pharmacol.*, 8, 353, 1979.
30. **Sowers, J. R., Rubenstein, L. Z., and Stern, N.,** Plasma norepinephrine responses to posture and isometric exercise increase with age in the absence of obesity, *J. Gerontol.*, 38, 315, 1983.
31. **Fleg, J. L., Tzankoff, S. P., and Lakatta, E. G.,** Age-related augmentation of plasma catecholamine during dynamic exercise in healthy males, *J. Appl. Physiol.*, 59, 1033, 1985.
32. **Ingall, T. J., McLeod, J. G., and O'Brien, P. C.,** The effect of ageing on autonomic nervous system function, *Aust. N.Z. J. Med.*, 20, 570, 1990.
33. **Miyazawa, K. and Yamaguchi, I.,** Cardiovascular response to exercise in healthy male septuagenarians: with reference to plasma norepinephrine, *Tohoku. J. Exp. Med.*, 143, 177, 1984.

34. **Minaker, K. L., Rowe, J. W., Young, J. B., Sparrow, D., Pallotta, J. A., and Landsberg, L.,** Effect of age on insulin stimulation of sympathetic nervous system activity in man, *Metabolism,* 31, 1181, 1982.
35. **Tonino, R. P., Minaker, K. L., Young, J. B., Landsberg, L., and Rowe, J. W.,** Splanchnic factors enhance the norepinephrine response to oral glucose in aged man, *Exp. Gerontol.,* 21, 413, 1986.
36. **McCarty, R.,** Age-related alterations in sympathetic-adrenal medullary responses to stress, *Gerontology,* 32, 172, 1986.
37. **Sachs, C., Hamberger, B., and Kaijser, L.,** Cardiovascular responses and plasma catecholamines in old age, *Clin. Physiol.,* 5, 553, 1985.
38. **Chen, M., Halter, J. B., and Porte, D., Jr.,** Plasma catecholamines, dietary carbohydrate, and glucose intolerance: a comparison between young and old men, *J. Clin. Endocrinol. Metab.,* 62, 1193, 1986.
39. **Van Brummelen, P., Buhler, F. R., Kiowski, W., and Amann, F. W.,** Age-related decrease in cardiac and peripheral vascular responsiveness to isoprenaline: studies in normal subjects, *Clin. Sci.,* 60, 571, 1981.
40. **Elam, H., Yao, T., Svensson, T. H., and Thoren, P.,** Regulation of locus coeruleus neurons and splanchnic, sympathetic nerves by cardiovascular afferents, *Brain Res.,* 290, 281, 1984.
41. **Ziegler, M. G., Lake, C. R., Wood, J. H., Brooks, B. R., and Ebert, M. H.,** Relationship between norepinephrine in blood and cerebrospinal fluid in the presence of a blood-cerebrospinal fluid barrier for norepinephrine, *J. Neurochem.,* 28, 677, 1977.
42. **Rubin, P. C., Scott, P. J. W., McLean, K., and Reid, J. L.,** Noradrenaline release and clearance in relation to age and blood pressure in man, *Eur. J. Clin. Invest.,* 12, 121, 1982.
43. **Hoeldtke, R. D. and Cilmi, K. M.,** Effects of aging on catecholamine metabolism, *J. Clin. Endocrinol. Metab.,* 60, 479, 1985.
44. **Veith, R. C., Featherstone, J. A., Linares, O. A., and Halter, J. B.,** Age differences in plasma norepinephrine kinetics in humans, *J. Gerontol.,* 41, 319, 1986.
45. **Schwartz, R. S., Jaeger, L. F., and Veith, R. C.,** The importance of body composition to the increase in plasma norepinephrine appearance rate in elderly men, *J. Gerontol.,* 42, 546, 1987.
46. **Poehlman, E. T., McAuliffe, T., and Danforth, E., Jr.,** Effects of age and level of physical activity on plasma norepinephrine kinetics, *Am. J. Physiol.,* 258, E256, 1990.
47. **Kjeldsen, S. E., Eide, I., Christensen, C., Westheim, A., and Müller, O.,** Renal contribution to plasma catecholamines — effect of age, *Scand. J. Clin. Lab. Invest.,* 42, 461, 1982.
48. **Meredith, I. T., Lambert, G. W., Jennings, G. L., and Esler, M. D.,** The effects of aging on the sympathetic nervous system, *J. Hypertens.,* 8 (Suppl. 3), s78, 1990.
49. **Stromberg, J. S., Linares, O. A., Supiano, M. A., Smith, M. J., Foster, A. H., and Halter, J. B.,** Effect of desipramine on norepinephrine metabolism in humans: interaction with aging, *Am. J. Physiol.,* 261, R1484, 1991.
50. **Esler, M., Skews, H., Leonard, P., Jackman, G., Bobik, A., and Korner, P.,** Age dependence of noradrenaline kinetics in normal subjects, *Clin. Sci.,* 60, 217, 1981.
51. **Hilsted, J., Christensen, N. J., and Larsen, S.,** Plasma clearance of noradrenaline does not change with age in normal subjects, *Clin. Physiol.,* 5(5), 443, 1985.
52. **Morrow, L. A., Linares, O. A., Hill, T. J., Sanfield, J. A., Supiano, M. A., Rosen, S. G., and Halter, J. B.,** Age differences in the plasma clearance mechanisms for epinephrine and norepinephrine in humans, *J. Clin. Endocrinol. Metab.,* 65, 508, 1987.
53. **Hoffmann, J. J. M. L., Lenders, W. M., Pijls, N. H. J., Benraad, Th. J., and Thien, Th.,** Vascular distribution of plasma catecholamines — pulmonary extraction, arterio-venous differences and effect of age, *Scand. J. Clin. Lab. Invest.,* 47, 179, 1987.
54. **Robinson, D. S.,** Changes in monoamine oxidase and monoamines with human development and aging, *Fed. Proc.,* 34, 103, 1975.

55. **Kuchel, O., Buu, N. T., Hamet, P., Larochelle, P., Bourque, M., and Genest, J.,** Catecholamine sulfates and platelet phenosulfotransferase activity in essential hypertension, *J. Lab. Clin. Med.,* 104, 238, 1984.
56. **Scarpace, P. J.,** Decreased receptor activation with age. Can it be explained by desensitization?, *J. Am. Geriatr. Soc.,* 36, 1067, 1988.
57. **Christensen, N. J.,** Catecholamines and sympathetic nervous activity in the elderly, *Acta Med. Scand.,* 676 (Suppl.), 52, 1983.
58. **Featherstone, J. A., Veith, R. C., Flatness, D., Murburg, M. M., Villacres, E. C., and Halter, J. B.,** Age and alpha-2 adrenergic regulation of plasma norepinephrine kinetics in humans, *J. Gerontol.,* 42, 271, 1987.
59. **Wallin, B. G., Sundlöf, G., Erikkson, B. M., Dominiak, P., Grobecker, H., and Lindblad, L. E.,** Plasma noradrenaline correlates to sympathetic muscle nerve activity in normotensive man, *Acta Physiol. Scand.,* 111, 69, 1981.
60. **Zigmond, R. E., and Ben-Ari, Y.,** Electrical stimulation of preganglionic nerve increases tyrosine hydroxylase activity in sympathetic ganglia, *Proc. Natl. Acad. Sci. U.S.A.,* 7, 3078, 1977.
61. **Zigmond, R. E., Schwarzschild, M. A., and Rittenhouse, A. R.,** Acute regulation of tyrosine hydroxylase by nerve activity and by neurotransmitters via phosphorylation, *Annu. Rev. Neurosci.,* 12, 415, 1989.
62. **Schalling, M., Stieg, P. E., Lindquist, C., Goldstein, M., and Hokfelt, T.,** Rapid increase in enzyme and peptide mRNA in sympathetic ganglia after electrical stimulation in humans, *Proc. Natl. Acad. Sci. U.S.A.,* 86, 4302, 1989.
63. **Kedzierski, W., and Porter, J. C.,** Quantitative study of tyrosine hydroxylase mRNA in catecholaminergic neurons and adrenals during development and aging, *Mol. Brain Res.,* 7, 45, 1990.
64. **Hervonen, A. A., Vaalasti, A., Partanen, M., Kanerva, L., and Hervonen, H.,** Effects of ageing on the histochemically demonstrable catecholamines and acetylcholinesterase of human sympathetic ganglia, *J. Neurocytol.,* 7, 11, 1978.
65. **Waterson, J. G., Frewin, D. B., and Soltys, J. S.,** Age-related differences in catecholamine fluorescence of human vascular tissue, *Blood Vessels,* 11, 79, 1974.
66. **Neubauer, B. and Christensen, N. J.,** The decrease in noradrenaline concentration in the posterior tibial artery with age, *Gerontology,* 24, 299, 1978.
67. **Kuchel, G. A. and Zigmond, R. E.,** Functional recovery and collateral neuronal sprouting examined in young and aged rats following a partial neural lesion, *Brain Res.,* 540, 195, 1991.
68. **Appenzeller, O. and Ogin, G.,** Myelinated fibers in the human paravertebral sympathetic chain; quantitative studies on white rami communicantes, *J. Neurol. Neurosurg. Psychiatry,* 36, 777, 1973.
69. **Low, P. A. and Dyck, P. J.,** Splanchnic preganglionic neurons in man. II. Morphometry of myelinated fibers of T7 ventral spinal root, *Acta Neuropathol.,* 40, 219, 1977.
70. **Low, P. A., Okazaki, H., and Dyck, P. J.,** Splanchnic preganglionic neurons in man. I. Morphometry of preganglionic cytons, *Acta Neuropathol.,* 40, 55, 1977.
71. **Bhatnagar, B. K. and Moore, K. E.,** Effects of electrical stimulation, α-methyltyrosine and desmethylimipramine on the norepinephrine contents of neuronal cell bodies and terminals, *J. Pharmacol. Exp. Ther.,* 178, 450, 1971.
72. **Bhatnagar, B. K. and Moore, K. E.,** Regulation of norepinephrine contents in neuronal cell bodies and terminals during and after cessation of preganglionic stimulation, *J. Pharmacol. Exp. Ther.,* 180, 265, 1972.
73. **Wang, P. C., Buu, N. T., Kuchel, O., and Genest, J.,** Conjugation patterns of endogenous plasma catecholamines in human and rat. A new specific method for analysis of glucuronide-conjugated catecholamines, *J. Lab. Clin. Med.,* 101, 141, 1983.
74. **Bell, C.,** Dopamine as a postganglionic autonomic neurotransmitter, *Neuroscience,* 7, 1, 1982.

75. **Goldstein, D. S., Stull, R., Eisenhofer, G., and Gill, J. R.,** Urinary excretion of dihydroxyphenylalanine and dopamine during alterations of dietary salt intake in humans, *Clin. Sci.*, 76, 517, 1989.
76. **Lokhandwala, M. F. and Hegde, S. S.,** Cardiovascular dopamine receptors. Role of renal dopamine and dopamine receptors in sodium excretion, *Pharmacol. Toxicol.*, 66, 237, 1990.
77. **Linares, O. A. and Halter, J. H.,** Sympathochromaffin system activity in the elderly, *J. Am. Geriatr. Soc.*, 35, 448, 1987.
78. **Amenta, F., Collier, W. L., Coppola, L., and De Michele, M.,** Senescence of the peripheral dopaminergic system, in *Peripheral Dopamine Pathophysiology*, Amenta, F., Ed., CRC Press, Boca Raton, FL, 1990, 243.
79. **Morgan, D. G. and Finch, C. E.,** Dopaminergic changes in the basal ganglia, a generalized phenomenon of aging in mammals, *Ann. N.Y. Acad. Sci.*, 515, 145, 1988.
80. **Clark, B. A., Young, J. B., West, C., Elahi, D., and Epstein, F. H.,** Decline in urinary excretion of dopamine and PGE_2 with age, *Clin. Res.*, 39 (Abstr.), 226A, 1991.
81. **Fukagawa, W. K., Bandini, L. G., Rowe, J. W., and Young, J. B.,** Effect of age on renal responses to protein ingestion in man, *Clin. Res.*, 37 (Abstr.), 490A, 1991.
82. **Garty, M., Stull, R., Kopin, I. J., and Goldstein, D. S.,** Skin color, aging, and plasma L-dopa levels, *J. Auton. Nerv. Syst.*, 26, 261, 1980.
83. **Freedman, L. S., Ohuchi, T., Goldstein, M., Axelrod, F., Fish, I., and Dancis, J.,** Changes in human serum dopamine-β-hydroxylase activity with age, *Nature (London)*, 236, 310, 1972.
84. **Annat, G., Vincent, M., Tourniaire, A., and Sassard, J.,** Relationships between blood pressure and plasma renin, aldosterone and dopamine-β-hydroxylase in the elderly. Studies in patients over 70 years of age, *Gerontology*, 27, 266, 1981.
85. **Bell, C.,** Anatomical aspects of growth and ageing in the autonomic nervous system, *Curr. Opin. Neurol. Neurosurg.*, 4, 524, 1991.
86. **Wurtman, R. J., Hefti, F., and Melamed, E.,** Precursor control of neurotransmitter synthesis, *Pharmacol. Rev.*, 32, 315, 1981.
87. **Johnston, J. L., Warsh, J. J., and Anderson, G. H. H.,** Obesity and precursor availability affect urinary catecholamine metabolite production in women, *Am. J. Clin. Nutr.*, 38, 356, 1983.
88. **Elwyn, D. H.,** The role of the liver in regulation of amino acid and protein metabolism, in *Mammalian Protein Metabolism*, Munro, H. N., Ed., Academic Press, New York, 1970, 523.
89. **Ikeda, M., Levitt, M., and Udenfriend, S.,** Hydroxylation of phenylalanine by purified preparations of adrenal and brain tyrosine hydroxylase, *Biochem. Biophys. Res. Commun.*, 18, 482, 1965.
90. **Goldstein, D. S., Udelsman, R., Eisenhofer, G., Stull, R., Keiser, H. R., Kopin, I. J., and Folio, C. J.,** Neuronal source of plasma dihydroxyphenylalanine, *J. Clin. Endocrinol. Metab.*, 64, 856, 1987.
91. **Banwart, B., Miller, T. D., Jones, J. D., and Tyce, G. M.,** Plasma dopa and feeding, *Proc. Soc. Exp. Biol. Med.*, 191, 357, 1989.
92. **Kvetnansky, R., Goldstein, D. S., Weise, V. K., Bagdy, G., Szemeredi, K., Holmes, C., and Kopin, I. J.,** Plasma dopa and tissue tyrosine hydroxylase during acute and repeated immobilization stress, *Soc. Neurosci.*, Abstr. 66.2, 16, 146, 1990.
93. **Cuche, J. L., Prinseau, J., Selz, F., Ruget, G., Tual, J. L., Reingeissen, L., Devoisin, M., Baglin, A., Guédon, J., and Fritel, D.,** Oral load of tyrosine or l-DOPA and plasma levels of free and sulfoconjugated catecholamines in healthy men, *Hypertension*, 7, 81, 1985.
94. **Masserano, J. M., Vulliet, P. R., Tank, A. W., and Weiner, N.,** The role of tyrosine hydroxylase in the regulation of catecholamine synthesis, in *Handbook of Experimental Pharmacology*, Vol. 90, *Catecholamines II*, Trendelenburg, U. and Weiner, N., Eds., Springer-Verlag, Berlin, 1988, 427.

95. **Shigetomi, S., Buu, N. T., and Kuchel, O.,** Dopaminergic abnormalities in borderline essential hypertensive patients, *Hypertension,* 17, 997, 1991.
96. **Kuchel, O. and Shigetomi, S.,** Deficient dopamine generation from L-DOPA in stable essential hypertension, *Hypertension,* 18 (Abstr.), 413, 1991.
97. **Kuchel, G. A., Avorn, J., Reed, M., and Fields, D.,** Cardiovascular responses to phlebotomy and sitting in middle-aged and elderly subjects, *Arch. Intern. Med.,* 152, 366, 1992.
98. **Judy, W. V., Watanabe, A. M., Henry, D. P., Besch, H. R., Jr., and Aprison, B.,** Effect of l-DOPA on sympathetic nerve activity and blood pressure in the spontaneously hypertensive rat, *Circ. Res.,* 43, 24, 1978.
99. **Soares-da-Silva, P. and Fernandes, M. H.,** A study on the renal synthesis of dopamine in aged rats, *Acta Physiol. Scand.,* 143, 287, 1991.
100. **Racz, K., Buu, N. T., Kuchel, O., and de Léan, A.,** Dopamine 3-sulfate inhibits aldosterone secretion in cultured bovine adrenal cells, *Am. J. Physiol.,* 247, E431, 1984.
101. **Ratge, D., Knoll, E., and Wisser, H.,** Plasma free and conjugated catecholamines in clinical disorders, *Life Sci.,* 39, 557, 1986.
102. **Schneider, E. L. and Rowe, J. W., Eds.,** *Handbook of the Biology of Aging,* 3rd ed., Academic Press, San Diego, 1990.
103. **Coleman, P. D., Finch, C. E., and Joseph, J. A.,** The need for multiple time points in aging studies, *Neurobiol. Aging,* 11, 1, 1990.
104. **Rowe, J. W.,** Plasma norepinephrine as an index of sympathetic activity in aging man, in *Banbury Report 27: Molecular Neuropathology of Aging,* Cold Spring Harbor Laboratory, Cold Spring Harbor, NY, 1987, 137.
105. **Wallin, B. G. and Fagius, J.,** The sympathetic nervous system in man — aspects derived from microelectrode recordings, *TINS,* February, 63, 1986.
106. **Dimsdale, J. E., Graham, R. M., Ziegler, M. G., Zusman, R. M., and Berry, C. C.,** Age, race, diagnosis, and sodium effects on the pressor response to infused norepinephrine, *Hypertension,* 10, 564, 1987.
107. **Shannon, R. P., Wei, J. Y., Rosa, R. M., Epstein, F. H. and Rowe, J. W.,** The effect of age and sodium depletion on cardiovascular response to orthostasis, *Hypertension,* 8, 438, 1986.
108. **Lipsitz, L. A., Pluchino, F. C., Wei, J. Y., Minaker, K. L., and Rowe, J. W.,** Cardiovascular and norepinephrine responses after meal consumption in elderly (older than 75 years) persons with postprandial hypotension and syncope, *Am. J. Cardiol.,* 58, 810, 1986.
109. **Shannon, R. P., Minaker, K. L., and Rowe, J. W.,** The influence of age on water balance in man, *Semin. Nephrol.,* 4, 346, 1984.
110. **Tonkin, A. L., Wing, L. M. H., Morris, M. J., and Kapoor, V.,** Afferent baroreflex dysfunction and age-related orthostatic hypotension, *Clin. Sci.,* 81, 531, 1991.
111. **Goldstein, D. S., Lake, C. R., Chernow, B., Ziegler, M. G., Coleman, M. D., Taylor, A. A., Mitchell, J. R., Kopin, I. J., and Keiser, H. R.,** Age-dependence of hypertensive-normotensive differences in plasma norepinephrine, *Hypertension,* 5, 100, 1983.
112. **Joossens, J. V.,** Sel et hypertension, dureté de l'eau et mortalité par affections cardiovasculaires, *Triangle,* 12, 9, 1973.
113. **Khaw, K. T. and Barrett-Connor, E.,** Increasing sensitivity of blood pressure to dietary sodium and potassium with increasing age, *Am. J. Hypertens.,* 3, 505, 1990.
114. **Kuchel, G. A., et al.,** unpublished data.

Chapter 5

SENSORY NEUROPEPTIDES AND AGING

Pierangelo Geppetti

TABLE OF CONTENTS

0-8493-6981-9/93/$0.00 + $.50

I. INTRODUCTION

We owe the introduction to physiopharmacology of capsaicin (8-methyl-*N*-vanillyl-6-nonenamide), the pungent principle contained in the plants of the genus *Capsicum,* to the late N. Jancso. He and his co-workers first recognized that this drug acted selectively on a subpopulation of unmyelinated or thinly myelinated primary sensory afferents.[1] Most of the neurons excited by capsaicin include small, dark cell-type neurons, which can be classified as C-fiber nociceptors. However, these neurons cannot be grouped according to any morphological, electrophysiological, or neurochemical criterion other than their sensitivity to capsaicin. This drug has the unique ability to excite these neurons in a time- and concentration-dependent manner and to desensitize them to a variety of pharmacological or physiological stimuli, including capsaicin itself.[2]

Capsaicin-sensitive primary afferents (CSPA) contain several peptides derived from larger polypeptide molecules. Some of these peptides are transported to the central and peripheral terminals of the neuron and then are released in a Ca^{2+}-dependent manner by a variety of stimuli at the level of both the spinal cord and peripheral tissues. Research during the past 20 years has disclosed the property of this subpopulation of sensory neurons which, in addition to conveying afferent impulses to the brain (sensory function), is also able, by local release of neuropeptides, to exert a variety of actions on peripheral tissues (''efferent'' function).[3] Various excellent reviews,[4,5] to which the reader is referred, have recently covered this topic. This chapter, after a brief overview on the general aspects of the physiopharmacology of CSPA, focuses on the morphological and functional aspects of these neurons and their peptide transmitters in aging and development. Particular attention is paid to the efferent function exerted by these neurons, since increasing evidence indicates that sensory neuropeptides released in peripheral organs may have a role in certain functions particularly relevant in aging, such as control of vascular tone, trophism, tissue response to injury, and tissue repair.

II. GENERAL ASPECTS OF CAPSAICIN-SENSITIVE AFFERENTS

Nerve fibers excited by capsaicin have been described in most mammalian species, including human, while, for instance, other animals classes such as birds are insensitive to this drug.[6] Recent studies have clarified the molecular basis of the action of capsaicin on sensory nerves. A ''vanilloid receptor'' has been described by using resiniferatoxin,[7] an ultrapotent analog of capsaicin. Binding to the putative receptor leads to the opening of a nonselective channel for cations, which allows the entry of Ca^{2+} and Na^+ and the efflux of K^+.[8-10] The biological action of capsaicin appears to be related to this ion flux, since generation of other intracellular messengers is secondary to this primary event. Capsaicin, by opening this channel, induces two kinds of events on CSPA. The first is a short-lasting stimulation that can be followed by reversible desensitization to capsaicin itself and other stimuli of sensory neurons. The second, obtained with suprathreshold doses, expresses itself in a long-term desensitization and morphological damage or even death of the neuron. It is easy to understand that these features, similar to those of guanetidine for sympathetic neurons, made capsaicin a critical tool that has contributed greatly to disclosing the functions exerted by CSPA.[11] Both the early excitatory effect and the delayed desensitization are Ca^{2+}-dependent events. However, while the absence of Ca^{2+} totally prevents peptide release, the influx of Na^{+} and the consequent passive entrance of H_2O into the nerve fibers plays a relevant role in its neurotoxic effect.[12]

Following stimulation of a peripheral ending of CSPA, the generated action potential is transmitted not only ortodromically to the brain, but also invades antidromically collateral fibers, with consequent depolarization-evoked peptide release. This mechanism is inhibited by tetrodotoxin and by voltage-sensitive Ca^{2+} channel (VSCC) blockers.[13,14] Inhibition by local anesthetics of the capsaicin-sensitive neurogenic flare produced by intradermal injection of histamine or by electrical stimulation[1] is a striking example of this phenomenon observed in humans. In contrast, the release evoked by capsaicin is not affected by local anesthetics, tetrodotoxin, or VSCC blockers, but is selectively inhibited by ruthenium red,[15,16] an inorganic dye with Ca^{2+} antagonist properties, which, at certain concentrations, has been shown to block selectively the cation channel operated by capsaicin.[17] Several exogenous and endogenous stimuli and autacoids have the ability to release peptides from CSPA. Some of these, including hydrogen ions[18,19] and prostacyclin,[20] appear to act at least in part by the same mechanism as that of capsaicin, while others such as bradykinin[21] are sensitive to VSCC inhibitor and still others, including veratridine[22] or hypertonicity,[81] may act through an as yet undefined mechanism. Studies on the modalities by which peptides are released from central and peripheral terminals underscore the complexity of this event and the increasing number of pharmacological tools that can intervene in its modulation.[23]

III. PEPTIDE TRANSMITTERS OF CAPSAICIN-SENSITIVE AFFERENTS

Among the large variety of peptides which have been described in the neuronal bodies of CSPA in the spinal, trigeminal, and vagal ganglia (see References 4, 5, 11, and 23), those for which Ca^{2+}-dependent release from both central and peripheral endings has been extensively demonstrated[4,5,11,23] are the tachykinins, substance P (SP) and neurokinin A (NKA), and calcitonin gene-related peptide (CGRP). From human tissue, release of vasoactive intestinal polypeptide (VIP) was also shown. SP and NKA are the products of larger precursors (α-, β-, and γ-preprotachykinin) which are the products of mRNAs originated by differential splicing of the tachykinin gene.[24] Neurokinin B, the third mammalian tachykinin, is not present in sensory neurons. The larger molecule from which CGRP derives is encoded in the calcitonin gene, and its mRNA is produced by an alternative splicing specific for the nervous tissue.[25] Upon release, tachykinin, by binding to NK-1, NK-2, and NK-3 receptors, and CGRP, by binding to CGRP-1 and CGRP-2 receptors, produce a large array of responses in effector cells.[4,5,11,23] Some of these effects, including vasodilation, plasma exudation, and leukocyte adhesion and diapedesis, have been collectively referred to as neurogenic inflammation, a phenomenon particularly important in the tissue response to injury. Regarding white blood cells, the action of tachykinins is not only directed to promote neutrophil extravasation, but also involves a variety of other different neutrophil functions, including chemokinesis,[26] oxidative metabolism,[27] enhancement of response to other mediators,[28] effects on T-lymphocytes,[29] and release of cytokines from monocytes.[30]

IV. "EFFERENT" FUNCTION OF CAPSAICIN-SENSITIVE AFFERENTS

Neurogenic inflammation, although a major function exerted by CSPA at the peripheral level, cannot be considered the only action exerted by peptides released from the peripheral terminals of these neurons. These peptides can produce many other responses on effector cells. Some of them are brought about in organs in which neurogenic inflammation takes place, and in this case, these responses can be considered part of the defense reaction. Miosis and changes in motility in the ciliary muscle, and contraction of the bronchial and tracheal smooth muscle may fall in this category. However, other effects have been reported that can barely be included in the host responses to noxious agents. These include the inotropic and chronotropic effect in the heart and relaxation in the vas deferens by CGRP, and the contraction and relaxation of the small intestine, and ureter and urinary bladder by tachykinins and CGRP, respectively.[4,5,11,23] In these cases, these responses may be evoked in response not only to pathological agents, but they may also be elicited in physiological conditions. Thus, the hypothesis can be put forward that the efferent function of CPSA may not only be considered as an emergency system against agents threatening the integrity of the tissue, but can also participate in the modulation of physiological functions such as peristalsis in the gastrointestinal and urinary tract.

V. PERIPHERAL EFFECTS RELEVANT IN AGING

The aforementioned effects produced by CPSA underscore the large array of functions in which changes of these neurons with age may affect organisms in the elderly. However, certain functions may have a specific and relevant importance in aging. A factor controlled by sensory neuropeptides and potentially relevant in aging is blood flow. Regional blood flow is increased by SP[31] or CGRP[32] in the airways and by CGRP in the stomach[33] and skin.[34]

Increased blood flow in the stomach and skin seems to be relevant for protection against the formation of gastric[35] and cutaneous ulcers,[36] respectively. A possible role of CSPA in the host response to noxious agents is also inferred by the observation that capsaicin-evoked depletion of sensory nerves reduced antibody production after exposure to antigen.[37] On the other hand, tachykinins have been shown to enhance the functions of various immunocompetent cells[38] and CGRP was found to promote proliferation of human polymorphonuclear cells.[39]

Recent evidence supports the view that tachykinins and CGRP may have a role in the maintenance of tissue trophism. Rats pretreated with a dose of capsaicin that ablated sensory nerves evidenced spontaneous lesions around the nose, eyes, and ears, a larger incidence of acid-evoked skin ulcerations, and a decrease in regrowth of abdominal skin hair.[36] The adhesion of cutaneous flaps is severely reduced in capsaicin-pretreated rat[40] and, conversely, enhanced by CGRP.[41] Even at the cellular level, evidence emerges indicating sensory neuropeptides as agents potentially capable of maintaining tissue integrity. Tachykinins were found to promote proliferation of human fibroblasts[42] and to be a mytogenic factor for human endothelial cells and to favor angiogenesis.[43]

The actions of sensory neuropeptides are also directed to modulate certain functions that may be involved in diseases particularly frequent in aging, such as type II diabetes, hypertension, and cerebrovascular diseases. The nonadrenergic, noncholinergic vasodilation produced by electrical stimulation in the rat mesenteric arteries, which is thought to be due to CGRP released from sensory nerves,[44,45] is markedly lowered in spontaneously hypertensive rats.[46] Vasodilation evoked by sensory neuropeptides in the cerebral circulation may play a role in stroke.[47] CGRP, as well as amylin, exerts a hyperglycemic effect, since it induces inhibition in the glucose uptake evoked by insulin.[48]

VI. AGE-RELATED SENSITIVITY TO TROPHIC FACTORS

The role of growth factors in the development of many cell types has been clearly established. Capsaicin-sensitive sensory nerves do not differ from this general rule. In fact, as shown for sympathetic neurons, sensory neurons also require nerve growth factor (NGF) for survival.[49] However, there is a striking age-related difference between sensory neurons taken from newborn or adult rat regarding the response to NGF. NGF is necessary for survival of sensory neurons taken from newborn rats which, in its absence, can degenerate and die.[50] In contrast, sensory neurons obtained from adult rats can be maintained in culture without an NGF supply, but these neurons show downregulation of SP and the CGRP level. This action does not seem to be due to a higher rate of peptide degradation or release, but, rather, to be the result of the action of the growth factor on protein synthesis, since it was associated with a decrease in mRNA for CGRP and tachykinins.[50] Also, sensitivity to capsaicin, and possibly expression of the molecular structure, now recognized as the putative capsaicin receptor and/or the channel operated by this drug, is downregulated in adult sensory neurons in cultures not supplemented with NGF.[51] Therefore, CSPA exhibit a dramatic sensitivity to NGF that expresses itself in an age-related fashion, being crucial for survival in the early phase of development, and still pivotal for their functions in the mature neurons.

VII. DEVELOPMENT

Considering aging as a continuum from the early phases of fetal and perinatal development to maturity and senescence, it seems appropriate to summarize studies that have addressed morphofunctional aspects of CSPA during ontogenesis. There is a broad consensus in the literature that from fetal life throughout the perinatal period, there is an increased expression of nerve fibers positive for sensory neuropeptides. With few exceptions, the maximal increase in sensory neuropeptide content or in fibers stained for these peptides is reported around the early postnatal period. In rat skin, both immunoreactivity for SP and CGRP underwent a time-related increase in the fetus, reaching a maximal level a few days after birth.[52,53] Likewise, the number of CGRP-positive nerve fibers, which were more abundant in the atria than in the ventricle, peaked in the perinatal period in the dog heart.[54] The nerves containing SP and CGRP in the mesenteric vascular bed, and those containing CGRP in the intracranial arterial vessels,[55,56] completed their pattern of distribution, resembling that of the adult animal, around the first week of age. Even in the inner ear of the guinea pig, the content of immunoreactive SP was higher at 2 d of life than in the adult animal.[57] In contrast with these reports, the CGRP content of the rat stomach which is mostly contained in CSPA was found to be decreased in the perinatal period.[58]

Morphological aspects of CSPA and their peptide transmitters during ontogenesis are not abundant in the literature. However, even fewer studies report functional data on the efferent function exerted by these nerves. These studies suggest, as shown for morphological development, that an age-related change in the response produced by release of sensory neuropeptides can also occur in the early phases of life. Responses comparable to those obtained in the adult animal were observed relatively later compared to the time course of maturation of fibers from a morphological point of view. For instance, the contractile effect of capsaicin in the rat urinary bladder peaked at 20 d of age. Since the response to SP did not change with age, the conclusion was made that changes in the response to capsaicin were due to an age-related maturation of sensory fibers and not of postjunctional factors on smooth muscle, including tachykinin receptors.[59] Similar results were reported for the potentiation of the electrically evoked contraction of the rabbit airways by SP.[60] In this model, the response showed an age-related increase until 11 weeks after birth. These findings do not allow any firm conclusion. However, considering that a higher content of peptides is found in the perinatal period, while their function appears to mature in a later phase, it is tempting to advance the hypothesis that the remarkable increase in sensory neuropeptide content observed in the perinatal period might be related to the fact that these peptides in the early phase of life may exert predominantly a developmental role rather than the actions commonly observed in the adult animals.

VIII. AGING

Few studies have investigated the morphological aspects of CSPA and the responses that they produce at the peripheral level in aging. With certain exceptions, a common finding emerges from these studies, indicating a decrease in peptide content and function of CSPA with aging. The binding of ^{125}I-CGRP was found to be increased in old rats,[61] and this finding may be interpreted as a compensatory mechanism to a possible decline of the peptide level in the spinal cord. Interestingly, CGRP-positive fibers were found to be decreased in the dorsal horn of spontaneously hypertensive rats (SHR) compared to age-matched normotensive Wistar-Kyoto (WKY) animals.[62] In the rat, dilation in response to electrical stimulation of mesenteric artery was due to CGRP release from CSPA.[44,45] This response decreased slightly with age in normotensive rats, but was greatly decreased in SHR.[46] Likewise, the release of immunoreactive CGRP in response to electrical stimulation was markedly decreased in SHR, but not in normotensive rats. In contrast, vasodilation induced by exogenous CGRP increased with age in SHR, whereas a decrease was found in WKY.[46] Altogether, these studies underscore the possible role of vasodilatory sensory neuropeptides in the development of hypertension. Furthermore, aging appears to be a crucial concurrent factor for the worsening of this pathology. In contrast with findings in the mesenteric vessels, at the level of the rat cerebral circulation, fibers positive for CGRP and VIP increased strikingly in old animals, while vasoconstrictor agents such as neuropeptide Y decreased.[63]

A decrease in peptides with age has been generally reported in other organs such as the gastrointestinal tract or the airways. SP in the jejunum only and VIP in all compartments of the small intestine were lowered in senescent rats.[64] In another study, the same pattern of decrease was observed in the small intestine for these two peptides.[65] In guinea pig bronchi, the contractile response to nicotine, due to SP released from sensory nerves, decreased with age.[66] In the same study, a lowering of SP release by nicotine in aged animals was also found.[66] In rat lung, the amount of immunoreactive SP was enhanced in aged animals.[67] However, this study did not investigate whether the immunoreactive SP found in the lung was contained in sensory nerves.

IX. STUDIES IN HUMANS

The pungent property of capsaicin has been extensively described in humans. This drug, if applied to human skin or mucosal surfaces,[1,68,69] can produce a painful sensation which undergoes desensitization after repeated administrations. This effect is due to excitation of CSPA. The possibility that in humans these neurons also exert an efferent function is due to the observation that the neurogenic flare produced in human skin by various stimuli is blocked by desensitization to capsaicin of the cutaneous area.[1] However, there is still great uncertainty regarding which molecules could be released from CSPA and play a role in such effects. Ruthenium red-sensitive release of immunoreactive CGRP by capsaicin has been reported from human coronary artery *in vitro*.[70] More recently, release by capsaicin of immunoreactive CGRP was reported from slices of human iris and ciliary body.[71] In this latter study, the capsaicin-evoked release underwent rapid tachyphilaxis and high-performance liquid chromatography (HPLC) analysis showed that authentic CGRP was present in the immunoreactive material released. Altogether, these studies suggest that CGRP can be released from peripheral endings of sensory neurons by the same mechanisms that have been shown previously in experimental animals. Capsaicin is also able to release VIP from the human small intestine *in vitro*.[72] Since this release was not blocked by tetrodotoxin,[73] and was inhibited by ruthenium red,[74] the hypothesis was advanced that VIP is released from sensory nerves in this human tissue. On the other hand, there is no report of secretion of SP from any human preparation. The results of these few *in vivo* and *in vitro* studies are sufficient to support the notion that in human beings there is a population of primary sensory neurons that, being excited by capsaicin, not only lead to the generation of sensory impulses and consequently of peripheral and central reflexes, but also to peripheral release of peptides basically in the same manner already shown in experimental animals.

Obviously, in human tissue, it is not possible to equate immunoreactivity for SP, CGRP, or, even more, VIP and nerve fibers positive for these peptides with CSPA. Keeping this point in mind, an overview of the literature indicates an early occurrence of fibers and neurons showing positive reactions for antisera directed to these peptides. Nerve fibers showing SP immunoreactivity were present in the dorsal horn of the fetal spinal cord as early as week 5 of gestation[74] as well as in the newborn and adult brainstem.[75] Cell bodies positively stained for this peptide were, however, found only in the dorsal ganglia of 28- to 40-week-old fetuses.[76] A dense immunoreactive SP innervation was present within the spinal trigeminal nucleus of the human infant.[77] Similar results were reported in another study in which an age-related increase in SP-immunoreactive nerve fibers was observed in the first postnatal year.[78] In agreement with data obtained in the rat, it seems that after the perinatal period, a decrease in sensory neuropeptide also takes place in the human. The percentage of perikaria positive for SP in the trigeminal ganglion in the human newborn was 23.6%, and dropped to 16.7% in adult subjects.[79]

From a functional point of view, only one study reported data on the effect of activation of CSPA at the peripheral level in aging.[80] This study reported a decrease in neurogenic flare reaction to topical capsaicin in the skin of aged men. This finding led the authors to advance the hypothesis that a loss of peptide in afferent nerve fibers may play a role in the morphological and functional changes observed in the skin of the elderly. Further and more numerous studies are needed in both experimental animals and humans to define the extent to which aging of CSPA may be relevant in the pathophysiology of the elderly. Recent knowledge on the broad array of functions exerted by these neurons in different tissues and organs, potentially important in aging, will certainly favor this promising field of investigation.

ACKNOWLEDGMENTS

This work was supported by grants from CNR (NATO-Senior) and CNR (Progetto bilaterale), Rome, Italy. I am grateful to Dr. C. Bertrand for his helpful comments on the manuscript.

REFERENCES

1. **Jancso, N., Jancso-Gabor, A., and Szolcsanyi, J.,** The role of sensory nerve endings in neurogenic inflammation in the human skin and in the eye and paw of the rat, *Br. J. Pharmacol.,* 33, 32, 1968.
2. **Jancso, N., Jancso-Gabor, A., and Szolcsanyi, J.,** Direct evidence for neurogenic inflammation and its prevention by denervation and by pretreatment with capsaicin, *Br. J. Pharmacol.,* 31, 138, 1967.
3. **Szolcsanyi, J.,** Capsaicin-sensitive chemoceptive neural system with dual sensory-efferent function, in *Antidromic Vasodilatation and Neurogenic Inflammation,* Chahl, L. A., Szolcsanyi, J., and Lembeck, F., Eds., Akademiai Kiado, Budapest, 1984, 26.
4. **Maggi, C. A.,** The sensory-efferent function of capsaicin-sensitive sensory neurons, *Gen. Pharmacol.,* 19, 43, 1988.
5. **Holzer, P.,** Local effector functions of capsaicin-sensitive sensory nerve endings: involvement of tachykinins, calcitonin gene-related peptide and other neuropeptides, *Neuroscience,* 24, 739, 1988.
6. **Pierau, F.-K., Sann, H., Harti, G., and Gamse, R.,** Neuropeptides in sensory neurons in pigeons and the insensitivity of avians to capsaicin, in *Fine Nerve Fibres and Pain,* Schmidt, R. F., Schaible, H.-G., and Vahle-Hinz, C., Eds., VCH Publishers, Weinheim, West Germany, 1987, 213.
7. **Szallasi, A. and Blumberg, P. M.,** Specific binding of resiniferatoxin, an ultrapotent analog of capsaicin, the irritant constituent of red pepper, *Brain Res.,* 524, 106, 1990.
8. **Marsh, S. J., Stansfeld, C. E., Brown, D. E., Davey, R., and McCarthy, D.,** The mechanism of action of capsaicin on sensory C-type neurons and their axons in vitro, *Neuroscience,* 23, 275, 1987.
9. **Wood, J. N., Winter, J., James, I. F., Rang, H. P., Yeats, J. C., and Bevan, S.,** Capsaicin induced ion fluxes in dorsal root ganglion cells in culture, *J. Neurosci.,* 8, 3208, 1988.
10. **Bleakman, D., Borson, J. R., and Miller, R. J.,** The effect of capsaicin on voltage-gated calcium currents and calcium signal in cultured dorsal root ganglion cells, *Br. J. Pharmacol.,* 101, 423, 1990.
11. **Holzer, P.,** Capsaicin: cellular targets, mechanism of action, and selectivity for thin sensory neurons, *Pharmacol. Rev.,* 43, 143, 1991.

12. **Bevan, S. and Szolcsanyi, J.,** Sensory neuron-specific actions of capsaicin-mechanisms and applications, *Trends Pharmacol. Sci.,* 11, 330, 1990.
13. **Maggi, C. A., Patacchini, R., Santicioli, P., Lippe, I. Th., Giuliani, S., Geppetti, P., Del Bianco, E., Selleri, S., and Meli, A.,** The effect of ω-conotoxin GVIA, a peptide modulator of N-type voltage-sensitive calcium channels on motor responses produced by activation of efferent and sensory nerves in mammalian smooth muscle, *Naunyn-Schmiedeberg's Arch. Pharmacol.,* 338, 107, 1988.
14. **Maggi, C. A., Tramontana, M., Cecconi, R., and Santicioli, P.,** Neurochemical evidence for the involvement of N-type calcium channels in transmitter secretion from peripheral nerve endings of sensory nerves in guinea-pigs, *Neurosci. Lett.,* 114, 203, 1990.
15. **Maggi, C. A., Patacchini, R., Santicioli, P., Giuliani, S., Del Bianco, E., Geppetti, P., and Meli, A.,** The 'efferent' function of capsaicin-sensitive nerves: ruthenium red discriminates between different mechanisms of action, *Eur. J. Pharmacol.,* 170, 167, 1989.
16. **Amann, R. and Lembeck, F.,** Ruthenium red selectively prevents capsaicin induced nociceptor stimulation, *Eur. J. Pharmacol.,* 161, 227, 1989.
17. **Dray, A., Forbes, C. A., and Burgess, G. M.,** Ruthenium red blocks the capsaicin-induced increase in intracellular calcium and activation of membrane currents in sensory neurons as well as the activation of peripheral nociceptors *in vitro, Neurosci. Lett.,* 110, 52, 1990.
18. **Bevan, S. and Yeats, J.,** Protons activated a cation conductance in a sub-population of rat isolated dorsal root ganglion neurones, *J. Physiol. (London),* 443, 145, 1991.
19. **Geppetti, P., Del Bianco, E., Patacchini, R., Santicioli, P., Maggi, C. A., and Tramontana, M.,** Low pH induced release of calcitonin gene-related peptide from capsaicin-sensitive sensory nerves: mechanism of action and biological response, *Neuroscience,* 41, 295, 1991.
20. **Mapp, C. E., Fabbri, L. M., Boniotti, A., and Maggi, C. A.,** Prostacyclin activates tachykinin release from capsaicin-sensitive afferents in guinea-pig bronchi through a ruthenium red-sensitive pathway, *Br. J. Pharmacol.,* 104, 49, 1991.
21. **Geppetti, P., Tramontana, M., Santicioli, P., Del Bianco, E., Giuliani, S., and Maggi, C. A.,** Bradykinin-induced release of calcitonin gene-related peptide from capsaicin-sensitive nerves in guinea-pig atria: mechanism of action and calcium requirements, *Neuroscience,* 38, 687, 1990.
22. **Tramontana, M., Del Bianco, E., Cecconi, R., Maggi, C. A., and Geppetti, P.,** Veratridine evokes release of CGRP from capsaicin-sensitive nerves of rat urinary bladder, *Eur. J. Pharmacol.,* 212, 137, 1992.
23. **Maggi, C. A.,** The pharmacology of the efferent function of sensory nerves, *J. Auton. Pharmacol.,* 11, 173, 1991.
24. **Nawa, H., Kotani, H., and Nakanishi, S.,** Tissue-specific generation of two preprotachykinin mRNAs from one gene by alternative RNA splicing, *Nature (London),* 312, 729, 1984.
25. **Rosenfeld, M. G., Memod, J. J., Amara, S. G., Swanson, L. W., Sanchenko, P. E., Rivier, J., Vale, W. W., and Evans, R. M.,** Production of a novel neuropeptide encoded by the calcitonin gene via tissue-specific RNA processing, *Nature (London),* 304, 129, 1983.
26. **Wiendermann, C. J., Wiendermann, F. J., Apperl, A., Kieselbach, G., Konwalinka, G., and Braunsteiner, H.,** In vitro human polymorphonuclear leukocyte chemokinesis and human monocyte chemotaxis are different activities of aminoterminal and carboxyterminal substance P, *Naunyn-Schmiedeberg's Arch. Pharmacol.,* 340, 185, 1989.
27. **Hafstrom, I., Gyllenhammar, H., Palmblad, J., and Ringertz, B.,** Substance P activates and modulates neutrophil oxidative metabolism and aggregation, *J. Rheumatol.,* 16, 1033, 1989.

28. **Perianin, A., Snyderman, R., and Malfroy, B.,** Substance P primes human neutrophil activation: a mechanism for neurological regulation of inflammation, *Biochem. Biophys. Res. Commun.,* 161, 520, 1989.
29. **Payan, D. G., Brewster, D. R., Missirian-Bastan, A., and Goetzl, E. J.,** Substance P recognition by a subset of human T lymphocytes, *J. Clin. Invest.,* 74, 1532, 1984.
30. **Lotz, M., Vaughan, J. H., and Carson, D. A.,** Effect of neuropeptides on production of inflammatory cytokines by human monocytes, *Science,* 241, 1218, 1988.
31. **Piedimonte, G., Hoffman, J. I., Husseini, W. K., and Nadel, J. A.,** Effect of sensory neuropeptides on blood flow in the rat airway microcirculation, *J. Appl. Physiol.,* in press.
32. **Matran, R., Alving, K., Martling, C.-R., Lacroix, J. S., and Lundberg, J. M.,** Effects of neuropeptides and capsaicin on tracheobronchial blood flow of the pig, *Acta Physiol. Scand.,* 135, 335, 1989.
33. **Li, D.-S., Raybould, H. E., Quintero, E., and Guth, P. H.,** Role of calcitonin gene-related peptide in gastric hyperemic response to intragastric capsaicin, *Am. J. Physiol.,* 261, G657, 1991.
34. **Brain, S. D., Williams, T. J., Tippins, J. R., Morris, H. R., and Macintyre, I.,** Calcitonin gene-related peptide is a potent vasodilator, *Nature (London),* 313, 54, 1985.
35. **Holzer, P., Pabst, M. A., Lippe, I. Th., Peskar, B. A., Livingstone, E. H., and Guth, P. H.,** Afferent nerve-mediated protection against deep mucosal damage in the rat stomach, *Gastroenterology,* 98, 838, 1990.
36. **Maggi, C. A., Borsini, F., Santicioli, P., Geppetti, P., Abelli, L., Evangelista, S., Manzini, S., Theodorsson-Norheim, E., Somma, V., and Amenta, F.,** Cutaneous lesions in capsaicin-pretreated rats. A trophic role of capsaicin sensitive afferents, *Naunyn-Schmiedeberg's Arch. Pharmacol.,* 336, 358, 1987.
37. **Stanitz, A. M., Befus, S., and Bienenstock, J.,** Differential effects of vasoactive intestinal peptide, substance P, and somatostatin on immunoglobulin synthesis and proliferations by lymphocytes from Peyer's patches, mesenteric lymph nodes, and spleen, *J. Immunol.,* 136, 152, 1986.
38. **Payan, D. G. and Goetzl, E. J.,** Modulation of lymphocyte function by sensory neuropeptides, *J. Immunol.,* 135, 783, 1985.
39. **Casini, A., Geppetti, P., Maggi, C. A., and Surrenti, C.,** Effects of calcitonin gene-related peptide (CGRP), neurokinin A and neurokinin A (4-10) on the mitogenic response of human peripheral blood mononuclear cells, *Naunyn-Schmiedeberg's Arch. Pharmacol.,* 339, 354, 1989.
40. **Kjatansson, J., Dalsgaard, G. J., and Janssen, C. E.,** Decreased survival of experimental critical skin flaps in rats after sensory denervation with capsaicin, *Plast. Reconstr. Surg.,* 79, 218, 1987.
41. **Dalsgaard, C. J., Halgardh-Nilsson, A., Haegerstrand, A., and Nilsson, J.,** Neuropeptides as growth factors: possible roles in human diseases, *Regul. Pept.,* 25, 1, 1989.
42. **Nilsson, J., von Euler, A. M., and Dalsgaard, C.-J.,** Stimulation of connective tissue cell growth by substance P and substance K, *Nature (London),* 315, 61, 1985.
43. **Ziche, M., Morbidelli, L., Pacini, M., Geppetti, P., Alessandri, G., and Maggi, C. A.,** Substance P stimulates neovascularization in vivo and proliferation of cultured endothelial cells, *Microvasc. Res.,* 40, 264, 1990.
44. **Manzini, S. and Perretti, F.,** Vascular effects of capsaicin in isolated perfused rat mesenteric bed, *Eur. J. Pharmacol.,* 148, 153, 1988.
45. **Kawasaki, H., Takasaki, K., Saito, A., and Goto, K.,** Calcitonin gene-related peptide acts as a novel neurotransmitter in mesenteric resistance vessels, *Nature (London),* 77, 1072, 1988.
46. **Kawasaki, H., Saito, A., and Takasaki, K.,** Age-related decrease of calcitonin gene-related peptide-containing vasodilator innervation in the mesenteric resistance vessel of the spontaneous hypertensive rat, *Circ. Res.,* 67, 733, 1990.

47. **Moskowitz, M. A., Macfarlane, R., Tasdemiroglu, E., Wei, E. P., and Kontos, H. A.,** Neuroeffector functions of sensory nerve fibers in the cerebral circulation after global cerebral ischemia, *Arzeim. Forsh.*, 41, 315, 1991.
48. **Young, A. A., Mott, D. M., Stone, K., and Cooper, G. J. S.,** Amilyn activates glycogen phosphorylase in the isolated soleous muscle of the rat, *FEBS Lett.*, 28, 149, 1991.
49. **Kessler, J. A. and Black, I. B.,** Nerve growth factor stimulates the development of substance P in sensory ganglia, *Proc. Natl. Acad. Sci. U.S.A.*, 77, 649, 1980.
50. **Lindsay, R. M. and Harmar, A. J.,** Nerve growth factor regulates expression of neuropeptide gene in adult sensory neurons, *Nature (London)*, 337, 362, 1989.
51. **Winter, J., Forbes, A. C., Sternberg, J. M., and Lindsay, R. M.,** Nerve growth factor (NGF) regulates adult rat cultured dorsal root ganglion neuron responses to the excitotoxin capsaicin, *Neuron*, 1, 973, 1988.
52. **Eedy, D. J., Shaw, C., Johnston, C. F., Armstrong, E. P., and Buchanan, K. D.,** Neuropeptides of the primary sensory neurons in rat skin: an ontogenic study, *Regul. Pept.*, 33, 175, 1991.
53. **Schotzinger, R. J. and Landis, S. C.,** Postnatal development of autonomic and sensory innervation of thoracic hairy skin in the rat. A histochemical, immunocytochemical and radioenzymatic study, *Cell. Tissue Res.*, 260, 575, 1990.
54. **Ursell, P. C., Ling Ren, C., and Danilo, P.,** Anatomic distribution of autonomic neural tissue in the developing dog heart. II. Nonadrenergic noncholinergic innervation by calcitonin gene-related peptide-immunoreactive tissue, *Anat. Rec.*, 230, 531, 1991.
55. **Scott, T. M., Robinson, J., and Foote, J.,** The peptidergic innervation of the developing mesenteric vascular bed in the rat, *J. Anat.*, 162, 177, 1989.
56. **Tsai, S.-H., Tew, J. M., and Shipley, M.,** Cerebral arterial innervation. II. Development of calcitonin-gene related peptide and norepinephrine in the rat, *J. Comp. Neurol.*, 279, 1, 1989.
57. **Nowak, R., Zelck, U., Rathsack, R., Oehme, P., Scholtz, H.-J., Koitschev, A., and Beleites, B.,** Postnatal development of substance P in the inner ear of the guinea-pig, *Eur. Arch. Otorhinolaryngol.*, 247, 147, 1990.
58. **Flatt, P. R., Swanston-Flatt, S. K., Bailey, C. J., McGregor, G. P., and Conlon, J. M.,** Substance P, neurokinin A and calcitonin gene-related peptide during development of the rat gastrointestinal tract, *Regul. Pept.*, 33, 313, 1991.
59. **Maggi, C. A., Santicioli, P., Geppetti, P., Frilli, S., Spillantini, M. G., Nediani, C., Hunt, S. P., and Meli, A.,** Biochemical, anatomical and functional correlates of postnatal development of the capsaicin-sensitive innervation of the rat urinary bladder, *Dev. Brain Res.*, 43, 183, 1988.
60. **Tanaka, D. T. and Grunstein, M. M.,** Maturation of neuromodulatory effect of substance P in rabbit airways, *J. Clin. Invest.*, 85, 345, 1990.
61. **Guidobono, F., Netti, C., Bettica, P., Sibilia, V., Pagani, F., Cazzamalli, E., and Pecile, A.,** Effect of age on binding sites for calcitonin gene-related peptide in the rat central nervous system, *Neurosci. Lett.*, 102, 20, 1989.
62. **Westlund, K. N., DiPette, D. J., Carson, J., and Holland, O. B.,** Decreased spinal cord content of calcitonin gene-related peptide in the spontaneously hypertensive rat, *Neurosci. Lett.*, 131, 183, 1991.
63. **Mione, M. C., Dhital, K. K., Amenta, F., and Burnstock, G.,** An increase in the expression of neuropeptidergic vasodilator, but not vasoconstrictor, cerebral nerves in aging rats, *Brain Res.*, 103, 113, 1988.
64. **Ferrante, F., Geppetti, P., and Amenta, F.,** Age-related changes in substance P and vasoactive intestinal polypeptide immunoreactivity in the rat stomach and small intestine, *Arch. Gerontol. Geriatr.*, 13, 81, 1991.
65. **Feher, E. and Penzes, L.,** Density of substance P, vasoactive intestinal polypeptide and somatostatin-containing nerve fibers in the ageing small intestine of the rat, *Gerontology*, 33, 341, 1987.

66. **Takayanagi, I., Moriya, M., Kurata, R., and Koile, K.,** Effects of ageing on nicotine-induced contraction and substance P-like materials release in guinea-pig bronchus, *Gen. Pharmacol.,* 22, 783, 1991.
67. **Amenta, F., Spillantini, M. G., Cavallotti, C., and Geppetti, P.,** Increase in substance P levels in the lung of old rats, *Arch. Gerontol. Geriatr.,* 7, 215, 1988.
68. **Geppetti, P., Fusco, B. M., Marabini, S., Maggi, C. A., Fanciullacci, M., and Sicuteri, F.,** Secretion, pain and sneezing induced by the application of capsaicin to the nasal mucosa in man, *Br. J. Pharmacol.,* 93, 509, 1988.
69. **Carpenter, S. E. and Lynn, B.,** Vascular and sensory responses of human skin to mild injury after topical treatment with capsaicin, *Br. J. Pharmacol.,* 73, 755, 1981.
70. **Franco-Cereceda, A.,** Calcitonin gene-related peptide and human epicardial coronary arteries: presence, release and vasodilator effects, *Br. J. Pharmacol.,* 102, 506, 1991.
71. **Geppetti, P., Del Bianco, E., Cecconi, R., Tramontana, M., Romani, A., and Theodorsson, E.,** Capsaicin releases calcitonin gene-related peptide from the human iris and ciliary body in vitro, *Regul. Pept.,* 41, 83, 1992.
72. **Maggi, C. A., Santicioli, P., Del Blanco, E., Geppetti, P., Barbanti, G., Turini, D., and Meli, A.,** Release of VIP, but not CGRP-like immunoreactivity by capsaicin from the human isolated small intestine, *Neurosci. Lett.,* 98, 317, 1989.
73. **Maggi, C. A., Giuliani, S., Santicioli, P., Patacchini, R., Said, S. I., Theodorsson, E., Turini, D., Barbanti, G., Giachetti, A., and Meli, A.,** Direct evidence for the involvement of vasoactive intestinal polypeptide in the motor response of the human isolated ileum to capsaicin, *Eur. J. Pharmacol.,* 185, 169, 1990.
74. **Renzi, D., Mantellini, P., Maggi, C. A., Fucini, C., and Surrenti, C.,** Ruthenium red selectively antagonizes capsaicin-induced release of vasoactive intestinal polypeptide (VIP) from the human colon, *Neurosci. Lett.,* 126, 134, 1991.
75. **Del Fiacco, M., Dessi, M., Atzori, M. G., and Levanti, M. C.,** Substance P in the human postmortem brainstem, an immunohistochemical study in the newborn and adult tissue, *Neuroscience,* 12, 591, 1984.
76. **Luo, C. B., Zheng, D. R., Guan, Y. L., and Yew, D. T.,** Localization of substance P and enkephalin by immunohistochemistry in the spinal cord of the human fetus, *Neuroscience,* 27, 989, 1988.
77. **Richard-Bell, G. C., Tork, I., Sullivan, C., and Scheiber, T.,** Distribution of substance P-like immunoreactive fibres and terminals in the medulla oblongata of the human infant, *Neuroscience,* 34, 133, 1990.
78. **Chigr, F., Najimi, M., Leduque, P., Chayvialle, J. A., Bouvier, R., and Kopp, N.,** Anatomical distribution of substance P-immunoreactive neurons in human brainstem during the first postnatal year, *Brain Res. Bull.,* 26, 515, 1991.
79. **Del Fiacco, M., Quartu, A., Floris, A., and Diaz, G.,** Substance P-like immunoreactivity in the trigeminal ganglion, *Neurosci. Lett.,* 110, 16, 1990.
80. **Helme, R. D. and McKernan, S.,** Neurogenic flare responses following topical application of capsaicin in humans, *Ann. Neurol.,* 18, 505, 1985.
81. **Geppetti, P.,** unpublished observation.

Chapter 6

EFFECT OF AGE ON THE RESPONSE OF TARGET ORGANS TO AUTONOMIC NEUROTRANSMITTERS

James R. Docherty

TABLE OF CONTENTS

0-8493-6981-9/93/$0.00 + $.50

I. INTRODUCTION

The progress from birth to old age is marked by change, due to both alterations in maturation and the process of senescence. With the increasing numbers of elderly people in Western societies, it becomes crucial that we learn more about how the elderly differ in terms of the physiological function of various organ systems, how differently they respond to drugs, and how these problems can be minimized. This chapter looks at how aging affects target organ response to neurotransmitters and neuromodulators (substances released by nerves), hormones (substances released into the bloodstream by endocrine glands), and other autacoids (endogenous substances not easily defined as either of the former). Hence, it is important to define the scope of this chapter.

This chapter concerns itself only with aging in the sense of senescence, the progress from maturity to old age, and, with few exceptions, does not address the process of maturation from birth to adulthood. Disease processes which become more prevalent in the elderly are not discussed, as the focus of this chapter is normal aging in the absence of overt pathology.

Autacoids cause target organ responses usually by activation of cell surface receptors. Aging can affect the response to autacoids at several levels: by altering the handling of the autacoid, and by alterations in cell surface receptor number or affinity for the agonist, the coupling of receptor activation to cellular events, the ability of the system to respond, and by alterations in reflex control of function.

Studies of the effects of aging on target tissues can be classified into three major groups of investigation which, for convenience, I will call physiological, pharmacological, and biochemical (Figure 1). Physiological studies are those which look at overall tissue response such as contraction of isolated smooth muscle or changes in blood pressure *in vivo*. Pharmacological studies, for the purpose of this review, are those which look at receptor number and affinity, usually using the ligand binding technique. Biochemical studies are defined as those which look at the events after receptor occupation such as second-messenger systems. Some overlap in the terminology may be unavoidable, but some advantage can be obtained. A physiological response is a combination of pharmacological changes in receptors, biochemical alterations, and other factors such as alterations in contractile machinery, etc.

This review considers how aging affects peripheral tissue responsiveness, looking at several of the major autacoids. Moreover, this review reflects the fact that one of the most studied areas in aging research is the cardiovascular system.

A. MAJOR CONTROL SYSTEMS

The peripheral nervous system consists of the somatic system to skeletal muscle and the autonomic nervous system which innervates almost every other type of cell. The somatic nervous system is not considered since this

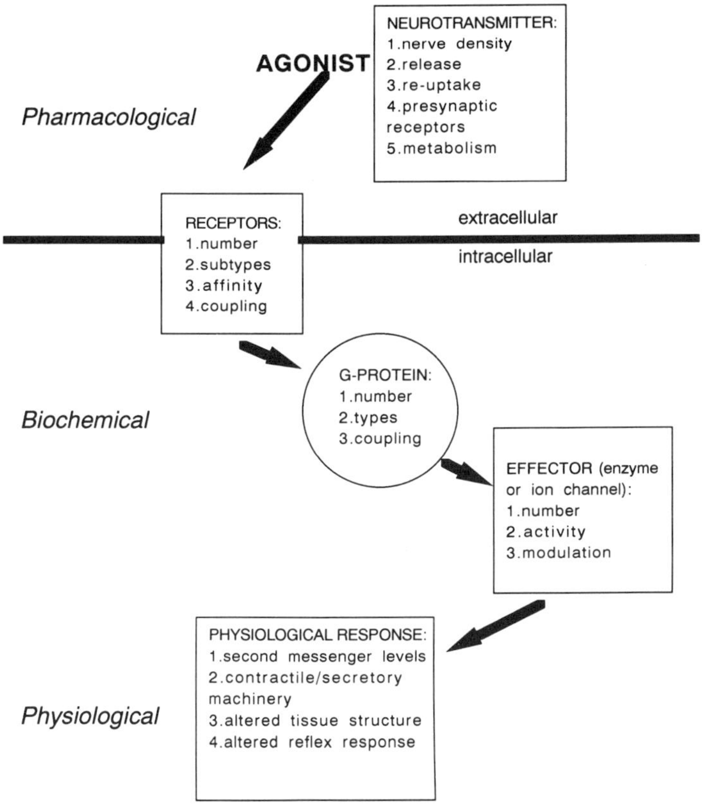

FIGURE 1. Illustration of various steps at which the effector tissue response to autonomic neurotransmitters can be modulated.

volume restricts itself to the autonomic nervous system. The autonomic nervous system is divided into the sympathetic and parasympathetic systems, with the major neurotransmitters noradrenaline and acetylcholine, respectively (Figure 2). Many of the results presented here concern these two neurotransmitters. However, in recent years, it has become clear that the peripheral nervous system contains many amine neurotransmitters such as dopamine, 5-hydroxytryptamine, Adenosine triphosphate (ATP), and many peptide neurotransmitters (or neuromodulators) such as neuropeptide Y and vasoactive intestinal polypeptide (VIP). However, neuropeptides are considered elsewhere in this volume.

In addition to neurotransmitter systems, peripheral tissues are under the influence of hormones such as adrenaline from the adrenal medulla (considered in detail elsewhere in this volume), angiotensin II, endothelin, and locally released autacoids such as prostaglandins and EDRF (endothelium-derived relaxant factor). Autacoids which have been extensively studied with respect to aging are dealt with individually.

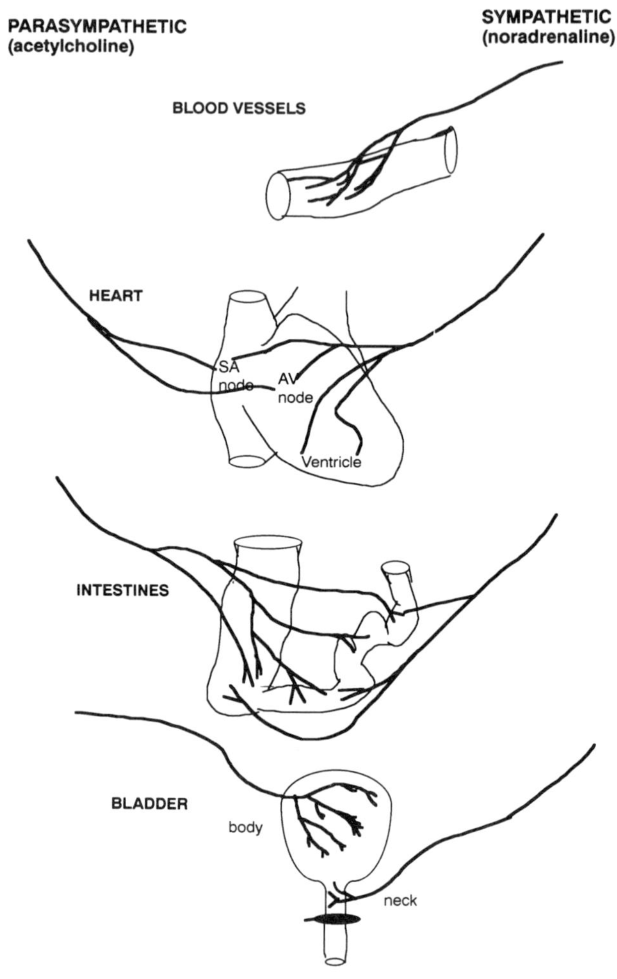

FIGURE 2. Illustration of some autonomically innervated tissues. Blood vessels are largely controlled by noradrenaline released from sympathetic nerves. The heart is innervated by the sympathetic and vagus (parasympathetic) nerves. The latter releases acetylcholine, but the cardiac ventricles are controlled only by the sympathetics. In the bladder, the major innervation is parasympathetic, but sympathetic fibers control the bladder neck.

Figure 2 summarizes some of the major autonomically controlled systems discussed in this article. This review looks at how responses to various agents are altered by aging, and tries to decide at what level the alteration is occurring. The major sites at which aging may alter responses to pharmacological agents are summarized in Figure 1. Ignoring pharmacokinetic factors (which may be important in studies *in vivo*), altered availability of neurotransmitter may be due to altered release (due to altered nerve density, nerve activity, or presynaptic inhibition), reuptake, or metabolism. At the pharmacological level, altered receptor-mediated response to an agonist may be due to alterations in receptor affinity for the agonist, in the number of receptors or proportions of different subtypes or states of receptor present, or in coupling to G-protein. At the biochemical level, altered activation of the effector (enzyme or ion channel) could be due to alterations in the relative proportions of receptors to G-protein and effector, in coupling, or in the activity or modulation of the effector. Altered physiological response can be due to any of these factors as well as alterations in the contractile/secretory machinery, altered tissue content or structure, and altered reflex response to activation of the organ system (Figure 1).

II. ADRENERGIC SYSTEM

A. NORADRENALINE AND ADRENALINE

The adrenergic system constitutes the major part of the sympathetic system and acts through the neurotransmitter noradrenaline and the hormone adrenaline. The pharmacological differences between these two amines are in terms of selectivity for subtypes of adrenoceptor: adrenaline has major actions at all types of adrenoceptor (α_1, α_2, β_1, and β_2), whereas noradrenaline acts predominantly α_1, α_2, and β_1. Molecular biology has revealed a much wider range of subtypes of adrenoceptor with the interesting symmetry of three subtypes of β(β_1, β_2, and β_3), α_1 (α_{1A}, α_{1B}, and α_{1C}), and α_2 (α_{2A}, α_{2B}, and a third subtype which we will term α_{2C}).[1-3] However, since little is known of the distribution of these additional subtypes and aging studies have not generally been concerned with subtypes, very little mention of the nine subtypes will appear, and we will subdivide this topic into α_1, α_2, and β.

B. α_1-ADRENOCEPTORS

α_1-Adrenoceptors are found on many types of smooth muscle, including vascular and intestinal, as well as cardiac muscle, and mediate contraction, normally by phosphoinositol (PI) hydrolysis.[4] In physiological studies, the majority of investigators have examined the contractile response of isolated smooth muscle and looked for alterations in either the potency of the agonist or in the maximum contractile response. The results of these studies are shown in Table 1.

Most studies have failed to find any change with age in contractile responses mediated by α_1-adrenoceptors in human and dog arteries, rat and rabbit blood vessels, and rat urinary bladder. However, studies employing the rat aorta, a tissue widely used in aging research, have generally found that aging causes a decreased contractile responsiveness or a decrease in receptor reserve. However, the rat aorta may not be the most useful model of the effects of aging on vascular tissue.

An increased α-adrenoceptor-mediated contractile response has been reported with increasing age in rat urinary bladder and rabbit urethra in terms

TABLE 1
Effects of Aging on α_1-Adrenoceptors

Species and tissue	Response	Effect of aging	Ref.
Human arteries	Contraction to NA	No change	5, 6
Dog arteries	Contraction	No change	7
Rat aorta	Contraction	No change	8
		Decrease	9–12
Rat blood vessels	Contraction	No change	13, 14
		Decrease	15
Rat urinary bladder	Contraction	Increase (body)	16
		No change (neck)	16
Rabbit urethra	Contraction	Slight increase in potency	17
Rat vas deferens	Contraction	No change	18
Rat liver	PI formation	No change	4
	Ligand binding	Decrease	19
Rat atria	Inotropic response	Decrease	20
	PI formation	Decrease	21
	mRNA	Decrease	20
Rat heart	Ligand binding	No change	22
		Decrease	23, 24
	PI formation	Decrease	23
Rabbit heart	Ligand binding	Decrease	25
Rat parotid gland	Ligand binding	No change	26
Rabbit spleen	Ligand binding	No change	25
Human systemic	Pressor	Decrease	27
		No change	28
Rabbit systemic	Pressor	No change	25
Rat systemic	Pressor	No change	29
Rat parotid gland	Glucose oxidation	Decrease	30
Human eye	Mydriasis	No change	31, 32

of an increased potency, and a decreased maximum inotropic response has been reported in rat heart with increasing age.[20]

In the eye, α_1-adrenoceptors mediate mydriasis by contraction of the smooth muscle of the dilator pupillae. The diameter of the dark-adapted pupil is reduced in the elderly,[31,32] perhaps due to reduced sympathetic dilator tone. Certainly, the α-adrenoceptor antagonist thymoxamine produced a smaller percentage reduction in the pupil diameter of elderly subjects, and differences in pupil diameter between young and old were eliminated by thymoxamine.[31] The dilation of the pupil produced by the α_1-adrenoceptor agonist phenylephrine is reported to be greater in the elderly[33] in terms of the percentage increase in diameter, but the maximum diameter reached was similar in young and old;[31] hence, differences in the effect of phenylephrine were mainly due to differences in resting pupil diameter. Since the potency of phenylephrine was unchanged,[31] there is no evidence to suggest that a change occurs with age in α_1-adrenoceptors of the pupil: apparent differences are due to differences in pupil diameter.

TABLE 2
Effects of Aging on Responses Mediated by α_2-Adrenoceptors

Species and tissue	Response	Effect of aging	Ref.
Human saphenous vein	Contraction	Decrease	34
Human platelet	Ligand binding	No change	35, 36
		Decrease	37
		Increase	38
Human platelet	Aggregation	No change	35
		Increase	38
Rat systemic	Pressor	Decrease	39

Pharmacological studies of ligand binding sites have reported a decrease in the number of α_1-ligand binding sites in rat liver and rabbit heart and spleen, a decrease or no change in rat heart, but no change in rat parotid gland (Table 1). These changes in ligand binding site number are accompanied by a decreased biochemical response in terms of PI response in rat heart and liver (Table 1).

C. α_2-ADRENOCEPTORS

α_2-Adrenoceptors are involved in two major types of response at different locations. On peripheral nerve terminals, α_2-adrenoceptors function as inhibitory prejunctional receptors and mediate an inhibition of transmitter release as both autoreceptors (on adrenergic nerves) and heteroceptors (on other types of nerve). Prejunctional (or presynaptic receptors) are dealt with in more detail later. α_2-Adrenoceptors are also present postjunctionally on smooth muscle, where, like α_1-receptors, they mediate a contraction, and on glands such as the islets of Langerhans, where they mediate inhibition of secretion. Very few studies have looked at the effects of aging on α_2-adrenoceptor function, mainly since α_2-adrenoceptor-mediated responses are generally difficult to find. These studies are listed in Table 2. In pharmacological studies of ligand binding sites, the majority of studies have found no change with age in the number or affinity of α_2-adrenoceptors on human platelets. In the current classification of α_2-adrenoceptors, these receptors belong to the α_{2A}-subtype.[2] In line with the reports of no consistent change in the number of α_2-ligand binding sites with age, there is no consistent evidence for any change in the aggregatory response mediated by these receptors on platelets, although two studies report an increased aggregation. An increased aggregatory response with age also appears to occur for other agents such as collagen and arachidonic acid.

Studies of the effects of aging on α_2-adrenoceptors mediating contractions of smooth muscle have been hindered by the lack of suitable isolated tissues containing α_2-adrenoceptors. However, the human saphenous vein is an example of a tissue in which contractions to noradrenaline are mediated pre-

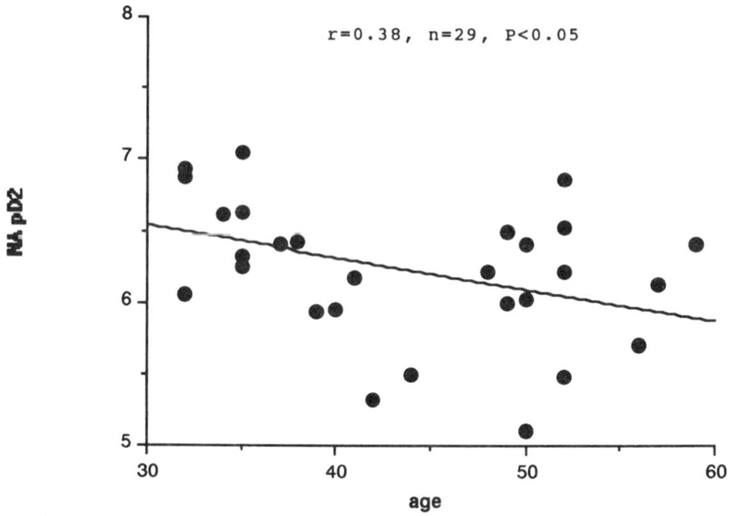

FIGURE 3. Correlation between noradrenaline potency (expressed as pD_2, the negative log of the concentration producing 50% of maximal contraction) and age for contractions of human isolated saphenous vein obtained from patients (predominantly female) undergoing varicose vein surgery. There was a significant negative correlation between noradrenaline potency and age (*p* <0.05).

dominantly, if not exclusively, by α_2-adrenoceptors. In a study of vessels obtained from varicose vein surgery of both male and female (predominantly female) patients, we found a significant age-related correlation between noradrenaline potency and age, such that noradrenaline became less potent with increasing age (Figure 3). These postjunctional α_2-adrenoceptors of human saphenous vein are different from those on human platelets in that they are not α_{2A}-receptors.[160]

Vascular α_2-adrenoceptors mediating contraction can be relatively easily identified *in vivo,* and indeed these postjunctional α_2-adrenoceptors were first identified in the pithed rat preparation. In the pithed rat preparation, the pressor potency of the α_2-adrenoceptor agonist xylazine is reduced by aging, but no such age-related difference can be seen for the α_1-adrenoceptor agonist amidephrine.[39] However, this age-related difference in the pressor potency of xylazine is abolished by the angiotensin-converting enzyme inhibitor captopril, suggesting that an age-related decline in angiotensin levels (or responsiveness) causes the change in α_2-responsiveness. It is well known that, *in vitro,* the demonstration of an α_2-adrenoceptor-mediated contraction can be dependent on an ongoing contraction to other vasoconstrictors such as angiotensin, and so alterations in levels of angiotensin *in vivo* may also have implications for α_2-responsiveness.

TABLE 3
Effects of Aging on β-Adrenoceptor-Mediated Responses

Species and tissue	Response	Effect of aging	Ref.
Human cardiovasculature	Tachycardia to isop	Decrease	40–42
Dog cardiovasculature	Tachycardia to isop	Decrease	43
Rat cardiovasculature	Tachycardia to isop	Decrease	10
	Tachycardia to NA	No change	44
Rat heart	Ligand binding	No change	45–47
		Decrease	48
		Increase	49
Human lymphocyte	Ligand binding	No change	50, 51
		Decrease	52, 53
		Increase	54
Rat lung	Relaxation	No change	55
	cAMP production	Decrease	56
	Ligand binding	Increase	51
		Decrease	52
Rat erythrocyte	cAMP production	Decrease	57
	Ligand binding	Decrease	57
Rat submandibular gland	Salivary protein	Little change	58
Rat liver	Ligand binding	Increase	19
Rat fat cells	Ligand binding	Decrease	59
Rat aorta, pulmonary artery	Relaxation	Decrease	10, 60, 61
Rat jugular, portal vein	Relaxation	No change	62, 63
Rabbit aorta, pulmonary artery	Relaxation	Decrease	64
Rabbit portal vein	Relaxation	No change	64
Dog mesenteric artery	Relaxation	Decrease	65
Human saphenous, hand vein	Relaxation	Decrease	66, 67
Human vasculature	Depressor	No change	41
Rat vasculature	Depressor	No change	68

D. β-ADRENOCEPTORS

β-Adrenoceptors mediate three major types of response: cardiac stimulation, smooth muscle dilation, and metabolic actions. Cardiac actions are predominantly β_1, whereas β_2-receptors mediate most other actions.

In most studies of the effects of aging on tachycardia to the β-adrenoceptor agonists noradrenaline or isoprenaline, a decreased response has been reported with increasing age (Table 3). For instance, in man the positive chronotropic response to isoprenaline has been reported to decline with age in many studies. However, a component of this isoprenaline-induced tachycardia may be due to a diminished baroreflex tachycardia to the vasodepressor action of isoprenaline. In the pithed rat preparation, in which the central nervous system is destroyed so that no reflexes are possible, the direct tachycardia to isoprenaline can be examined in isolation without any complications due to vasodilator actions. In the pithed rat, there was a small, twofold difference between young and old in the potency of isoprenaline at producing tachycardia, indicating that isoprenaline was only slightly less potent in old than in young rats. Furthermore, in the pithed rat preparation, there was no significant difference between young and old in the potency of noradrenaline at producing tachycardia; hence, the cardiac response to the physiological agonist noradrenaline is unchanged by aging in the pithed rat despite a decrease in the potency of isoprenaline. This is probably due to a decline in the noradrenaline reuptake process (isoprenaline is not a substrate), as explained in Section VI.

In addition to mediating cardiac responses, β-adrenoceptors (particularly β_2) mediate relaxation of various types of smooth muscle, including vascular, bronchial, and intestinal. Despite many reports of reduced vasodilator responses to β-adrenoceptor agonists in isolated tissues, particularly arteries (Table 3), no change in potency or maximum depressor response to β-adrenoceptor agonists was found during aging in the human and rat vasculature *in vivo* (Table 3). Hence, differences found in isolated tissues are not always reflected in changes in overall cardiovascular responsiveness, especially when other factors such as reflexes come into play.

III. CHOLINERGIC SYSTEM

A. MUSCARINIC RECEPTORS

Acetylcholine is the major neurotransmitter of the parasympathetic nervous system, and its effects on smooth muscle, cardiac muscle, and glands are mediated by muscarinic receptors. Although molecular biology has revealed the existence of five separate genes coding for muscarinic receptors, this diversity has not yet been studied in research on aging. The response to activation of muscarinic receptors is to produce bradycardia and negative inotropic effects (particularly in atria) in cardiac muscle, contraction of smooth muscle (although acetylcholine will dilate vascular smooth muscle indirectly by releasing EDRF from the endothelium; see later), and secretion from exocrine glands.

Bradycardia to muscarinic agonists is reported variously to be decreased, unchanged, or even increased, and negative inotropic effects to be reduced, by aging in rat atria (Table 4). The increased potency of acetylcholine, but not carbachol, in one of the above studies of rat atria appears to be due to an age-related reduction in acetylcholinesterase activity,[73] since differences are abolished by cholinesterase inhibitors. Similarly, in rabbit bronchus, an increased contractile potency of acetylcholine, but not carbachol, is due to an age-related decrease in acetylcholinesterase activity.[74] However, the reverse was true in rat jejunum, where an increased activity of acetylcholinesterase could explain why the response to acetylcholine was unchanged by aging, whereas the response to methacholine was reduced.[70] In the rat, vagotomy causes a significant tachycardia in young, but not old, animals,[75] and the tachycardia to blockade of muscarinic receptors by atropine is reported to be decreased[76] or unchanged[75] with age in the rat and decreased with age in man.[77] However, the reduced effects of vagotomy and atropine with age need not necessarily be due to a reduced responsiveness of cardiac muscle to

TABLE 4
Effects of Aging on Responses Mediated by Muscarinic Receptors

Species and tissue	Response	Effects of aging	Ref.
Rat heart	Ligand binding	No change	69
		Increase	46
Rat jejunum	Contraction	No change (ACh)	70
		Increase (methacholine)	70
		Increase (AChE)	70
Mouse/rat ileum	Contraction	Decrease	71
Rat vas deferens	Contraction	No change	72
Rabbit bronchus	Contraction	Increase (ACh)	74
		No change (carbachol)	74
		Decrease (AChE)	74
		Decrease (Ch Acet)	74
Rat heart	Heart rate	Increase (ACh)	73
		No change (carbachol)	73
		Decrease (AChE)	73
		No change	69
		Decrease	75

Note: ACh, acetylcholine; AChE, acetylcholinesterase; Ch Acet, choline acetylase.

acetylcholine: a decline in nerve activity with age could also be a factor. Any reduced responsiveness to muscarinic receptor activation does not seem to be due to a decrease in the number of receptors, since the density of muscarinic sites appears to be unchanged or even increased in rat atria with increasing age (Table 4).

IV. ENDOTHELIUM-DEPENDENT RELAXATIONS

Whereas certain vasodilators such as isoprenaline and adrenaline produce vasodilation by action at receptors on vascular smooth muscle (in this case, β-adrenoceptors), evidence has accumulated in recent years that the endothelium plays a major role in the modulation of vascular tone. Various physiological and pharmacological stimuli cause release of a substance from the endothelium, termed endothelium-dependent relaxant factor (EDRF), which is thought to be nitric oxide (NO) or some derivative.[78] EDRF acts on smooth muscle to activate the enzyme guanylate cyclase and so cause vascular relaxations. Several groups have looked at the effects of aging on vascular relaxations mediated by EDRF, but it should be remembered that since these studies look only at the ability of a substance such as histamine to cause vascular relaxation, age-related changes may occur not only in the effects of EDRF, but also in the ability of the substance to release EDRF or in the tissue response to activation of guanylate cyclase. Endothelium-dependent relaxa-

TABLE 5
Effects of Aging on Responses Mediated by 5-HT

Species and tissue	Response	Effects of aging	Ref.
Human saphenous vein	Contraction	Decrease	34
Human forearm	Pressor	No change	85
Rat aorta	Contraction	Decrease	68
		Increase	63
Dog coronary artery	Contraction	No change	84
Rat coronary artery	Contraction	Increase affinity	86
		Decrease number	86
Rat carotid artery	Contraction	No change	63

tions to histamine are reported to decline with age in the rat mesenteric artery;[79] relaxations to acetylcholine are unchanged[8] or decreased[13] by age in the rat mesenteric artery, and are generally reported to decrease with age[80-82] (or be unchanged[83]) in rat aorta, and to decrease with age in dog mesenteric[7] and coronary arteries.[84] In one investigation of rat aorta, it was suggested that the decreased relaxation with age to acetylcholine was not due to any change in the ability of acetylcholine to cause a relaxation, but rather to an increase in acetylcholine-induced endothelium-dependent contractions, which in this case could be prevented by indomethacin.[80] Since EDRF and nitrovasodilators are thought to act through the same mechanism, i.e., activation of guanylate cyclase, the fact that the response to nitrovasodilators is unchanged by aging in rat aorta suggests that changes in receptors on endothelium cells complicate the study of changes in the vascular effects of EDRF.

V. RESPONSES TO OTHER VASOACTIVE AGENTS

A. 5-HYDROXYTRYPTAMINE

5-Hydroxytryptamine (5-HT), although present as a neurotransmitter in some peripheral nerves, is located mainly in platelets and chromaffin cells in the periphery. Most studies that investigated the effects of aging on smooth muscle contractions mediated by 5-HT failed to examine which subtype of 5-HT was involved. However, since the predominant 5-HT receptor mediating contractions in many tissues is the 5-HT_2 receptor, it can be assumed, except if noted otherwise, that 5-HT_2 receptors are being examined. During aging, most studies report no change in the contractile potency of 5-HT (Table 5), but a decreased or even increased potency has been reported in the rat aorta. We have recently looked at how aging affects the potency of 5-HT at producing contractions of human saphenous vein, and found that the potency of 5-HT declined with age in a group of predominantly female patients undergoing varicose vein surgery.[34]

Since platelets are a major source of 5-HT, it is of interest to examine how aging affects platelet 5-HT function. The 5-HT content of platelets is reported to be decreased in aged female (but not male) rats,[87] and in man,[88] but the aggregatory response to a variety of stimuli is reported to be increased or unchanged by aging[89] (effects on the response to 5-HT were not reported).

B. ANGIOTENSIN

Angiotensin, or more correctly the renin-angiotensin-aldosterone system, is of importance in both the short-term control of blood pressure and the long-term control of blood pressure and blood volume. In the short-term control of blood pressure, angiotensin acts as a direct vasoconstrictor and facilitates vascular contractions, particularly those involving vascular α_2-adrenoceptors.[90] As previously mentioned, pressor responses to α_2-adrenoceptor agonists in the pithed rat preparation are critically dependent on angiotensin: blocking angiotensin formation with the converting enzyme inhibitor captopril markedly reduces the pressor response to the α_2-agonist xylazine.[39] Interestingly, while this is true of young rats, the pressor response to xylazine is much reduced in pithed old rats compared with young rats, and captopril has no effect.[39] Hence, the difference between young and old rats appears to be due to a decreased responsiveness to angiotensin in old rats. Indeed, there are several reports of an age-related decline in the renin-angiotensin-aldosterone system in terms of renin and aldosterone levels.[91,92]

In rat aorta, although responses to angiotensin II, like those to α_1-adrenoceptor agonists and 5-HT, decrease with increasing age, this change occurs mainly in maturation.[68,93] The vasoconstrictor response to angiotensin is reported to be unchanged by aging in the human vasculature,[28] but the number of binding sites on human platelets for angiotensin is reported to increase with age.[94]

C. ATRIAL NATRIURETIC PEPTIDE

Atrial natriuretic peptide (atriopeptin) is released from cardiac atrial cells in response to stretch and is a hormone involved in the control of blood volume by causing sodium and water excretion. At physiological concentrations, atriopeptin increases sodium excretion with little effect on blood pressure in man,[95] although it will decrease blood pressure in higher concentrations. Despite the increased resting plasma level of atriopeptin,[96-98] the elderly have a reduced ability to concentrate urine and excrete a saline load,[99] at least partly due to the 50% decrease in renal blood flow which occurs during aging.[100] The higher plasma levels of atriopeptin found in the elderly may be caused by a decreased clearance, a decreased renal responsiveness to atriopeptin, or to the increased atrial pressure found in the elderly.

There have been a few investigations of how aging affects responsiveness to atriopeptin. In man, there is a report of an age-related increase in skin blood flow to atriopeptin,[101] and in conscious rats, the depressor response to atriopeptin is prolonged in aged animal.[112] In rat aorta and rat and monkey renal arteries, there is a decreased vasodilator response to atriopeptin in tissues from aged animals, although in rat aorta, such changes may occur more in maturation than in aging. The platelet atriopeptin binding site number is reported to be unchanged by aging.[94]

D. ENDOTHELIN

Endothelin is a newly discovered peptide[106] released from the endothelium which acts as a very potent vasoconstrictor. In studies of rat aorta, it has been reported that the contractile response decreases during maturation,[107,108] with no further change in aging.[108] The potency of endothelin is reported to decrease with age in rat perfused blood vessels.[13]

VI. NEUROTRANSMISSION

So far, we have looked at how cells and tissues respond particularly to exogenously applied pharmacological agents. However, while some of these agents may reach target cells via the bloodstream as hormones or by acting locally as autacoids, some substances affect tissues by being released from nerves as neurotransmitters. Tissue response to neurotransmitters is likely to be affected by changes in nerve activity, density of innervation, presynaptic control of neurotransmission, and neurotransmitter disposition mechanisms in addition to the target organ factors outlined in Figure 1. Since the two most important neurotransmitters in the autonomic system are noradrenaline and acetylcholine, these have been investigated in most detail.

A. ADRENERGIC NEUROTRANSMISSION

Although we have looked in detail at how aging affects target organ adrenergic receptor number and response to agonists, the agonists chosen for these studies were not always the physiological agonists noradrenaline or adrenaline, and even where these agents were used, the response was not always examined under physiological conditions. The adrenergic control of peripheral tissues is largely mediated by noradrenaline released by adrenergic nerves to act locally at postjunctional receptors in the vicinity of the nerve terminal. The overall adrenergic control of a tissue, and how it changes in aging, depends not only on changes in the responsiveness of postjunctional receptors, but also on alterations in the central sympathetic drive to the periphery, in the level of innervation, in presynaptic control of transmission, and in transmitter disposition mechanisms. These factors are examined separately (Figure 4).

Little is known about how central control of the sympathetic nervous system is altered in aging, but many studies have looked at the catecholamine content of peripheral tissues as an indication of how innervation is changed, with conflicting results. Catecholamine content is reported to decline with age in male rat heart,[109] but not female, and to increase in the tail artery of the F344,[110] but decrease in the Ivanos,[15] strain. Other reports suggest no change or a decrease with age in other rat arteries and veins.[110] However, there are reports of an increased synthesis of catecholamines in tissues from aged rats, with an increased activity of tyrosine hydroxylase[111] and dopamine-β-hydroxylase[112] reported. Studies investigating the release of tritium from tissues preloaded with tritium-labeled noradrenaline indicate that the stimu-

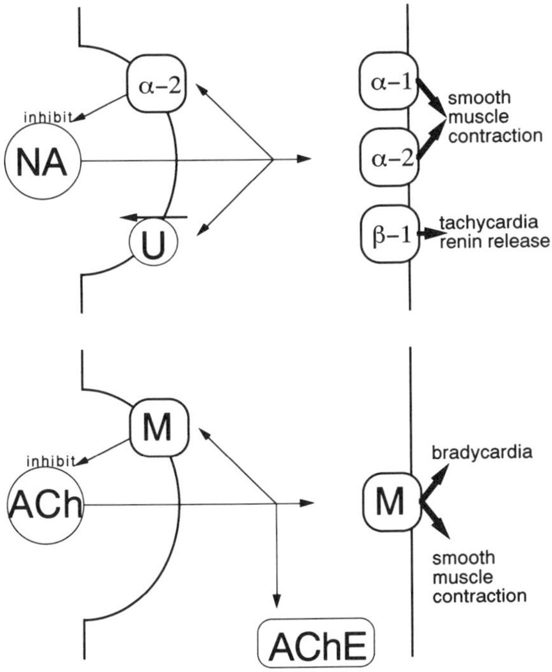

FIGURE 4. Illustration of neuromuscular transmission in adrenergic and cholinergic nerves and possible sites of age-related changes. NA, noradrenaline; ACh, acetylcholine; U, noradrenaline uptake process; M, muscarinic acetylcholine receptor; AChE, acetylcholinesterase.

lation-evoked release of noradrenaline as a fraction of tissue tritium is decreased in rat atria[44] and rat heart,[113] suggesting a decreased adrenergic nerve density with age at least in the male rat heart, in agreement with the catecholamine studies. Other studies report age-related decreases in adrenergic innervation of guinea pig,[114] rat,[115,116] and rabbit[117] blood vessels and an age-related decrease in the proportion of nerves in sympathetic ganglia innervated by enkephalin-containing nerve fibers.[118] Levels of peptides[114] and 5-HT,[117] presumably in nerves, are also reported to fall with age.

The level of adrenergic nerve activity can also be gauged by the effects of adrenergic antagonists. β-Adrenoceptor antagonists are reported to be less effective at lowering blood pressure in elderly hypertensives compared to younger hypertensives,[119] and propranolol is reported to be slightly less effective in older subjects at reducing cardiac output during exercise,[120] but produced larger falls in systolic blood pressure.[121] Propranolol eliminated the age-related difference in cardiac output,[120] suggesting that the decreased cardiac output in the elderly group was due to decreased sympathetic drive to the heart. Other studies have reported that the maximum cardiac output during exercise declines with age,[122] but this may be at least partly due to the increased incidence of heart disease in the elderly.[123] In a study of subjects without

TABLE 6
Effects of Aging on Neuronal Uptake of Noradrenaline

Species and tissue	Response	Effects of aging	Ref.
Rat heart	Tachycardia	Decrease	44
		Increase	113
Rat heart	Release of tritium	Decrease	44
		Increase	113
Rat vasculature	Pressor	Decrease	44
Rat arteries	Contraction	Decrease	129
		Increase	130
Rat vas deferens	Release of tritium	Decrease (prostatic)	131
		No change (epididymal)	131

overt heart disease, cardiac output at rest and during exercise was unaffected by age,[124] and β-adrenoceptor antagonists have been reported to be effective at lowering blood pressure in the elderly.[125]

In terms of α-adrenoceptor antagonists, no change with age was found in the effects of the nonselective α-antagonist phentolamine in the human forearm,[126] nor in the hypotensive action of the α_1-selective antagonists prazosin and terazosin in man.[27,127] In the human saphenous vein where stimulation-evoked contractions are mediated predominantly by α_2-adrenoceptors, the potency of the α_2-adrenoceptor antagonist yohimbine at inhibiting nerve stimulation-evoked contractions declined with age.[128] Since there was no evidence of an age-related decline in the affinity of yohimbine for α_2-adrenoceptors (as judged by its ability to antagonize exogenous noradrenaline), this negative correlation between yohimbine potency and age may reflect the fact that nerve stimulation-evoked contractions of the human saphenous vein become more and more non-α_2-adrenergic with age. Further investigation is required to establish what happens with age to nerve responses in the human saphenous vein, especially as these will have bearing on the ability of the vein to constrict and thus take part in the response to changes in posture. It is well established that the elderly have an increased incidence of postural hypotension, although the causes of this are still open to debate.

B. NORADRENALINE REUPTAKE

The major route of disposition of noradrenaline released from adrenergic nerves is by reuptake into the nerve terminals (Figure 4). The effectiveness of the neuronal reuptake system can be assessed using the neuronal uptake blocker cocaine: a decreased ability of cocaine to increase transmission should indicate a decline in the effectiveness of the reuptake process. The effects of age on the uptake process have been studied in various tissues, with conflicting results (Table 6). In rat heart, the effects of aging on neuronal uptake appears to be strain dependent, with reports of a reduced functioning of the reuptake process in old Sprague Dawley and Wistar, but an increased functioning in old F344, rats (Table 6). Similarly, in blood vessels, there are reports of both decreased and increased functioning of the uptake system in aging (Table 6). Other studies of central neurones report a decreased accumulation of noradrenaline with increasing age in rat hypothalamus.[132] A decreased functioning of the reuptake system could be due to biochemical changes or simply to a reduction in the number of nerve terminals with increasing age, in line with evidence for decreased tissue content of noradrenaline.

C. PLASMA LEVELS OF NORADRENALINE

In man, one of the most consistently reported changes during aging is an increased plasma level of noradrenaline, both at rest[133-136] and in response to physiological stress.[136-138] While some authors have sought to equate increased plasma levels of noradrenaline with increased adrenergic nerve activity, it must be stressed that plasma levels are a poor indicator of this. Increased plasma levels of noradrenaline can be due to an increased rate of appearance in the plasma, a decreased plasma clearance, or both. An age-related increased rate of appearance of noradrenaline in the plasma has been reported,[135,139,140] and while this could be due to increased neurotransmission with age, this seems unlikely as the catecholamine content of tissues tends to decrease with age (see above). A decreased functioning of the disposition mechanisms for noradrenaline, particularly the reuptake process, is the most likely explanation of the increased plasma levels. Hence, it is important to establish whether reuptake declines with age in the majority of tissues or is selective to tissues such as the heart. A diminished reuptake could simply be due to decreased density of the adrenergic nerve endings, which has been reported to occur with age (see above).

D. PREJUNCTIONAL RECEPTORS

In the 1970s, it became clear that peripheral nerve terminals contained receptors involved in modulation of transmitter release, the so-called presynaptic (prejunctional) receptors. The best-known prejunctional receptor on adrenergic nerve terminals is the inhibitory α_2-autoreceptor (i.e., a receptor on a nerve for its own transmitter), but there are also inhibitory muscarinic autoreceptors on cholinergic nerves as well as a wide variety of heteroceptors (receptors on a nerve for another transmitter). The effects of aging on prejunctional α_2-adrenoceptors have been studied in the greatest detail, and responsiveness is reported to either decrease or remain unchanged in aging (Table 7). A decreased sensitivity of prejunctional α_2-adrenoceptors should result in an increased release of neurotransmitter, and this coupled with a reported decline in the reuptake of noradrenaline with age should serve to maintain transmission even if postjunctional responsiveness declines with age.

In the rat vas deferens, there is an age-related decline in both prejunctional α_2- and 5-HT_1-receptor responsiveness, suggesting that the decline in response occurs at a postreceptor level, either at the level of the G-protein or beyond.

TABLE 7
Effects of Aging on the Responsiveness of Prejunctional Inhibitory Receptors

Species and tissue	Response	Effects of aging	Ref.
	α_2-Adrenoceptors		
Rat heart	Inhibit tachycardia	Decrease	141
Rat vas deferens	Inhibit twitch	Decrease	18, 142
Rat heart	Inhibit overflow	Decrease	143
Rat vas deferens	Inhibit overflow	No change	142
Human saphenous vein	Inhibit overflow	No change	144
	5-HT Receptors		
Rat vas deferens	Inhibit twitch	Decrease	145

E. CHOLINERGIC NEUROTRANSMISSION

In the cholinergic system, actions of the neurotransmitter acetylcholine are terminated not by reuptake into nerve terminals, but by enzymatic destruction of acetylcholine by acetylcholinesterase which is present in the synaptic cleft (Figure 4). Alterations with age in the activity of this enzyme should have differential effects on the response to acetylcholine (substrate for the enzyme) and agonists which are not substrates for the enzyme (carbachol, methacholine). As previously mentioned, there is evidence that the activity of acetylcholinesterase decreases with age in rat heart and rabbit bronchus, but increases with age in rat jejunum (Table 5).

Although cholinergic neurotransmission is, like adrenergic neurotransmission, subject to presynaptic modulation involving muscarinic receptors (autoreceptors) and receptors for other neurotransmitters (heteroceptors), there is very little information on the effects of aging on this control. The level of cholinergic nerve activity can also be gauged by the effects of the muscarinic receptor antagonist atropine. The ability of atropine to produce tachycardia is reduced with age in both rat[75] and man,[77] as is the bradycardia to vagal stimulation.[75]

F. AUTONOMIC REFLEX FUNCTION

In man, baroreflex function is normally measured in terms of the tachycardia to falls in blood pressure produced by a vasodilator, or in terms of the bradycardia to rises in blood pressure produced by a vasoconstrictor, and so involves both sympathetic and vagal functions. Baroreflex sensitivity is reported to be reduced by aging in man[146-148] and rat,[149] but not in anesthetized dog.[150] Increased arterial stiffness occurs with aging, which could explain diminished baroreflexes, but the baroreflex could equally well be altered by diminished central integration, diminished efferent nerve activity, of diminished target organ response to transmitter.

Reflex vasoconstriction to cold is reported to be less in old than in young humans[147] and rats.[151] With increasing age, there is a decrease in the vasodilatory response to direct heat on the forearm in man.[152] The diminished response to heat or cold stress may be at least partly due to reduced responsiveness of cutaneous microvessels, since it has been shown that capillary blood flow is reduced in the nailfold circulation of the elderly.[153] Furthermore, during acute cold exposure, systemic blood pressure rises more slowly at first in the elderly, but at 2 h, the rise in systolic pressure is greater than in the young;[154] such increases in systolic pressure due to cold may partly contribute to the increased winter mortality among the elderly.[155] Hence, the elderly have a diminished control of cutaneous blood flow coupled with relatively large blood pressure changes to cold exposure, presumably due to diminished baroreflex function.

As a further manifestation of diminished baroreflex function, heart rate variability[156] and respiratory sinus arrhythmia[157] decrease with age.

Postural hypotension is reported to be common in the elderly,[156,158] and a failure to increase peripheral resistance sufficiently[159] or arterial rigidity[156] has been implicated. Venous dysfunction, as reported above, as well as diminished baroreflex function are likely to contribute to postural hypotension.

VII. SUMMARY

Although it is difficult to generalize from so much apparently conflicting data, several important facts emerge concerning target organ response in the autonomic nervous system. Target organ response to administration of exogenous adrenergic or cholinergic agonists is generally reported to be unchanged or decreased by aging. A decreased response can occur due to reduced numbers of receptors on the target organ or to diminished second-messenger formation, as has been demonstrated particularly for the β-adrenergic system. An increased response to agonists with age has generally been due to a decline in the function of agonist disposition mechanisms, as in the case of acetylcholine, where an increased response can be shown to be due to the decreased activity of acetylcholinesterase. These studies of pharmacological responses to exogenously applied agents tell us little about how the system responds to autonomic neurotransmitters: we must also examine how aging changes autonomic neurotransmission. Studies of innervation and the neurotransmitter content of tissues tend to suggest that there is a loss of nerves with aging, at least in some tissues, and investigations of the effects of neurotransmitter antagonists generally suggest a reduced or unchanged neural control of tissues. Certain factors conspire to maintain transmission despite this decline: diminished acetylcholinesterase activity, noradrenaline reuptake, and presynaptic inhibition of neurotransmission all tend to make full use of the available transmitter, presumably at the expense of fine control and at the risk of overstimulation. One manifestation of altered handling of neurotransmitters is the increased plasma level of noradrenaline in the elderly. Diminished neurotransmitter disposition mechanisms may largely counteract diminished target organ response to the neurotransmitter, leaving the function of the organ largely unchanged at rest. However, the autonomic nervous system must respond reflexively to changes from the resting state, and the elderly are generally reported to have diminished reflex control systems. Hence, an increased incidence of postural hypotension, diminished temperature regulation, and disturbances of bladder function may be manifestations of diminished reflex control of autonomic function.

ACKNOWLEDGMENTS

Work carried out in the author's laboratory was supported by the Royal College of Surgeons in Ireland, the Health Research Board (Ireland), and the Irish Heart Foundation.

REFERENCES

1. **Libert, F., Parmentier, M., Lefort, A., Dinsart, C., Van Sande, J., Maenhaut, C., Simmons, M.-J., Dumont, J.-E., and Vassart, G.,** Selective amplification and cloning of four new members of the G protein-coupled receptor family, *Science,* 244, 569, 1989.
2. **Lorenz, W., Lomasney, J. W., Collins, S., Regan, J. W., Caron, M. G., and Lefkowitz, R. J.,** Expression of three alpha-2 adrenergic receptor subtypes in rat tissues: implications for alpha-2 receptor classification, *Mol. Pharmacol.,* 38, 599, 1990.
3. **Emorine, L. J. and Strosberg, D. A.,** Molecular analysis of the three human beta-adrenergic receptors, in *Adrenoceptors: Structure, Mechanisms, Function,* Szabadi, E. and Bradshaw, C. M., Eds., Birkhauser, Basel, 1991, 79.
4. **Borst, S. E. and Scarpace, P. L.,** Alpha-1 adrenergic stimulation of inositide hydrolysis in liver of senescent rats, *Mech. Ageing Dev.,* 56, 275, 1990.
5. **Scott, P. J. W. and Reid, J. L.,** The effect of age on the response of human isolated arteries to noradrenaline, *Br. J. Clin. Pharmacol.,* 13, 237, 1982.
6. **Stevens, M. J., Lipe, S., and Moulds, R. F. W.,** The effect of age on the response of human isolated arteries and veins to noradrenaline, *Br. J. Clin. Pharmacol.,* 14, 750, 1982.
7. **Toda, N. and Shimizu, I.,** Neuroeffector function in mesenteric arteries isolated from beagles of different ages, *J. Pharmacol. Exp. Ther.,* 240, 223, 1987.
8. **Hynes, M. R. and Duckles, S. P.,** Effect of increasing age on the endothelium-mediated relaxation of rat blood vessels in vitro, *J. Pharmacol. Exp. Ther.,* 241, 387, 1987.
9. **Tuttle, R. S.,** Age related changes in the sensitivity of rat aortic strips to norepinephrine and associated chemical and structural alterations, *J. Gerontol.,* 21, 510, 1966.
10. **Simpkins, J. W., Field, F. P., and Ress, R. J.,** Age-related decline in adrenergic responsiveness of the kidney, heart and aorta of male rats, *Neurobiol. Aging,* 4, 233, 1983.
11. **Hyland, L., Warnock, P., and Docherty, J. R.,** Age-related alterations in alpha-1 and beta-adrenoceptor mediated responsiveness of rat aorta, *Naunyn-Schmiedeberg's Arch. Pharmacol.,* 335, 50, 1987.
12. **Wanstall, J. C. and O'Donnell, S. R.,** Influence of age on calcium entry blocking drugs in rat aorta is spasmogen-dependent, *Eur. J. Pharmacol.,* 159, 241, 1989.

13. **Dohi, Y. and Luscher, T. F.,** Aging differentially affects direct and indirect actions of endothelin-1 in perfused mesenteric resistance arteries of the rat, *Br. J. Pharmacol.*, 100, 889, 1990.
14. **Duckles, S. P., Carter, B. J., and Williams, C. L.,** Vascular adrenergic neuroeffector function does not decline in aged rats, *Circ. Res.*, 56, 109, 1985.
15. **Fouda, A.-K. and Atkinson, J.,** Sensitivity to noradrenaline and electrical stimulation decreases with age in the rat tail artery, *Naunyn-Schmiedeberg's Arch. Pharmacol.*, 334, 37, 1986.
16. **Ordway, G. A., Kolta, M. G., Gerald, M. C., and Wallace, L. J.,** Age-related change in alpha-adrenergic responsiveness of the urinary bladder of the rat is regionally specific, *Neuropharmacology*, 25, 1335, 1986.
17. **Yoshida, M., Latifpour, J., Nishimoto, T., and Weiss, R. M.,** Pharmacological characterization of alpha adrenergic receptors in the young and old female rabbit urethra, *J. Pharmacol. Exp. Ther.*, 257, 1100, 1991.
18. **Pagenkopf, C. W., Bourreau, J.-P., and Challice, C. E.,** Changes in prejunctional potency of B-HT 933 during aging in the rat vas deferens, *Clin. Exp. Pharmacol. Physiol.*, 17, 557, 1990.
19. **Borst, S. E. and Scarpace, P. L.,** Reduced high-affinity alpha-1 adrenoceptors in liver of senescent rats: implications of assessment at various temperatures, *Br. J. Pharmacol.*, 101, 650, 1990.
20. **Kimball, K. A., Cornett, L. E., Seifen, E., and Kennedy, R. H.,** Aging: changes in cardiac α-1 adrenoceptor responsiveness and expression, *Eur. J. Pharmacol.*, 208, 231, 1991.
21. **Borst, S. E., Narang, N., Crews, F. T., and Scarpace, P. J.,** Reduced alpha-1 adrenergic receptor-mediated inositide hydrolysis in cardiac atria of senescent rats, *J. Cardiovasc. Pharmacol.*, 16, 444, 1990.
22. **Schaefer, W. and Williams, R. S.,** Age dependent changes in expression of alpha-1 adrenergic receptors in rat myocardium, *Biochem. Biophys. Res. Commun.*, 138, 387, 1986.
23. **Miyamoto, A. and Ohshika, H.,** Age-related changes in [^{3}H]-prazosin binding and phosphoinositide hydrolysis in rat ventricular myocardium, *Gen. Pharmacol.*, 20, 647, 1989.
24. **Partilla, J. S., Hoopes, M. T., Ito, H., Dax, E. M., and Roth, G. S.,** Loss of rat ventricular alpha-1 adrenergic receptors during aging, *Life Sci.*, 31, 2507, 1982.
25. **Dalrimple, H. W., Hamilton, C. A., and Reid, J. L.,** The effect of age on peripheral alpha-adrenoceptors in vivo and in vitro in the rabbit, *Br. J. Pharmacol.*, 77, 322P, 1982.
26. **Ishikawa, Y., Gee, M. V., Baum, B. J., and Roth, G. S.,** Decreased signal transduction in rat parotid cell aggregates during aging is not due to loss of alpha-1 adrenergic receptors, *Exp. Gerontol.*, 24, 25, 1989.
27. **Elliott, H. L., Sumner, D. J., McLean, K., and Reid, J. L.,** Effect of age on the responsiveness of vascular alpha-adrenoceptors in man, *J. Cardiovasc. Pharmacol.*, 4, 388, 1982.
28. **Rosendorff, C., Kalliatakis, B., Radford, H. M., and Patton, J.,** Age dependence of the pressor sensitivity to noradrenaline and angiotensin II during calcium channel blockade in hypertensive patients, *J. Cardiovasc. Pharmacol.*, 12, S69, 1988.
29. **Docherty, J. R. and Hyland, L.,** Alpha-adrenoceptor responsiveness in the aged rat, *Eur. J. Pharmacol.*, 126, 75, 1986.
30. **Gee, M. V., Ishikawa, Y., Baum, B. J., and Roth, G. S.,** Impaired adrenergic stimulation of rat parotid cell glucose oxidation during aging: the role of calcium, *J. Gerontol.*, 41, 331, 1986.
31. **Buckley, C., Curtin, D. McA., Docherty, J. R., and Eustace, P.,** Ageing and alpha adrenoceptors in the iris, *Eye*, 1, 211, 1987.
32. **Korcyn, A. D., Laor, N., and Nemet, P.,** Sympathetic pupillary tone in old age, *Arch. Ophthalmol.*, 94, 1905, 1976.

33. **Smith, S. A. and Fasler, J. J.,** Age-related changes in autonomic function: relations with postural hypotension, *Age Ageing,* 12, 205, 1983.
34. **Docherty, J. R.,** Effects of ageing on contractions to noradrenaline and 5-HT in human saphenous vein, *Br. J. Pharmacol.,* 105, 305P, 1992.
35. **Elliott, J. M. and Grahame-Smith, D. G.,** The binding characteristics of [^{3}H]-dihydroergocryptine on intact human platelets, *Br. J. Pharmacol.,* 76, 121, 1982.
36. **Buckley, C., Curtin, D., Walsh, T., and O'Malley, K.,** Ageing and platelet alpha-2 adrenoceptors, *Br. J. Clin. Pharmacol.,* 21, 721, 1986.
37. **Brodde, O.-E., Anlauf, M., Graben, N., and Bock, K. D.,** Age dependent decrease of alpha-2 adrenergic receptor number in human platelets, *Eur. J. Pharmacol.,* 81, 345, 1982.
38. **Yokoyama, M., Kusui, A., Sakamoto, S., and Fukazaki, H.,** Age-associated increments in human platelet alpha-adrenoceptor capacity: possible mechanism for platelet hyperactivity to epinephrine in aging man, *Thromb. Res.,* 34, 287, 1984.
39. **Docherty, J. R.,** The effects of ageing on vascular alpha-adrenoceptors in pithed rat and rat aorta, *Eur. J. Pharmacol.,* 146, 1, 1988.
40. **Vestal, R. E., Wood, A. J. J., and Shand, D. G.,** Reduced beta-adrenoceptor sensitivity in the elderly, *Clin. Pharmacol. Ther.,* 26, 181, 1979.
41. **Kendall, M. J., Woods, K. L., Wilkins, M. R., and Worthington, D. J.,** Responsiveness to beta-adrenergic receptor stimulation: the effects of age are cardioselective, *Br. J. Clin. Pharmacol.,* 14, 821, 1982.
42. **Montamat, S. C. and Davies, A. O.,** Physiological response to isoproterenol and coupling of beta-adrenoceptors in young and elderly subjects, *J. Gerontol.,* 44, M100, 1989.
43. **Yin, F. C. P., Spurgeon, H. A., Greene, H. L., Lakatta, E. G., and Weisfeldt, M. L.,** Age-associated decrease in heart rate response to isoproterenol in dogs, *Mech. Ageing Dev.,* 10, 17, 1978.
44. **Borton, M. and Docherty, J. R.,** The effects of ageing on neuronal uptake of noradrenaline in the rat, *Naunyn-Schmiedeberg's Arch. Pharmacol.,* 340, 139, 1989.
45. **Guarnieri, T., Filburn, C. R., Zitnik, G., Roth, G. S., and Lakatta, E. G.,** Contractile and biochemical correlates of beta-adrenergic stimulation of the aged heart, *Am. J. Physiol.,* 239, H501, 1980.
46. **Narayanan, N. and Derby, J. A.,** Alterations in the properties of beta-adrenergic receptors of myocardial membranes in aging: impairments in agonist-receptor interactions and guanine nucleotide regulation accompany diminished catecholamine responsiveness of adenylate cyclase, *Mech. Ageing Dev.,* 19, 127, 1982.
47. **Tumer, N., Houck, W. T., and Roberts, J.,** Upregulation of adrenergic beta-receptor subtypes in the senescent rat heart, *Mech. Ageing Dev.,* 49, 235, 1989.
48. **Fan, T. H.-M. and Banerjee, S. P.,** Age-related reduction of beta-adrenoceptor sensitivity in rat heart occurs by multiple mechanisms, *Gerontology,* 31, 373, 1985.
49. **Kusiak, J. W. and Pitha, J.,** Decreased response with age of the cardiac catecholamine sensitive adenylate cyclase system, *Life Sci.,* 33, 1679, 1983.
50. **Abrass, I. B. and Scarpace, P. J.,** Human lymphocyte beta-adrenergic receptors are unaltered with age, *J. Gerontol.,* 36, 298, 1981.
51. **Doyle, V., O'Malley, K., and Kelly, J. G.,** Human lymphocyte beta-adrenoceptor density in relation to age and hypertension, *J. Cardiovasc. Pharmacol.,* 4, 738, 1982.
52. **Schocken, D. D. and Roth, G. S.,** Reduced beta-adrenergic receptor concentrations in ageing man, *Nature (London),* 267, 856, 1977.
53. **Montamat, S. C. and Davies, A. O.,** Physiological response to isoproterenol and coupling of beta-adrenergic receptors in young and elderly human subjects, *J. Gerontol.,* 44, M100, 1989.
54. **Middeke, M., Remein, J., and Holzgreve, H.,** The influence of sex, age, blood pressure and physical stress on beta-2 adrenoceptor density of mononuclear cells, *J. Hypertens.,* 2, 261, 1984.

55. **Vanscheeuwijck, P., Van de Velde, E., and Fraeyman, N.,** Effect of aging on properties and function of beta-adrenoceptors in rat lung, *Eur. J. Pharmacol.*, 172, 373, 1989.
56. **Scarpace, P. J. and Yu, B. P.,** Diet restriction retards the age-related loss of beta-adrenergic receptors and adenylate cyclase activity in rat lung, *J. Gerontol.*, 42, 442, 1987.
57. **Bylund, D. B., Tellez-Inon, M. T., and Hollenberg, M. D.,** Age-related decline in beta-adrenergic receptors, adenylate cyclase and phosphodiesterase activity in rat erythrocyte membranes, *Life Sci.*, 21, 403, 1977.
58. **Tanaka, E., Habu, T., Letic-Gavrilovic, A., and Abe, K.,** Protein secretion by rat submandibular glands in response to isoproterenol, α-methylnoradrenaline and clonidine during aging, *Mech. Ageing Dev.*, 54, 221, 1990.
59. **Giudicelli, Y. and Pecquery, R.,** Beta-adrenergic receptors and catecholamine-sensitive adenylate cyclase in rat fat-cell membranes: influence of growth, cell size and aging, *Eur. J. Biochem.*, 90, 413, 1978.
60. **Parker, R. J., Berkowitz, B. A., Lee, C.-H., and Denckla, W. D.,** Vascular relaxation, aging and thyroid hormones, *Mech. Ageing Dev.*, 8, 397, 1978.
61. **O'Donnell, S. R. and Wanstall, J.C.,** Beta-1 and beta-2 adrenoceptor-mediated responses in preparations of pulmonary artery and aorta from young and aged rats, *J. Pharmacol. Exp. Ther.*, 228, 733, 1984.
62. **Duckles, S. P. and Hurlbert, J. S.,** Effects of age on beta-adrenergic relaxation of the rat jugular vein, *J. Pharmacol. Exp. Ther.*, 236, 71, 1986.
63. **Emmick, J. T. and Cohen, M. L.,** Ageing and vasodilatation to atrial peptides, *Clin. Exp. Hypertens.*, A8, 75, 1986.
64. **Fleisch, J. H. and Hooker, C. S.,** The relationship between age and relaxation of vascular smooth muscle in the rabbit and rat, *Circ. Res.*, 38, 243, 1976.
65. **Shimizu, I. and Toda, N.,** Alterations with age of the response to vasodilator agents in isolated mesenteric arteries of the beagle, *Br. J. Pharmacol.*, 89, 769, 1986.
66. **Ikezono, K., Zerkowski, H.-R., Beckeringh, J. J., Michel, M. C., and Brodde, O.-E.,** Beta-2 adrenoceptor mediated relaxation of the isolated human saphenous vein, *J. Pharmacol. Exp. Ther.*, 241, 294, 1987.
67. **Pan, H. Y.-M., Hoffman, B. B., Pershi, R. A., and Blaschke, T. F.,** Decline in beta-adrenergic receptor-mediated vascular relaxation with aging in man, *J. Pharmacol. Exp. Ther.*, 228, 802, 1986.
68. **Docherty, J. R.,** Changes in adrenoceptor function with age, in *Vascular Neuroeffector Mechanisms,* Bevan, J. A., Majewski, H., Maxwell, R. A., and Story, D. F., Eds., IRL Press, Oxford, 1988, 281.
69. **Elfellah, M. S., Johns, A., and Shepherd, A. M. M.,** Effect of age on responsiveness of isolated rat atria to carbachol and on binding characteristics of atrial muscarinic receptors, *J. Cardiovasc. Pharmacol.*, 8, 873, 1986.
70. **Kobashi, Y. L., Breung, E. P., and Markus, R. P.,** Age-related changes in the reactivity of the rat jejunum to cholinoceptor agonists, *Eur. J. Pharmacol.*, 115, 133, 1985.
71. **Ochillo, R. F. and Cheung, S. T.,** A comparative study of the effects of aging on the responsiveness of the cholinergic receptor of the isolated ileum of mouse and rat, *Res. Commun. Chem. Pathol. Pharmacol.*, 60, 261, 1988.
72. **Huang, J.-T. and Nakajima, T.,** Effect of some smooth muscle stimulants and trypsin on the contraction of vas deferens of aging rats, *Jpn. J. Pharmacol.*, 38, 321, 1985.
73. **Kennedy, R. H. and Seifen, E.,** Aging: effects on chronotropic actions of muscarinic agonists in isolated rat atria, *Mech. Ageing Dev.*, 51, 81, 1990.
74. **Takayanagi, I., Moriya, M., Kizawa, Y., Koike, K., Shinkai, M., and Murakami, H.,** Effect of ageing on response to nicotine in rabbit bronchial preparation, *Gen. Pharmacol.*, 22, 787, 1991.
75. **Kelliher, G. J. and Conahan, S. T.,** Changes in vagal activity and response to muscarinic receptor agonists with age, *J. Gerontol.*, 35, 842, 1980.

76. **Rothbaum, D. A., Shaw, D. G., Angell, C. S., and Shock, N. W.,** Age differences in the baroreceptor response of rats, *J. Gerontol.*, 29, 488, 1974.
77. **Dauchot, P. and Gravenstein, J. S.,** Effects of atropine on the electrocardiogram in different age groups, *Clin. Pharmacol. Ther.*, 12, 274, 1971.
78. **Furchgott, R. F.,** Studies on endothelium-dependent vasodilation and the endothelium-derived relaxing factor, *Acta Physiol. Scand.*, 139, 257, 1990.
79. **Moritoki, H., Hosoki, E., and Ishida, Y.,** Age-related decrease in endothelium-dependent dilator response to histamine in rat mesenteric artery, *Eur. J. Pharmacol.*, 126, 61, 1986.
80. **Koga, T., Takata, Y., Kobayashi, K., Takishita, S., Yamashita, Y., and Fujishima, M.,** Age and hypertension promote endothelium-dependent contractions to acetylcholine in the aorta of the rat, *Hypertension*, 14, 542, 1989.
81. **Kawai, Y. and Ohhashi, T.,** Age-related changes in relaxant response of vascular smooth muscles to atrial natriuretic peptide, *J. Pharmacol. Exp. Ther.*, 252, 1234, 1990.
82. **Ishihata, A., Katano, Y., Morinobu, S., and Endoh, M.,** Influence of aging on the contractile response to endothelin of rat thoracic aorta, *Eur. J. Pharmacol.*, 200, 199, 1991.
83. **Kazanietz, M. G. and Enero, M. A.,** Decreased beta-adrenoceptor-mediated vasodilation in aorta from aged rats: possible involvement of a stimulatory GTP-binding protein, *Eur. J. Pharmacol.*, 198, 177, 1991.
84. **Toda, N., Bian, K., and Inoue, S.,** Age-related changes in the response to vasoconstrictor and dilator agents in isolated beagle coronary arteries, *Naunyn-Schmiedeberg's Arch. Pharmacol.*, 336, 359, 1987.
85. **Blauw, G. J., Van Brummelen, P., Chang, P. C., Vermeij, P., and Van Zwieten, P. A.,** Arterial dilatation and venous constriction induced by serotonin in the elderly, *Drugs*, 36 (Suppl. 1), 74, 1988.
86. **Nyborg, N. C. B.,** Ageing is associated with increased 5-Ht-2-receptor affinity and decreased receptor reserve in rat isolated coronary arteries, *Br. J. Pharmacol.*, 102, 282, 1991.
87. **Yonezawa, Y., Kondo, H., and Nomaguchi, T. A.,** Age-related changes in serotonin content and its release reaction of rat platelets, *Mech. Ageing Dev.*, 47, 65, 1989.
88. **Shuttleworth, R. D. and O'Brien, J. R.,** Intraplatelet serotonin and plasma 5-hydroxytryptamine in health and disease, *Blood*, 57, 505, 1981.
89. **Johnson, M., Ramey, E., and Ramwell, P. W.,** Sex and age differences in human platelet aggregation, *Nature (London)*, 253, 355, 1975.
90. **Palluck, R., Hoefke, W., and Graida, W.,** Postjunctional alpha-adrenoceptor and influence of angiotensin II in different isolated blood vessels, *Arzneim. Forsch.*, 35, 395, 1985.
91. **James, G., Sealey, J., Mueller, B., Alderman, M., Madhavan, S., and Laragh, J.,** Renin relationship to sex, race and age in a normotensive population, *J. Hypertens.*, 4 (Suppl. 5), S387, 1986.
92. **Cugini, P., Murano, G., Lucia, P., Letizia, C., Scavo, D., Halberg, F., and Schramm, H.,** The gerontological decline of the renin-aldosterone system: a chronobiological approach extended to essential hypertension, *J. Gerontol.*, 42, 461, 1987.
93. **Wakabayashi, I., Sakamoto, K., Hatake, K., Yoshimoto, S., and Kurahashi, M.,** Effect of age on contractile response to angiotensin II in rat aorta, *Life Sci.*, 47, 771, 1990.
94. **Duggan, J., Kilfeather, S., Lightman, S., and O'Malley, K.,** The effect of age on platelet angiotensin II and ANP receptor density, *Br. J. Clin. Pharmacol.*, 26, 629P, 1988.
95. **Anderson, J. V., Donckier, J., Payne, N. N., Beacham, J., Slater, J. H. D., and Bloom, S. R.,** Atrial natriuretic peptide: evidence of action as a natriuretic hormone at physiological plasma concentrations in man, *Clin. Sci.*, 72, 305, 1987.

96. **Haller, B. G. D., Zust, H., Shaw, S., Gnadinger, M. P., Uehlinger, D. E., and Weidmann, P.,** Effects of posture and ageing on circulating atrial natriuretic peptide levels in man, *J. Hypertens.,* 5, 551, 1987.
97. **Montorsi, P., Tonolo, G., Polonia, J., Hepburn, D., and Richards, A. M.,** Correlates of plasma atrial natriuretic factor in health and hypertension, *Hypertension,* 10, 570, 1987.
98. **Singer, D. R. J., Sagnella, G. A., Markandu, N. D., Buckley, M. G., and MacGregor, G. A.,** Atrial natriuretic peptide, blood pressure, and age, *Lancet,* 2, 1394, 1987.
99. **Davis, P. J. and Davis, F. B.,** Water excretion in the elderly, *Endocrinol. Aging,* 16, 867, 1987.
100. **Schmucker, D. L.,** Ageing and drug disposition: an update, *Pharmacol. Rev.,* 37, 133, 1985.
101. **Jansen, T. L. Th. A., Tan, A. C. I. T. L., Wollersheim, H., Benraad, Th. J., and Thien, Th.,** Age-dependent vasodilation of the skin microcirculation by atrial natriuretic factor, *J. Cardiovasc. Pharmacol.,* 18, 622, 1991.
102. **Grimaldi, R., Zini, I., Agnati, L. F., Fuxe, K., and Harfstrand, A.,** Intravenous administration of atriopeptin III affects blood pressure differentially in young and old male rats, *Acta Physiol. Scand.,* 129, 525, 1987.
103. **Kawai, Y. and Ohhashi, T.,** Age-related changes in relaxant response of vascular smooth muscles to atrial natriuretic peptide, *J. Pharmacol. Exp. Ther.,* 252, 1234, 1990.
104. **Carvajal, A., Shafii, M., and Docherty, J. R.,** Effects of maturation and ageing on the response to vasodilators in the rat aorta, *Br. J. Pharmacol.,* 96, 190P, 1989.
105. **Duckles, S. P.,** Influence of age on relaxation of rat blood vessels to atriopeptin II, *J. Pharmacol. Exp. Ther.,* 200, 697, 1987.
106. **Randall, M. D.,** Vascular actions of the endothelins, *Pharmacol. Ther.,* 50, 73, 1991.
107. **Cargnelli, G., Rossi, G.-P., Bova, S., and Pessina, A. C.,** In vitro vascular reactivity to endothelin: a comparison between young and old normotensive and hypertensive rats, *Clin. Exp. Hypertens.,* A12, 1437, 1990.
108. **Ishihata, A., Katano, Y., Morinobu, S., and Endoh, M.,** Influence of aging on the contractile response to endothelin of rat thoracic aorta, *Eur. J. Pharmacol.,* 200, 199, 1991.
109. **Martinez, J. L., Vasquez, B. J., Messing, R. B., Jensen, R. A., Liang, K. C., and McGaugh, J. L.,** Age-related changes in the catecholamine content of peripheral organs in male and female F344 rats, *J. Gerontol.,* 36, 280, 1981.
110. **Handa, R. K. and Duckles, S. P.,** Influence of age on norepinephrine content in arteries and veins of Fischer 344 rats, *Neurobiol. Aging,* 8, 511, 1987.
111. **Partanen, M., Waller, S. B., London, E. D., and Hervonen, A.,** Indices of neurotransmitter synthesis and release in aging nervous system, *Neurobiol. Aging,* 6, 227, 1985.
112. **Banerji, T. K., Parkening, T. A., and Collins, T. J.,** Adrenomedullary catecholaminergic activity increases with age in male laboratory rodents, *J. Gerontol.,* 39, 264, 1984.
113. **Daly, R. N., Goldberg, P. B., and Roberts, J.,** Effects of age on neurotransmission at the cardiac sympathetic neuroeffector junction, *J. Pharmacol. Exp. Ther.,* 245, 798, 1988.
114. **Dhall, U., Cowen, T., Haven, A. J., and Burnstock, G.,** Perivascular noradrenergic and peptide-containing nerves show different patterns of changes during development and ageing in the guinea-pig, *J. Auton. Nerv. Syst.,* 16, 109, 1986.
115. **Koistinaho, J., Wadhwani, K. C., and Rapoport, S. I.,** Adrenergic innervation of the tibial and vagus nerves in rats of different ages, *Mech. Ageing Dev.,* 52, 195, 1990.
116. **Vega, J. A., Ricci, A., and Amenta, F.,** Age-dependence of the sympathetic innervation of the rat kidney, *Mech. Ageing Dev.,* 54, 185, 1990.

117. **Gale, J. D., Alberts, J. C. J., and Cowen, T.,** A quantitative study of changes in old age of 5-hydroxytryptamine-like immunoreactivity in perivascular nerves of the rabbit, *J. Auton. Nerv. Syst.*, 28, 51, 1989.
118. **Jarvi, R., Helen, P., Pelto-Huikko, M., Rapoport, S., and Hervonen, A.,** Age-related changes of enkephalinergic innervation of human sympathetic neurons, *Mech. Ageing Dev.*, 44, 143, 1988.
119. **Buhler, F. R.,** Age and cardiovascular response adaptation: determinants of an antihypertensive treatment concept primarily based on beta-blockers and calcium entry blockers, *Hypertension,* 5, III-94, 1983.
120. **Conway, J.,** Effect of age on the response to propranolol, *Int. Zeits. Klin. Pharm. Ther. Tox.*, 4, 148, 1970.
121. **Conway, J., Wheeler, R., and Sannerstedt, R.,** Sympathetic nervous activity during exercise in relation to age, *Cardiovasc. Res.*, 5, 577, 1971.
122. **Gerstenblith, G., Lakatta, E. G., and Weisfeldt, M. L.,** Age changes in myocardial function and exercise response, *Prog. Cardiovasc. Res.*, 19, 1, 1976.
123. **Tejada, C., Strong, J. P., Montenegro, M. R., Restrepo, C., and Solberg, L. A.,** Distribution of coronary and aortic atherosclerosis by geographical location, race, and sex, *Lab. Invest.*, 18, 509, 1968.
124. **Kostis, J. B., Moreyra, A. E., Amendo, M. T., Di Pietro, J., Cosgrove, N., and Kuo, P. T.,** The effect of age on heart rate in subjects free of heart disease, *Circulation,* 65, 141, 1982.
125. **Klein, C., Gerber, J. G., Gal, J., and Nies, A. S.,** Beta-adrenergic receptors in the elderly are not less sensitive to timolol, *Clin. Pharmacol. Ther.*, 40, 161, 1986.
126. **Buhler, F. R., Kiowski, W., Van Brummelen, P., Amann, F. W., Bertal, O., Landemann, R., Lutold, B. E., and Bolli, P.,** Plasma catecholamines and cardiac, renal and peripheral vascular adrenoceptor mediated responses in different age groups in normal and hypertensive subjects, *Clin. Exp. Hypertens.*, 2, 409, 1980.
127. **McNeil, J. J., Drummer, O. H., Conway, E. L., Workman, B. S., and Louis, W. J.,** Effect of age on pharmacokinetics of and blood pressure responses to prazosin and terazosine, *J. Cardiovasc. Pharmacol.*, 10, 168, 1987.
128. **Hyland, L. and Docherty, J. R.,** An investigation of age-related changes in pre- and postjunctional α-adrenoceptors in human saphenous vein, *Eur. J. Pharmacol.*, 114, 361, 1985.
129. **Duckles, S. P.,** Effect of norepinephrine uptake blockade on contractile responses to adrenergic nerve stimulation of isolated rat blood vessels: influence of age, *J. Pharmacol. Exp. Ther.*, 243, 521, 1987.
130. **Handa, R. K. and Duckles, S. P.,** Age-related changes in adrenergic vasoconstrictor responses of the rat hindlimb, *Am. J. Physiol.*, 253, H1566, 1987.
131. **Werneck de Avellar, M. C., Kobashi, Y. L., and Markus, R. P.,** Age-related changes in neuronal uptake of noradrenaline, *Naunyn-Schmiedeberg's Arch. Pharmacol.*, 341, 295, 1990.
132. **McIntosh, H. H. and Westfall, T. C.,** Influence of aging on catecholamine levels, accumulation, and release in F-344 rats, *Neurobiol. Aging,* 8, 233, 1987.
133. **Pedersen, E. B. and Christensen, N. J.,** Catecholamines in plasma and urine in patients with essential hypertension determined by double-isotope derivative techniques, *Acta Med. Scand.*, 198, 373, 1975.
134. **Zeigler, M. G., Lake, C. R., and Kopin, I. J.,** Plasma noradrenaline increases with age, *Nature (London),* 261, 33, 1976.
135. **MacGilchrist, A. J., Hawksby, C., Howes, L. G., and Reid, J. L.,** Rise in plasma noradrenaline with age results from an increase in spillover rate, *Gerontology,* 35, 7, 1989.
136. **Vargas, E., Lye, M., Faragher, E. B., Goddard, C., Moser, B., and Davies, I.,** Cardiovascular haemodynamics and the response of vasopressin, aldosterone, plasma renin activity and plasma catecholamines to head-up tilt in young and old healthy subjects, *Age Ageing,* 15, 17, 1986.

137. **Palmer, G. J., Zeigler, M. G., and Lake, C. R.,** Response of norepinephrine and blood pressure to stress increases with age, *J. Gerontol.*, 33, 482, 1978.
138. **Young, J. B., Rowe, J. W., Pallotta, J. A., Sparrow, D., and Landsberg, L.,** Enhanced plasma norepinephrine responses to upright posture and oral glucose administration in elderly human subjects, *Metabolism,* 29, 532, 1980.
139. **Rubin, P. C., Scott, P. J. W., McKean, K., and Reid, J. L.,** Noradrenaline release and clearance in relation to age and blood pressure in man, *Eur. J. Clin. Invest.*, 12, 121, 1982.
140. **Veith, R. C., Featherstone, J. A., Linares, O. A., and Halter, J. B.,** Age differences in plasma norepinephrine kinetics in humans, *J. Gerontol.*, 41, 319, 1986.
141. **Docherty, J. R. and Hyland, L.,** α-Adrenoceptor responsiveness in the aged rat, *Eur. J. Pharmacol.*, 126, 75, 1986.
142. **Hyland, L. and Docherty, J. R.,** Further examination of the effects of ageing on the adrenoceptor responsiveness of the rat vas deferens, *Eur. J. Pharmacol.*, 110, 241, 1985.
143. **Daly, R. N., Goldberg, P. B., and Roberts, J.,** The effect of age on presynaptic $alpha_2$ adrenoceptor autoregulation of norepinephrine release, *J. Gerontol.*, 44, 859, 1989.
144. **Hyland, L. and Docherty, J. R.,** An investigation of age related changes in pre- and postjunctional α-adrenoceptors in human saphenous vein, *Eur. J. Pharmacol.*, 114, 361, 1985.
145. **Borton, M. and Docherty, J. R.,** The effects of ageing on prejunctional 5-hydroxytryptamine receptors in the rat vas deferens, *Naunyn-Schmiedeberg's Arch. Pharmacol.*, 342, 130, 1990.
146. **McGarry, K., Laher, M., Fitzgerald, D., Horgan, J., O'Brien, E., and O'Malley, K.,** Baroreflex function in elderly hypertensives, *Hypertension,* 5, 763, 1983.
147. **Collins, K. J., Exton-Smith, A. N., James, M. H., and Oliver, D. J.,** Functional changes in autonomic nervous responses with ageing, *Age Ageing,* 9, 17, 1980.
148. **Simpson, D. M. and Wicks, R.,** Spectral analysis of heart rate indicates reduced baroreceptor-related heart rate variability in elderly persons, *J. Gerontol.*, 43, M21, 1988.
149. **Docherty, J. R., Fitzgerald, D., and O'Malley, K.,** Age-related reduction of baroreflex tachycardia without loss of β-adrenoceptor-mediated tachycardia in Sprague-Dawley rats, *J. Cardiovasc. Pharmacol.*, 8, 376, 1986.
150. **Cox, R. H., Bagshaw, R. J., and Detweiler, D. K.,** Arterial baroreceptor reflexes in young and old racing greyhounds, *Am. J. Physiol.*, 240, H383, 1981.
151. **McCarty, R.,** Sympathetic-adrenal medullary and cardiovascular responses to acute cold stress in adult and aged rats, *J. Auton. Nerv. Syst.*, 12, 15, 1985.
152. **Richardson, D.,** Effects of age on cutaneous circulatory response to direct heat on the forearm, *J. Gerontol.*, 44, M189, 1989.
153. **Richardson, D. and Schwartz, R.,** Comparison of capillary blood flow in the nailfold circulation of young and elderly men, *Age,* 8, 70, 1985.
154. **Collins, K. J.,** Effects of cold on old people, *Br. J. Hosp. Med.*, 38, 506, 1987.
155. **Bull, G. M. and Morton, J.,** Relationships of temperature with death rate from all causes and from certain respiratory and arteriosclerotic diseases in different age groups, *Age Ageing,* 4, 232, 1975.
156. **MacLennan, W. J., Hall, M. R. P., and Timothy, J. I.,** Postural hypotension in old age: is it a disorder of the nervous system or of blood vessels?, *Age Ageing,* 9, 25, 1980.
157. **Hellman, J. B. and Stacy, R. W.,** Variation of respiratory sinus arrhythmia with age, *J. Appl. Physiol.*, 41, 734, 1976.
158. **Robbins, A. S. and Rubenstein, L. Z.,** Postural hypotension in the elderly, *J. Am. Geriatr. Soc.*, 32, 769, 1984.
159. **Williams, B. O., Caird, F. L., and Lennox, I. M.,** Haemodynamic response to postural stress in the elderly with and without postural hypotension, *Age Ageing,* 14, 193, 1985.
160. **Docherty, J. R.,** unpublished observations.

Age-Dependent Changes of Neuroeffector Mechanisms in Target Organs of the Autonomic Nervous System

Chapter 7

CHANGES IN THE EFFECT OF THE AUTONOMIC NERVOUS SYSTEM ON THE AGING HEART

Jay Roberts, David L. Snyder, Mark D. Johnson, and Joel Horwitz

TABLE OF CONTENTS

0-8493-6981-9/93/$0.00 + $.50

I. INTRODUCTION

While the aging process affects the function of many different organ systems, the heart is one of the most important organs that changes with the passage of time. Klausner and Schwartz[1] pointed out that the study of aging in any organism, particularly those organisms with a relatively long, complex life span, is complicated by disease or exposure to other stimulants which are extremely difficult to disassociate and define against the background of the aging process. Nevertheless, the role of pathology and environment in producing changes in the heart over a period of time is fundamental to our understanding of changes in heart function and adrenergic nervous system control of cardiac function. In the present chapter, we attempt whenever possible to comment on the relationship of pathology and environmental factors to age-related changes. Indeed, in many of the human studies in which cardiac output has been shown to decrease with age, much of this decrease has been due to latent pathological changes and does not necessarily represent the aging process.[2]

The autonomic nervous system plays a fundamental role in regulating many systems involved in homeostasis, such as the respiratory and cardiovascular systems. This chapter focuses on age-related changes in autonomic nervous system control of the heart in humans and animals. The autonomic nervous system affects many cardiac functions, including rate, force of contraction, and electrophysiological parameters. In this regard, much is known about age-related changes in sympathetic nervous system* function (the nerves which liberate noradrenaline, NA, at the neuroeffector synapse), while surprisingly little is known about the changes in parasympathetic function with age (the nerves which release acetylcholine, ACh, at the neuroeffector synapse). Any discussion of age-related changes in autonomic control of the heart should emphasize that the autonomic nervous system has many components which can be affected by age or other processes that may be parallel to or interact with the aging process. These include afferent systems such as baroreceptors and chemoreceptors which contribute to autonomic reflexes, i.e., the carotid sinus reflex and aortic sinus reflex. Changes may also occur in centers of the central nervous system which control autonomic nervous system function. On the efferent side, the preganglionic fibers, the ganglionic synapse, and the neuroeffector junction, consisting of the synapse between the nerve terminal and the end organ, make up the components. Much of the discussion of this chapter is on the efferent side. We discuss both presynaptic events (referring to the nerve terminal) and postsynaptic events (referring to the end organ, the heart itself) (Figure 1). It should be noted that while

* The terms "adrenergic" and "sympathetic", although used interchangeably in the present chapter and in the literature, are not precisely the same, i.e., "adrenergic" is based on the nature of the chemical transmitter, while "sympathetic" is based on anatomic and physiologic parameters.

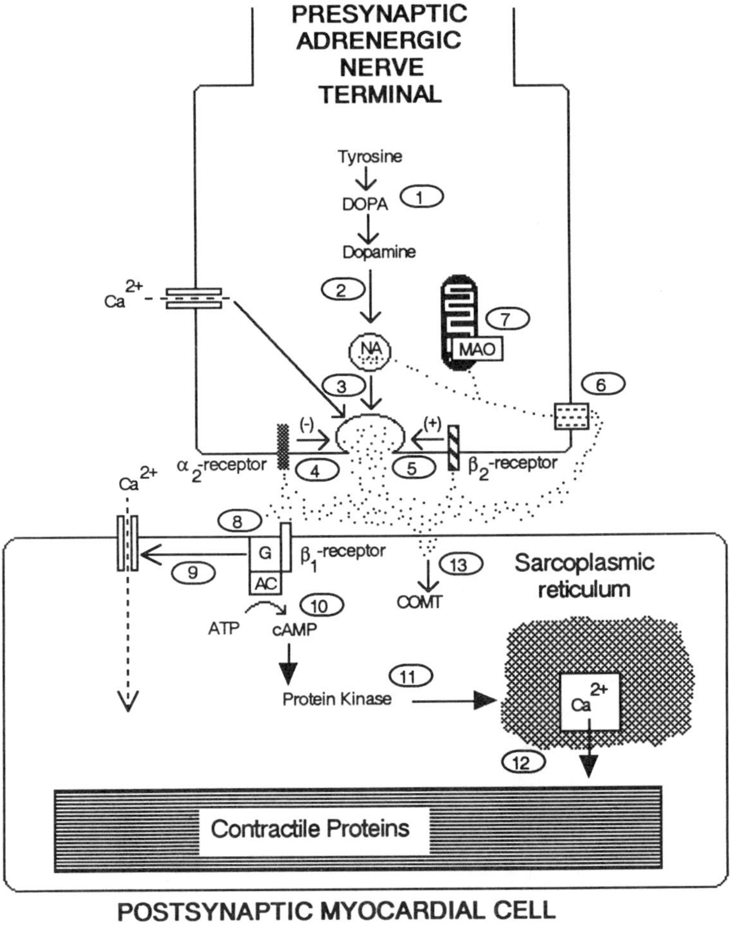

FIGURE 1. Cardiac adrenergic neuroeffector junction. (1) Synthetic pathway from tyrosine to 3,4-dihydroxyphenylalanine (DOPA) to dopamine to noradrenaline (NA); (2) transport and storage in vesicles; (3) membrane depolarization opens voltage-gated calcium channels, and allows Ca^{2+} to enter the nerve terminal and initiate exocytosis of NA vesicles; (4) binding of NA to presynaptic α_2-receptors inhibits further NA release; (5) binding of NA to presynaptic β_2-receptors promotes further NA release; (6) active neuronal uptake of NA; (7) degradation of NA by monoamine oxidase; (8) activation of postsynaptic β-receptors; (9) β_1-Receptor activation of G-proteins promotes Ca^{2+} entry via membrane calcium channels; (10) activated postsynaptic β_1-receptors and G-protein stimulate adenylate cyclase to produce cAMP from ATP; (11) cAMP activates protein kinase, which phosphorylates other enzymes; (12) phosphorylated enzymes stimulate Ca^{2+} release from the sarcoplasmic reticulum, which generates heart muscle contraction; (13) NA taken up into nonneuronal cells is degraded by catechol-*O*-methyltransferase (COMT). (Modified from Shepherd, G. M., *Neurobiology,* 2nd ed., Oxford University Press, New York, 1988, chap. 8. With permission.)

significant changes in the cardiac response to adrenergic influences have been observed in many different species (including the human), similar effects are not always seen in the vasculature (see Chapter 8 of this volume).[3-9] Age-related changes in autonomic control of the heart may not be representative of changes in other parts of the cardiovascular system. Sympathetic nervous system changes in the heart seem to be paralleled by those in the pupil,[10] but may not be paralleled by changes in other organ systems, such as the kidney,[11] muscle,[12] and sudomotor axon reflex.[13] Thus, changes in sympathetic nervous influences during aging seem to be organ dependent.

II. PHYSIOLOGY OF AGE-RELATED CHANGES IN AUTONOMIC CONTROL OF THE HEART

A. CHANGES IN HEART FUNCTION

Klausner and Schwartz[1] identified many of the difficulties involved in assessing the effect of age on the heart in the intact animal. They pointed out that the assessment of intrinsic myocardial performance is extremely difficult because of the interaction between the heart, vascular system, autonomic nervous system, and peripheral metabolic demands. Consequently, if one is to identify changes intrinsic to myocardial function, isolated preparations are suggested. The authors agree with this analysis and further suggest that the interactions between neuroeffector junction and myocardial tissue are best determined in isolated nerve-heart preparations so that the other influences cited above will not be major obstructions to observing age-related effects.

Weisfeldt[14] also pointed out that the effects of age on cardiovascular structure and function have been difficult to study. In cross-sectional studies, subjects should be free of cardiovascular disease so as not to confuse pathology with aging. Longitudinal studies pose other problems as the research must continue over an adequate span of time. Other problems complicating the attempt to define the impact of age on cardiac function relate to maturation, which must be differentiated from aging changes, although both occur as a function of time.

1. Cardiac Electrophysiology

Although there is considerable age-related change in the electrophysiology of the sinoatrial node, atrioventricular node, and ventricle,[1] the relationship of these changes to alterations in autonomic nervous system function has not been studied in detail and consequently will not be discussed further.

2. Heart Rate

In humans, measurement of the R-R interval variation (the interval between the R-wave of two consecutive electrocardiographic complexes) has been used to study heart rate changes during aging. In a study by O'Brien et al.,[15] the R-R interval was examined in 310 healthy subjects, age 18 to 85 years. Heart rate variation showed a skewed distribution and a significant

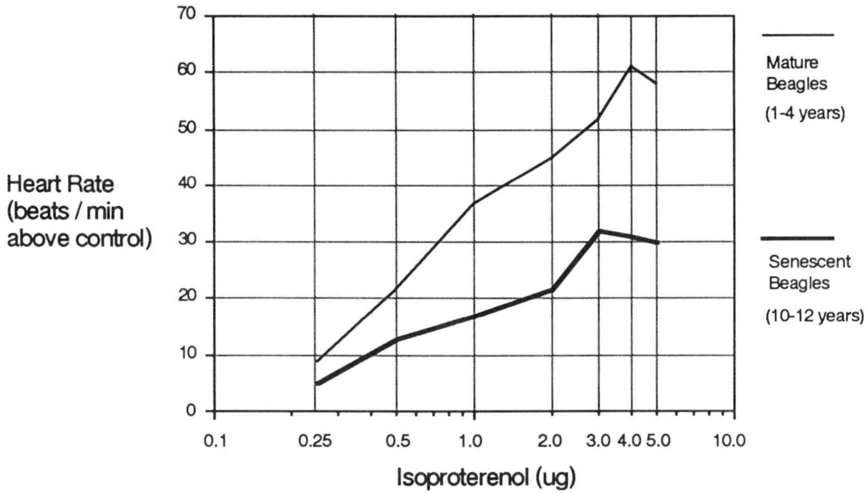

FIGURE 2. Effect of age on β-adrenergic stimulation of heart rate. Maximum heart rate increase during infusion of isoproterenol was diminished in older dogs. (From Weisfeldt, M., *Hosp. Pract.*, 20, 115, 1985. Illustrations by Albert Miller. With permission.)

negative correlation with age. The authors concluded that in view of the decline in heart rate variation with increasing age, normal ranges or tests of autonomic function must be related to the age of the subject. The intrinsic heart rate after autonomic blockade with atropine and propranolol also decreased with age with a slope similar to the maximal exercised heart rate.[16] In a cross-sectional study of over 2300 elderly subjects, change in heart rate was positively correlated with age for both men and women, but a longitudinal study showed a decrease in heart rate with age.[17] Numerous studies have shown that maximal heart rate during exercise declines with age (see Section II.A.4).

Age-related decreases in chronotropic responses to infusion of the β-adrenergic agonist isoproterenol have been found in rats, dogs, and humans (Figure 2).[18] Although there is an age-related reduction in β-receptor-mediated responses in the heart, this reduction appears to be due to age-related changes in postreceptor events (see Section III.B).

Changes during aging have also been observed in the cholinergic system. The negative chronotropic response to electrical stimulation and cholinergic agents decreases with age in rats.[19] The basis for this altered nerve response has not been elucidated. The decrease in heart rate produced by methacholine treatment, a parasympathomimetic agent, was greater than that of electrical stimulation at each age, indicating a defect in the presynaptic response. In addition to these findings, an age-related decrease in ACh synthesis and degradation[20-22] and an age-related increase in muscarinic receptors in the atria, but not the ventricle, occurs in the rat.[23] After vagotomy in senescent

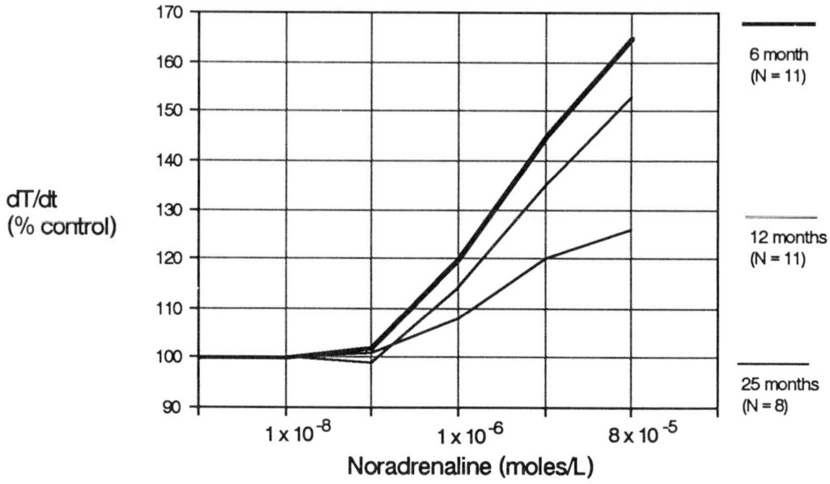

FIGURE 3. Effect of age on the inotropic response of cardiac muscle to noradrenaline. Maximum rate of tension rise (dT/dt) in response to increasing concentrations of noradrenaline was reduced in trabecular muscle from senescent rats. (From Weisfeldt, M., *Hosp. Pract.*, 20, 115, 1985. Illustrations by Albert Miller. With permission.)

rats[19] and elderly humans,[20] a small increase in heart rate occurs, indicating a reduction in vagal tone with age.

3. Contraction of the Heart

Studies have shown that the ability to develop tension and force is unimpaired in isolated cardiac muscle preparations from hearts of senescent animals.[14] With regard to the events leading to contraction, higher tensions have been shown to develop in older muscles when they are stretched to lengths greater than 115% of the length at which resting tension remains slightly above 0. However, the passive- and active-length tension curves determined from measurements of isometric contraction under the same preloaded conditions demonstrated no significant differences between the young and old rats; there is a prolongation of time to peak isometric tension in the older rats.

Experiments with isolated muscle preparations have shown that when the concentration of calcium is altered in the perfusion fluid while the preload is held constant, there is no difference in force development or maximum rate of force development measured in adult and senescent rat hearts.[24-26] Nevertheless, in older hearts, the inotropic effect of catecholamines (Figure 3) and cardiac glycosides are diminished.[24,26-28] This suggests that age-related changes in the inotropic response to calcium are not the basis for the reduced response to these drugs. Furthermore, since changes in response to cAMP, and changes in cAMP-dependent protein kinase (PKA) activity did not change with age, the decrease in inotropic response may be distal to activation of PKA (Figure 4).[28-30]

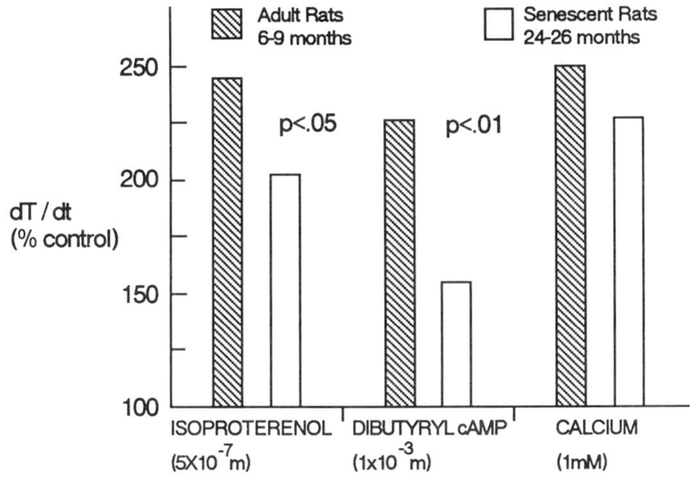

FIGURE 4. Effect of age on the inotropic response of cardiac muscle to isoproterenol, dibutyryl cAMP, and calcium. Maximum rate of tension rise (dT/dt) in response to isoproterenol and dibutyryl cAMP was reduced in interventricular septa muscle from senescent rats, but no age difference was found when calcium in the perfusate was increased from 0.3 to 1 m*M*. All muscles were contracting isometrically at the peak of their length-tension curve, stimulated 75/min at 29°C, and perfused with Krebs-Ringer bicarbonate solution. (From Lakatta, E. G., *J. Chron. Dis.*, 36, 15, 1983. With permission.)

There are at least four functions related to contraction that are altered with age in the heart.[14] The relaxation phase of contraction is delayed or prolonged and thus extends the duration of systole. The response to sympathetic stimulation is decreased. The left ventricular compliance decreases and impedance to left ventricular ejection increases. Weisfeldt[14] reported a prolonged relaxation in the dog and rat, which is probably caused by myocardial sequestration of calcium by the sarcoplasmic reticulum. Many studies have shown that as the heart ages, a degree of hypertrophy develops (for references, see Reference 14). There appears to be a definite similarity between aging myocardium and myocardium in which hypertrophy has been produced experimentally.

Amounts of connective tissue or precursors of connective tissue have been observed to increase in both experimental cardiac hypertrophy and during aging. Changes in collagen linkage rendered the tissues stiffer. An increase in stiffness during diastole does not have much influence on left ventricular function unless the Frank-Starling mechanism (increased stroke volume and end-diastolic volume) is called upon for compensation. Stiffness could have a functional advantage during the contraction cycle. This increase in stiffness may help explain the apparent paradox of an age-related decrease in the velocity of shortening while active tension and the rate of tension developed are maintained.[14]

The increase in collagen with age may also have consequences for neurochemical transmission. The increase in the amounts of collagen noted by McClean et al.[31] in the cardiac sympathetic neuroeffector junction may contribute to the reduction in sympathetic nervous influences on the heart with age, by either binding the transmitter or providing a physical barrier to receptor binding.

Weisfeldt[14] has reported that differences between his studies and those of others may reflect the failure to exclude subjects with ischemic heart disease, and he believes that aging does not affect cardiac output per se. The incidence of ischemic heart disease increases to about 50% by the end of the sixth decade. This could have influenced a study by Port et al.[32] In this study, a decline in left ventricular function, as reflected in altered wall motion and reduction in ejection fraction, was observed more frequently in older subjects. Lakatta[2] also pointed out that cardiovascular function may decline with age because of occult disease and physical activity. Lakatta[33] reported that in three separate studies:[34-36] "Cardiac output at rest was found to remain unchanged, decrease substantially, or even increase with age." The variation in results probably stems from the heterogeneity of selection criteria among studies. However, consensus does exist regarding age-related changes in some aspects of cardiac function.[33] Heart size and end-diastolic, diastole, or end-systole volume do not change substantially with age. Moderate myocardiohypertrophy was shown to occur with aging in several studies, regardless of how subjects were chosen. Many studies also found that the rate of left ventricular filling during early diastole declines markedly between 20 and 80 years of age. However, enhanced filling later in diastole in older persons and adaptive mechanism to maintain adequate filling volume resulted largely from the enhanced contribution of arterial contraction to ventricular filling.

4. Response to Physical Activity

With respect to exercise, the autonomic nervous system stimulates heart rate and muscle contraction and increases cardiac output and blood flow to skeletal muscles. Without the autonomic nervous system response, prolonged exercise would be impossible due to early muscle fatigue from lactic acid buildup during anaerobic metabolism. Does aging influence NA release in response to exercise? The following studies indicate that NA release during exercise is actually greater in older subjects (Figure 5). Tzankoff et al.[37] showed a large increase in circulating levels of NA among older healthy men during dynamic exercise. The authors repeated their study in healthy men screened to exclude the presence of cardiovascular disease. At maximum effort, plasma NA was significantly higher in subjects over 68 years old.[38] Even among competitive cyclists with training experience of 8 to 10 years, plasma NA was higher in older subjects at rest, at identical work loads, and at exhaustion.[39]

Consideration should be given to the fact that plasma NA levels during exercise may not be reflective of NA release from cardiac adrenergic nerve

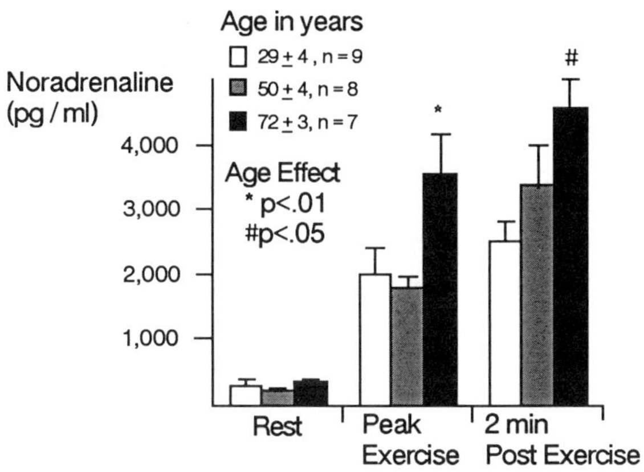

FIGURE 5. Effect of age on plasma noradrenaline levels during exercise. Plasma noradrenaline levels were measured at rest, during maximum treadmill exercise, and shortly after exercise. Subjects were participants in the Baltimore Longitudinal Study of Aging who were free of occult coronary artery disease. (From Lakatta, E. G., *Geriatrics,* 42, 84, 1987. With permission.)

terminals. Reuptake, degradation, clearance, and release of NA from organs innervated by the sympathetic adrenergic nervous system determine plasma NA levels. Studies with isolated rat hearts have shown an age-related decrease in NA release during electrical stimulation (Figure 6).[40,41] A more direct means of measuring NA release in the heart *in vivo* is needed to determine the effect of exercise and aging on adrenergic neurotransmission in the heart.

Physical work capacity declines with age, but the extent to which this decline is attributable to decreased cardiac reserve is uncertain. At least part of the age-related decline in maximum oxygen consumption appears to result from peripheral circulatory factors and decreased muscle mass.[33] A number of studies have shown age-related reductions in cardiac output, heart rate, and ejection fraction during exercise.[17,32,33,42-47] Given the reductions in both β-adrenergic response and/or release of NA in the old heart, a reduced ability to increase heart function during exercise is not surprising. Several studies have shown that during exercise, the aging heart utilizes increased stroke volume and end-diastolic volume to increase cardiac output rather than increased heart rate, as occurs in younger hearts.[34,48-50] The greater reliance of the older heart on the Frank-Starling mechanism to increase cardiac output during exercise underscores the older heart's inability to adequately increase heart rate in response to adrenergic nervous influences.

B. AGE-RELATED CHANGES IN THE BARORECEPTOR REFLEX AFFECTING THE HEART

The heart rate response to alterations in blood pressure and to the valsalva maneuver, each mediated by the baroreceptor reflex, is blunted by aging. Frey and Hoffler[51] found that in older men, there were greater increases in

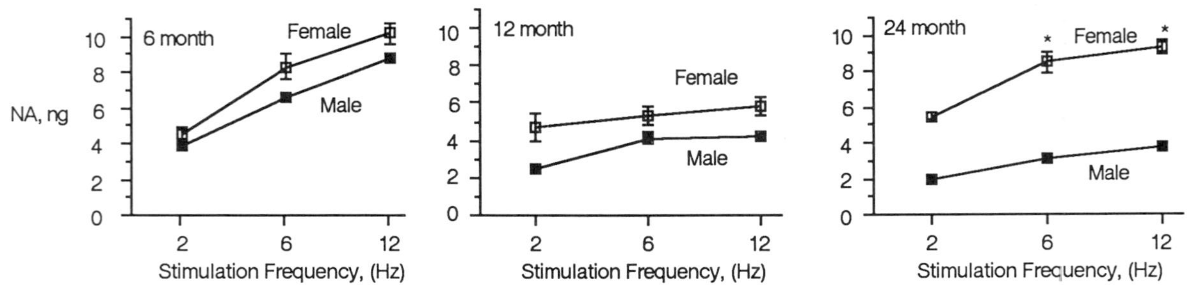

FIGURE 6. Cardiac sympathetic nerve noradrenaline release in male and female F344 rats. The cardiac sympathetic nerve was stimulated electrically and NA release was measured by high-performance liquid chromatography/electrochemical detection methods. All preparations were treated with cocaine (10^{-6} *M*) plus metanephrine (10^{-5} *M*). Results represent the means and standard errors of determinations from 4 to 10 rats from each age and gender. *p <0.05 using ANOVA. (From Tumer N., Mortimer, M. L., and Roberts, J., *Exp. Gerontol.*, 27, 301, 1992. With permission.)

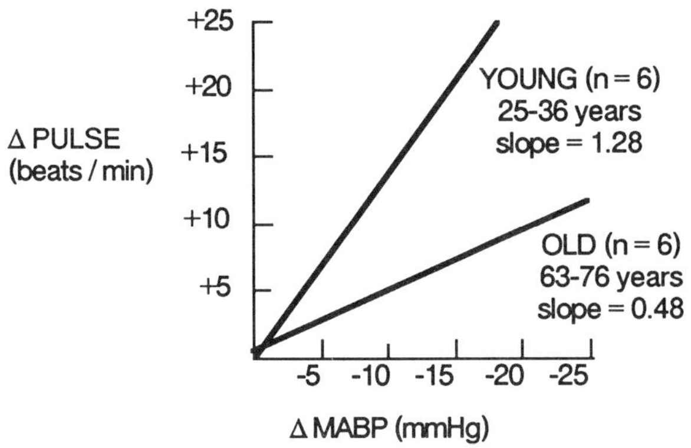

FIGURE 7. Effect of age on cardiovascular response to induced hypotension. Reductions in mean arterial blood pressure (MABP) were induced by nitroprusside infusion. The change in pulse for each unit decrease in MABP was significantly greater in the young (p <0.02). (From Minaker, K. L., Meneilly, G. S., Young, J. B., Landsberg, L., Staff, J. S., Robertson, G. L., and Rowe, J. W., *J. Gerontol.*, 46, M151, 1991. With permission.)

peripheral resistance and less heart rate elevation in response to lower body negative pressure. Studies with nitroprusside-induced hypotension in aging men showed that cardiac acceleration due to hypotension is impaired with age (Figure 7).[52] This reduction probably involves both a reduction in NA release and a reduction in β-adrenergic receptor response in the heart (see Section III.B.1).

Simpson and Wicks[53] reported that the supine resting heart rate is similar for all ages, whereas the erect resting heart rate decreases with age.[54,55] Although postural change produced significant increases in both heart rate and baroreceptor heart rate variation in both young and elderly participants, these increases were greater in the young group. The smaller heart rate response to standing observed in normotensive older persons has also been reported by Collins et al.[56] and Strandell.[57]

Ferrari et al.[58] reported that aging impairs baroreceptor control of heart rate in rats, but alters baroreceptor control of blood pressure, as assessed by response to common carotid artery occlusion, only in its fast-developing component, leaving its long-term component unaffected. They also noted that several studies have shown that reflex bradycardia induced by increasing arterial blood pressure via injections of vasopressor drugs and the reflex tachycardia induced by reducing arterial blood pressure via the injection of vasodepressor drugs or tilting are less pronounced in elderly subjects than in normo- or hypertensive younger subjects. These observations indicate that baroreceptor control of the cardiovascular system is affected adversely by aging.[59-61] However, aging is accompanied by a fall in blood volume,[62] depression of the renin angiotensin system,[63] reduced vascular compliance,[64] and impaired cardiopulmonary reflex.[65] These and other age-related changes may account for impaired blood pressure homeostasis in addition to alterations of baroreceptor mechanisms.

Goldstein and Shapiro[66] confirmed the blunted heart rate response to standing and the fact that when blood pressure is recorded by conventional means, the response exhibited during tilting does not differ in healthy groups of young and elderly subjects. However, the elderly exhibit a smaller immediate systolic and diastolic drop in response to change in upright position and less variability in beat-to-beat blood pressure changes. Shiraki et al.[67] showed that cardiac acceleration during a tilt test was greater in young men than in old. Other cardiovascular responses to tilting were qualitatively similar to that of the young except forearm blood flow and forearm vascular conduction. Old men did not show significant changes in forearm blood flow during tilting, suggesting a deterioration in sympathetic nervous reflex with aging. However, circulatory adaptations seem to overcome this deficiency, resulting in tilting tolerance similar to that of the young.

Storm et al.[68] showed that during the strain of the valsalva maneuver, aging was associated with diminished blood pressure and heart rate responses to sympathetic stimulation. In women, however, the predominant effect was a significant reduction in bradycardia response to systolic blood pressure overshoot following the release of strain. Shimada et al.,[69] using changes in R-R intervals during phase 4 of the valsalva maneuvers, reported that the baroreceptor sensitivity is correlated negatively with age. The decrease in reactivity of the baroreceptor reflex is also suggested by experiments in which phenylephrine was administered to elicit an increase in blood pressure and the effect measured was a decrease in cardiac rate. It was observed that in comparable hypertensive doses, phenylephrine produced smaller graded cardiac responses in old humans and animals than in younger counterparts, i.e., heart rate changes were less intense in response to alteration in blood pressure.[59,70,71]

Loss of baroreceptor sensitivity has been suggested as an explanation for the age-related increase in blood pressure and in plasma NA concentration in man (Figure 8).[72] Pfeifer et al.[10] concluded that in humans, the age-related decline of baroreceptor sensitivity may lead to compensatory autonomic responses which could mask underlying functional defects in the autonomic nervous system. As Phillips et al.[73] have pointed out, measurement of sympathetic nervous activity in humans is difficult. The plasma NA concentration may only indirectly reflect the rate of release of NA from postganglionic sympathetic noradrenergic neurons. Numerous studies have shown that the plasma NA concentration is positively correlated with age, but this correlation does not involve hypertension (Figure 8).[10,74-79] Plasma dopamine concentrations do not seem to be age dependent.[80] The age-related increase in plasma NA concentration has been reported to be due to reduced NA clearance with

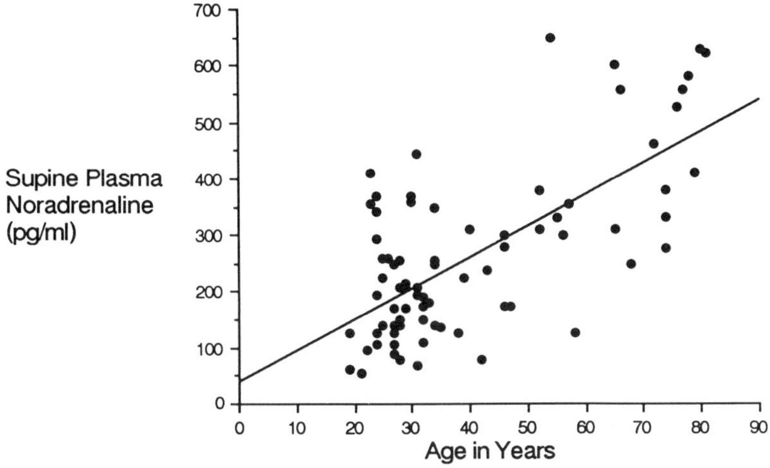

FIGURE 8. Effect of age on supine plasma noradrenaline levels. Seventy-seven male subjects were examined. The plasma levels of noradrenaline increased with age by 5.32 pg/ml/year. (From Pfeifer, M. A., Weinberg, C. R., Cook, D., Best, J. D., Reenan, A., and Halter, J. B., *Am. J. Med.*, 75, 249, 1983. With permission.)

age as well as an increase in NA release (for references, see References 81 and 82).

Abnormalities in baroreceptor function with age may also be important in the increased prevalence of hypertension with age and response to therapy. This is discussed in Chapter 8 of this volume.

Young et al.[83] reported that the increase in plasma NA while standing was greater in healthy old individuals than in healthy young individuals, despite similar changes in blood pressure and pulse. Of particular interest in this study was the finding that plasma NA persisted at significantly higher levels in older subjects than in younger subjects. Palmer et al.[84] studied plasma NA response to upright posture followed by an isometric exercise and cold pressor test in young and old individuals. In this study, basal NA levels were slightly higher in the older group than in the younger, and both groups exhibited equivalent increases in NA above basal after 5 min of standing. The addition of isometric exercise to postural stimulus resulted in a greater additional increase in NA in the old group than in the young. In the cold pressor test, plasma NA was obtained in five young and three old subjects before and after the test. Despite the small numbers of subjects, the old individuals had significantly higher basal NA levels than the young. Plasma NA rose after cold pressor test in both groups. The increase above basal in older subjects was greater, although not significantly so in this study. In studies in which basal and poststanding plasma NA were obtained in hypertensive subjects and age-matched normotensive controls, it was reported that increased plasma NA above basal, with upright posture, increases with age.

While numerous investigators have reported an increase in plasma NA levels in the basal condition as well as in response to various physiological challenges with age (upright posture, exercise, glucose ingestion, the cold pressor test, and mental stress testing) (see Reference 81 for references), others have not reported changes in basal NA levels in the elderly (see Reference 85 for references). In any case, it should be noted that plasma levels do not necessarily indicate the effects at any one particular neuroeffector junction. Indeed, Eisenhofer et al.[86] observed in the dog that when NA is released at the cardiac neuroeffector junction, 16 times more NA is taken up into the cardiac adrenergic neurons that reaches the plasma. This indicates that the relationships between NA release from the heart and plasma NA levels are not direct. With this major difference in the distribution of NA, it is difficult to consider plasma NA as being indicative of what is happening at the cardiac neuroeffector junction.

C. CHANGES IN CENTRAL VASOMOTOR FUNCTION THAT INFLUENCE CARDIAC ACTIVITY

Although there is considerable evidence that the major age-related defects in autonomic control of the heart occur primarily in the periphery, changes in the vasomotor functions of the brain during aging may also impair cardiac function. Severe emotional stress causes intense vagal stimulation, causing the heart rate to slow markedly.[87] Nyakas et al.[88] have shown that this bradycardia heart response to emotional stress, evoked by stimuli associated with a previous painful experience, is absent in 21- and 33-month-old Wistar rats. The bradycardia response remains intact after neonatal sympathectomy, indicating that the response is due to central vagal activation rather than a baroreceptor reflex mechanism. In a later study, low doses of amphetamine, an indirect acting sympathomimetic, were used to stimulate central adrenergic activity without increasing heart rate in the resting conditionl.[89] With amphetamine treatment, the reduction in heart rate after emotional stress was identical in young and old rats and enhanced compared to untreated rats.

Central activation of the vagus nerve appears to diminish with age, and Nyakas et al.[89] believe that this may contribute to the lack of a heart response after emotional stress. Support for this hypothesis comes from studies which show age-related changes in catecholaminergic systems in the brain. Welsh and Gold[90] showed that footshock reduced NA and dopamine in the forebrain and brainstem of 12-month-old, but not 24-month-old, male F344 rats. Reductions in NA levels, uptake, and release stimulated by potassium and high frequency were reported in several brain areas of 21- to 26-month-old F344 rats but not in rats younger than 14 months, but amphetamine induced greater NA release in older brain sections.[91] Direct electrical stimulation of the vagus and bolus injections of ACh have been shown to slow heart rate to a greater extent in 24-month-old Sprague-Dawley rats compared to 6-month-old rats.[92] Therefore, the hearts of old rats seem to be capable of responding to vagal stimulation, but an age-related defect in the brain of old rats prevents the response.

An extensive series of studies by Landsberg, Young, and colleagues has demonstrated a link between dietary intake, centrally mediated sympathetic activity, and the heart. Two days of fasting in rats significantly reduced NA turnover in the heart.[93] A similar reduction in turnover in fed animals can be achieved by ganglionic blockade, but blockade in fasted animals was without additional effect. Overfeeding sucrose stimulates sympathetic nervous system activity in rats.[94] Evidence suggests that insulin-mediated glucose metabolism is the link between nutritional status and centrally mediated sympathetic activity. 2-Deoxy-D-glucose, an inhibitor of intracellular glucose metabolism, suppresses cardiac NA turnover,[95] and diabetes induced in the rat by streptozotocin treatment also depresses sympathetic activity.[96] Rats treated with gold-thioglucose, which destroys insulin-glucose-sensitive neurons in the hypothalamus, fail to suppress sympathetic activity when fasted.[97]

These findings imply that fasting-induced suppression of sympathetic activity to the heart depends upon activation of an inhibitory pathway from hypothalamic neurons to tonically active brainstem centers that regulate sympathetic outflow.[98] Diet-induced changes in sympathetic activity persist in older rats. Rappaport et al.[99] have shown that sucrose overfeeding increases and fasting decreases cardiac NA turnover in male F344 rats to the same extent from 3 to 24 months of age.

Landsberg[98] has hypothesized that in man, hyperinsulinemia induced by obesity leads to increased sympathetic activity, greater cardiac output, vasoconstriction, and eventually hypertension. Normal human aging is associated with impairment of glucose metabolism, which results in a mild increase in fasting blood glucose, prolonged hyperglycemia after oral glucose challenge,[100] and insulin resistance.[101,102] Therefore, hyperinsulinemia associated with aging may alter cardiac function through the sympathetic nervous system. There is evidence that dietary intake and obesity can stimulate the human sympathetic nervous system. In a cross-sectional study of men in the Normative Aging Study, 24-h urinary NA excretion was used to assess sympathetic activity.[103] Increased body mass index and total caloric intake, hyperglycemia, and hyperinsulinemia were associated with increased urinary NA excretion. In another study, plasma NA was shown to increase more in elderly than young men after oral glucose ingestion.[83] Caloric restriction, weight loss, exercise, and high-fiber diets all reduce insulin secretion and decrease sympathetic activity.[104] More study is needed to clarify the possible links between aging, insulin, and central sympathetic nervous system activity, and if these factors actually alter heart function and contribute to hypertension.

III. MOLECULAR BASIS FOR AGE-RELATED CHANGES IN AUTONOMIC CONTROL OF THE HEART

As mentioned by Lakatta,[33] among the most notable age-related changes in the cardiovascular system is the diminution in β-adrenergic receptor-mediated responses. In the heart, this diminution results from both presynaptic and postsynaptic changes in adrenergic influences (Figure 1).

A. PRESYNAPTIC EVENTS

Alterations in sympathetic neurons leading to diminished NA release may contribute to the compromised regulation of cardiac function during aging. There is an age-related reduction in NA release from cardiac sympathetic nerves at a given impulse frequency. Isolated nerve-heart preparations from senescent rats release less NA upon nerve stimulation or potassium depolarization than do comparable preparations from adult rats.[40,41,105] This pattern of reduction in NA release is gender specific. NA release is reduced in both male and female hearts isolated from 12-month-old animals, but continues to decline only in the hearts isolated from 24-month-old males, while NA release in 24-month-old females returns to 6-month levels (Figure 6). Consistent with the observation of reduced release is the finding of Mazzeo and Horvath,[106] who studied the turnover of catecholamines after inhibition of tyrosine hydroxylase by α-methyl-*p*-tyrosine; there was a significant decrease in NA turnover with age.

As Daly et al.[40] pointed out, there are several possible explanations for the decline in NA overflow seen in heart preparations from 12- and 24-month-old animals. It might be due to reduced availability of the neurotransmitter in the neuronal storage vesicles during nerve stimulation. This is probably not caused by alterations in the enzymes responsible for biosynthesis and degradation of catecholamines, since the activity of these enzymes has been reported to increase with age (see Reference 81 for references). However, an age-related decrease in catecholamine content has been reported in the myocardium.[99,106,107] Using ultrastructural analysis, McLean et al.[31] demonstrated a significant sympathetic axonal degeneration between 3, 12, and 24 months of age. Thus, the decrease in catecholamine content with age seems to be due to axonal degeneration rather than reduced noradrenergic vesicles. In any case, Kreider et al.[108] found that perfusion with tyramine did not result in significant differences in the amount of NA appearing in the effluent of hearts from 6-, 12-, and 24-month-old animals, indicating that tyramine-sensitive stores of NA are not influenced by age. The amount released at each age represents more than 93% of total cardiac NA content, indicating that it is possible to cause almost 100% release of the available NA pools regardless of age. This agrees with the study of Duckles,[109] who concluded that the functional transmitter pool does not change with age. The reduced capacity of older preparations to release NA appears to be unrelated to changes in total content.

An autoreceptor component involving a relative increase in inhibitory influences may contribute to reduced release in the aging intact heart. However, both presynaptic α_2-inhibitory and β_2-facilitory influences decrease with age in the heart;[110,111] the loss of β_2-facilitory influences may possibly be more important physiologically. Further experiments are needed to establish these relationships.

The effect of cocaine on neuronal uptake has been variously reported. Potentiation of chronotropic responses by cocaine is reported to be enhanced in hearts from senescent rats, suggesting an age-related increase in neuronal uptake of catecholamines.[112] Similarly, potentiation of depolarization-induced NA release by cocaine and by desmethylimipramine is enhanced in hearts from senescent rats[40] and in perfused cardiac slices.[113] Similar findings relative to alterations associated with age in neuronal uptake mechanisms have been reported by other investigators. Gey et al.[114] reported that whereas uptake of ^{14}C-labeled NA is delayed in older heart preparations, the maximum level of ^{14}C-labeled NA is greater in older hearts. However, it should be noted that only 1-, 18-, and 24-month-old animals were compared, and the difference in uptake might be due to maturation rather than aging. Limas[115] also reported an increased uptake of NA in older rat hearts, but the capacity to retain this NA was greater in younger hearts. Hody et al.[116] reported that uptake of NA by myocardial slices from healthy C57BL/6J male mice is not impaired by aging (8 to 28 months). Doggrell and Vincent[117] did not observe any effect of cocaine (1 μM) on rate or force of contractions produced by NA, or field stimulation in ventricles from young adult Wistar rats, but Tessel et al.[118] found that in atria isolated from young Sprague-Dawley rats, cocaine (10 μM) did potentiate NA-induced tachycardia. Borton and Docherty[119] report decreased chronotropic effects and decreased stimulation of NA release by cocaine during aging.

Different results have also been reported with regard to the activity of neuronal uptake in organs other than the heart. In vascular smooth muscle from rabbits, a decline in neuronal uptake pump activity was found during maturational stages to adult ages.[9] In addition, cocaine has been reported to be less effective in potentiating NA-induced pressor response in old rats using the anesthetized-pithed rat preparation.[120] However, Handa and Duckles[121] reported that in the blood-perfused hindlimb preparation of the rat, cocaine was more potent in potentiating the effects of NA in older rats.

Release of catecholamines is thought to be directly linked to the influx of Ca^{2+} through voltage-sensitive channels. Tyramine-induced release of NA which is not dependent on Ca^{2+} is not affected by age, and the amount of NA released by ionomycin, an agent which carries Ca^{2+} through the membranes, is similar in both young and old animals.[41,108] Thus, the decrease in release of catecholamines with age may be due to alterations in Ca^{2+} channels. Roberts et al.[41] have shown that NA release in the hearts of 24-month-old rats was more sensitive to ω-conotoxin, a specific N-type Ca^{2+} channel-blocking agent, relative to release in the 6-month-old rat. This may be due to an age-related reduction in adrenergic nerve terminals of older hearts. Parikh et al.[122] have reported an age-related reduction in ω-conotoxin binding in cardiac synaptosome preparations from rats. Thus, fewer ω-conotoxin-sensitive Ca^{2+} channels seem to be present in adrenergic nerve terminals of older hearts. The difference between the pattern of male and female reduction in NA release may be due to age-related changes in serum sex hormones in the female rat.[123] Indeed, serum progesterone levels are increased in 24-month-old female rats, suggesting that progesterone may be influencing NA release from cardiac adrenergic nerve terminals in the female.

While there is less information on the parasympathetic nervous system, the indications are that there is also less input from this system in older subjects. Bradycardia produced by vagal activation or muscarinic stimulation is decreased and there is a reduced ability of atropine to produce tachycardia with age.[19,20] Consistent with these observations, there is a decrease in choline uptake, decreased ACh synthesis, and decreased ACh release in minced atrial slices.[124] The results with muscarinic receptor binding studies are inconsistent. There are no reported muscarinic receptor changes in the ventricle,[125] but an increase in the atria has been reported.[23]

B. POSTSYNAPTIC EVENTS

1. β-Receptors

Numerous studies have shown that cardiac responses to β-adrenergic receptor stimulation decline during aging. The myocardial chronotropic and ionotropic response to β-adrenergic agonists is diminished in older animals[24,28,126] (Figures 2 and 3) and in elderly humans.[127] Biochemical responses to β-agonists are diminished in myocardial membranes from senescent rats, including blunted stimulation of adenycyclase activity.[128] This is not unique to cardiac tissue, as similar findings have been reported in a number of organs (see Chapter 6 of this volume).

Decreased cardiac β-receptor-mediated responses could be due to alterations at several sites, including the receptors themselves and the postreceptor elements. Most investigators report no change in β-receptor number during aging, but several laboratories observe a modest decline of 15 to 30%.[30,126,129,130] The ability to upregulate[131,132] and downregulate[133] β-adrenergic receptor number in response to alterations in adrenergic tone is retained during aging, although the rate of upregulation may be diminished.[134] The proportion of β_1- and β_2-receptor subtypes does not change during aging, and both subtypes are able to upregulate.[131] Thus, most authors conclude that if β-receptor numbers do decline during aging, the magnitude of the reduction is not sufficient to account for the reduced responses. There is some evidence that the affinity of cardiac β-receptors for agonists decreases during aging.[128]

Not only do responses to β-agonists decrease with age, but it has also been reported that the action of propranolol to block cardiac β-receptor stimulation decreases with age.[123] The studies showed that the capacity of propranolol to block the effects of isoproterenol was decreased in older men. However, in a more recent study using timolol, another β-blocker, Klein et al.[135] concluded that although the elderly are less sensitive to isoproterenol, they are equally sensitive to timolol. Thus, their data do not implicate a change in the interaction of β-adrenergic receptor mechanisms with antagonists. The difference between these studies is not clear, but certainly requires further experimentation to come to a definitive conclusion regarding changes with age in response to β-adrenergic antagonists.

The decrease in senescent myocardial adenylate cyclase response persists when β-receptors are bypassed by stimulating cyclase with the GTP analog Gpp(NH)p, which is thought to activate G-proteins, or with forskolin, which is thought to act on the cyclase or the cyclase-G_s complex.[136] Additionally, there is some evidence for alterations distal to cyclic AMP formation (Figure 4).[18] There is evidence that an age-related decrease in calcium mobilization occurs possibly due to changes in the sarcoplasmic reticulum and a decrease in phosphorylation of proteins in the hormonal cascade (see Figure 1, number 11). Thus, it seems clear that postreceptor alterations do contribute to the reduction in β-receptor responsiveness. Further studies are needed to establish the nature and cause of these alterations.

2. α-Receptors

Postsynaptic α_1-adrenergic receptors are present in the heart and are involved in the positive ionotropic response to catecholamines.[137,138] A receptor-stimulated contraction is decreased with age in rat ventricle[139] and α_1-receptor-stimulated inositol phosphate formation is blunted.[140] α_1-Receptor density decreases with age in rat ventricle,[139,141,142] and this may contribute to the reduced α-responses.

3. Muscarinic Receptors

Age-related alterations in cardiac responses to muscarinic stimulation have been noted, but there is no general agreement as to their occurrence or direction. The negative chronotropic effects of muscarinic agonists are reported to be increased, decreased, or unchanged in aged rats.[19,21,92] The discrepancies may be due to the use of different ages and strains of rats and to the presence or absence of intact autonomic reflexes in the different preparations. The chronotropic response of isolated atria to the muscarinic agonist carbachol is reported to be unchanged,[143] but carbachol is both a nicotinic and muscarinic agent. The response to ACh is reported to increase during aging, but the age difference is not observed in the presence of a cholinesterase inhibitor, suggesting that it is secondary to reduced acetylcholinesterase activity.[144] The biochemical sequelae of cardiac muscarinic receptor stimulation have not been studied as a function of age. Muscarinic receptor density and antagonist affinity do not change in the heart during aging.[143,145,146] These same authors disagree as to whether agonist affinity is unaltered or decreased.

IV. SUMMARY

In animals and humans, sympathetic adrenergic influence on the heart decreases with age. There is considerable evidence indicating that there is a decrease in NA release from the adrenergic terminals as well as in response to NA by the myocardial cell in humans and animals. Nevertheless, there are reports that when more remote indicators of sympathetic neurochemical transmission are employed, i.e., an increase in plasma NA, there appears to be an increase in sympathetic nervous activity. However, the studies by Eisenhofer et al.,[86] which show that NA released at the cardiac neuroeffector junction is mostly taken up into the nerve, indicate that plasma NA levels are not a very good indicator of neural activity at the cardiac neuroeffector junction. Consequently, it would be extremely important to have a biological marker of adrenergic neurotransmission at the cardiac neuroeffector junction which can be used in humans and animals in a noninvasive manner. In this way, the status of cardiac adrenergic neurotransmission as a function of age can be more precisely determined in the clinical setting.

ACKNOWLEDGMENT

Studies by the authors were supported by funds from the National Institute of Aging, U.S. Public Health Service (AG 03362, AG 00131).

REFERENCES

1. **Klausner, C. S. and Schwartz, A. B.,** The aging heart, *Clin. Geriatr. Med.*, 1, 119, 1985.
2. **Lakatta, E. G.,** Why cardiovascular function may decline with age, *Geriatrics*, 42, 84, 1987.
3. **Tuttle, R. S.,** Age-related changes in the sensitivity of rat aortic strips to norepinephrine and associated chemical and structural alterations, *J. Gerontol.*, 21, 510, 1966.
4. **Fleisch, J. H., Maling, A. M., and Brodie, B. B.,** β-Receptor activity in aorta: variations with age and species, *Circ. Res.*, 26, 151, 1970.
5. **Fleisch, J. H.,** Further studies on the effect of aging on β-adrenoceptor activity of rat aorta, *Br. J. Pharmacol.*, 42, 311, 1971.
6. **Shibata, S., Hattori, K., Sakurai, I., Mori, J., and Fujiwara, M.,** Adrenergic innervation and cocaine-induced potentiation of adrenergic responses of aortic strips from young and old rabbits, *J. Pharmacol. Exp. Ther.*, 177, 621, 1971.
7. **Santer, R. M., Partanen, M., and Hervonen, A.,** Glyoxylic acid fluorescence and ultrastructural studies of neurons in the coeliac-superior mesenteric ganglion of the aged rat, *Cell Tissue Res.*, 211, 475, 1980.
8. **Cowen, T., Hoven, A. J., Wen-Qin, C., Gallen, D. D., Franc, F., and Brunstock, G.,** Development and ageing of perivascular adrenergic nerves in the rabbit. A quantitative fluorescence histochemical study using image analysis, *J. Auton. Nerv. Syst.*, 5, 317, 1982.
9. **Duckles, S. P.,** Age-related changes in adrenergic neuronal function of rabbit vascular smooth muscle, *Neurobiol. Aging*, 4, 151, 1983.
10. **Pfeifer, M. A., Weinberg, C. R., Cook, D., Best, J. D., Reenan, A., and Halter, J. B.,** Differential changes of autonomic nervous system function with age in man, *Am. J. Med.*, 75, 249, 1983.

11. **Hajduczok, G., Chapleau, M. W., and Abboud, F. M.,** Increase in sympathetic activity with age. II. Role of impairment of cardiopulmonary baroreflexes, *Am. J. Physiol.,* 260 (Heart Circ. Physiol. 29), H1121, 1991.
12. **Iwase, S., Mano, T., Watanabe, T., Saito, M., and Kobayashi, F.,** Age-related changes of sympathetic outflow to muscles in humans, *J. Gerontol.,* 46, M1, 1991.
13. **Low, P. A., Opfer-Gehrking, T. L., Proper, C. J., and Zimmerman, I.,** The effect of aging on cardiac autonomic and postganglionic sudomotor function, *Muscle Nerve,* 13, 152, 1990.
14. **Weisfeldt, M.,** The aging heart, *Hosp. Pract.,* 20, 115, 1985.
15. **O'Brien, I. A., O'Hare, P., and Corrall, R. J. M.,** Heart rate variability in healthy subjects: effect of age and the derivation of normal ranges for tests of autonomic function, *Br. Heart J.,* 55, 348, 1985.
16. **Jose, A. D. and Collison, D.,** The normal range and determinants of the intrinsic heart rate in man, *Cardiovasc. Res.,* 4, 160, 1970.
17. **Marks, R. G., Hale, W. E., Moore, M. T., May, F. E., and Stewart, R. B.,** Resting heart rates in an ambulatory elderly population: an evaluation of age, sex, symptoms, and medication, *Gerontology,* 35, 210, 1989.
18. **Lakatta, E. G.,** Determinants of cardiovascular performance: modification due to aging, *J. Chron. Dis.,* 36, 15, 1983.
19. **Kelliher, G. J. and Conahan, S. T.,** Changes in vagal activity and response to muscarinic receptor agonists with age, *J. Gerontol.,* 35, 842, 1980.
20. **Dauchot, P. and Gravenstein, J. S.,** Effects of atropine on the electrocardiogram in different age groups, *Clin. Pharmacol. Ther.,* 12, 274, 1971.
21. **Frolkis, V. V., Bezrukov, V. V., Duplenko, Y. K., Shchegoleva, I. V., Shevkchuk, V. G., and Verkhratsky, N. S.,** Ach metabolism and cholinergic regulation of functions in aging, *Gerontologia,* 19, 45, 1973.
22. **Frolkis, V. V., Shevtchu, V. G., Verkhratsky, N. S., Stupina, A. S., Karpova, S. M., and Lakiza, T. Y.,** Mechanisms of neurohumoral regulation of heart function in aging, *Exp. Aging Res.,* 5, 441, 1979.
23. **Narayanan, N. and Derby, J. A.,** Effects of age on muscarinic cholinergic receptors in rat myocardium, *Can. J. Physiol. Pharmacol.,* 61, 822, 1983.
24. **Lakatta, E. G., Gerstenblith, G., Angell, C. S., et al.,** Diminished inotropic response of aged myocardium to catecholamines, *Circ. Res.,* 36, 262, 1975.
25. **Spurgeon, H. A., Thorne, P. R., Yin, F. C. P., et al.,** Increased dynamic stiffness of trabeculae carnae from senescent rats, *Am. J. Physiol.,* 232, H373, 1977.
26. **Gerstenblith, G., Spurgeon, H. A., Froehlich, J. P., et al.,** Diminished inotropic responsiveness to ouabain in aged rat myocardium, *Circ. Res.,* 44, 517, 1979.
27. **Guarnieri, T., Spurgeon, H., Froehlich, J. P., et al.,** Diminished inotropic response but unaltered toxicity to acetylstrophanthidin in the senescent beagle, *Circulation,* 60, 1548, 1979.
28. **Guarnieri, T., Filburn, C. R., Zitnik, G., et al.,** Contractile and biochemical correlates of β-adrenergic stimulation of the aged heart, *Am. J. Physiol.,* 239, H501, 1980.
29. **Lakatta, E. G.,** Age-related alterations in the cardiovascular response to adrenergic mediated stress, *Fed. Proc.,* 39, 3173, 1980.
30. **Lakatta, E. G. and Yin, F. C. P.,** Myocardial aging: functional alterations and related cellular mechanisms, *Am. J. Physiol.,* 242, H927, 1982.
31. **McLean, M. R., Goldberg, P. B., and Roberts, J.,** An ultrastructural study of the effects of age on sympathetic innervation and atrial tissue in the rat, *J. Mol. Cell. Cardiol.,* 15, 75, 1983.
32. **Port, S., Cobb, F. R., Coleman, E., et al.,** Effect of age on the response of the left ventricular ejection fraction to exercise, *N. Engl. J. Med.,* 303, 1133, 1980.
33. **Lakatta, E. G.,** The aging heart, *Ann. Intern. Med.,* 113, 455, 1990.

34. **Rodenheffer, R. J., Gerstenblith, G., Becker, L. C., Fleg, J. L., Weisfeldt, M. L., and Lakatta, E. G.,** Exercise cardiac output is maintained with advancing age in healthy human subjects: cardiac dilatation and increased stroke volume compensate for a diminished heart rate, *Circulation,* 69, 203, 1984.
35. **Branfronbrener, M., Landowne, M., and Shock, N. W.,** Changes in cardiac output with age, *Circulation,* 12, 557, 1955.
36. **Julius, S., Amery, A., Whitlock, L. S., and Conway, J.,** Influence of age on the hemodynamic response to exercise, *Circulation,* 36, 222, 1967.
37. **Tzankoff, S. P., Fleg, J. L., Norris, A. H., and Lakatta, E. G.,** Age-related increase of plasma catecholamines in response to dynamic exercise in healthy adult men, *Physiologist,* 23, 50, 1980.
38. **Fleg, J. L., Tzankoff, S. P., and Lakatta, E. G.,** Age-related augmentation of plasma catecholamines during dynamic exercise in healthy males, *J. Appl. Physiol.,* 59, 1033, 1985.
39. **Lehmann, M. and Keul, J.,** Age-associated changes of exercise-induced plasma catecholamine responses, *Eur. J. Appl. Physiol.,* 55, 302, 1986.
40. **Daly, R. N., Goldberg, P. B., and Roberts, J.,** Effects of age on neurotransmission at the cardiac sympathetic neuroeffector junction, *J. Pharmacol. Exp. Ther.,* 245, 798, 1988.
41. **Roberts, J., Mortimer, M. L., Ryan, P. J., Johnson, M. D., and Tumer, N.,** Role of calcium in adrenergic neurochemical transmission in the aging heart, *J. Pharmacol. Exp. Ther.,* 253, 957, 1990.
42. **Gerstenblith, G., Lakatta, E. G., and Weisfeldt, M. L.,** Age changes in myocardial function and exercise response, *Prog. Cardiovasc. Dis.,* 19, 1, 1976.
43. **Kuikka, J. T. and Lansimies, E.,** Effect of age on cardiac index, stroke index and left ventricular ejection fraction at rest and during exercise, *Acta Physiol. Scand.,* 114, 339, 1982.
44. **Astrand, I.,** Aerobic work capacity in men and women with special reference to age, *Acta Physiol. Scand.,* 49 (Suppl. 169), 1, 1960.
45. **Kostis, J. B., Moreyra, A. E., Amendo, M. T., et al.,** The effect of age on heart rate in subjects free of heart disease, *Circulation,* 65, 141, 1982.
46. **Lester, M., Sheffield, L. T., Trammeil, P., et al.,** The effect of age and athletic training on the maximal heart rate during muscular exercise, *Am. Heart J.,* 76, 370, 1968.
47. **Robinson, S., Dill, P. B., Tzankoff, S. P., et al.,** Longitudinal studies of aging in 37 men, *J. Appl. Physiol.,* 38, 263, 1975.
48. **Mann, D. L., Denenberg, B. S., Gash, A. K., Makler, P. T., and Bove, A. A.,** Effects of age on ventricular performance during graded supine exercise, *Am. Heart J.,* 111, 108, 1986.
49. **Gerstenblith, G., Renlund, D. G., and Lakatta, E. G.,** Cardiovascular response to exercise in younger and older men, *Fed. Proc.,* 46, 1834, 1987.
50. **Konishi, T., Koyama, T., Aoki, T., Makino, K., Yamamuro, M., Nakai, K., Nakamura, M., and Nakano, T.,** Influence of age on left ventricular performance during exercise in normal Japanese subject: assessment by radionuclide ventriculography, *Ann. Nucl. Med.,* 4, 19, 1990.
51. **Frey, M. A. B. and Hoffler, G. W.,** Association of sex and age with responses to lower-body negative pressure, *J. Appl. Physiol.,* 65, 1752, 1988.
52. **Minaker, K. L., Meneilly, G. S., Young, J. B., Landsberg, L., Stoff, J. S., Robertson, G. L., and Rowe, J. W.,** Blood pressure, pulse, and neurohumoral responses to nitroprusside-induced hypotension in normotensive aging men, *J. Gerontol.,* 46, M151, 1991.
53. **Simpson, D. M. and Wicks, R.,** Spectral analysis of heart rate indicates reduced baroreceptor-related heart rate variability in elderly persons, *J. Gerontol.,* 43, M21, 1988.
54. **Waddington, J. L., MacCulloch, M. J., and Sambrooks, J. E.,** Resting heart rate variability in man declines with age, *Experientia,* 35, 1197, 1979.

55. **Sato, I., Hasegawa, Y., Takahashi, N., Hirata, Y., Shimomura, K., and Hotta, K.,** Age related changes of cardiac control function in man, *J. Gerontol.*, 36, 564, 1981.
56. **Collins, K. J., Exton-Smith, A. N., James, M. H., and Oliver, D. J.,** Functional changes in autonomic nervous responses with aging, *Age Ageing*, 9, 17, 1980.
57. **Strandell, T.,** Circulatory studies in healthy old men, *Acta Med. Scand.*, 175 (Suppl.), 414, 1964.
58. **Ferrari, A. U., Dafonchio, A., Albergati, F., and Mancia, G.,** Differential effects of aging on the heart rate and blood pressure influences of arterial baroreceptors in awake rats, *Hypertension*, 9, 615, 1991.
59. **Gribbin, B., Pickering, T. G., Sleight, P., and Peto, R.,** Effect of age and high blood pressure on baroreflex sensitivity in man, *Circ. Res.*, 29, 424, 1971.
60. **Korner, P. I., West, M. J., Shaw, J., and Uther, J. B.,** Steady state properties of the baroreceptor-heart rate reflex in essential hypertension in man, *Clin. Exp. Pharmacol. Physiol.*, 1, 65, 1974.
61. **Duke, P. C., Wade, J. G., Hickey, R. F., and Larson, C. P.,** The effects of age on baroreceptor reflex function in man, *Can. J. Anaesth.*, 23, 111, 1976.
62. **Epstein, M. and Oster, J. R.,** General pathophysiological considerations, in *Handbook of Hypertension*, Vol. 12, *Hypertension in the Elderly*, Amery, A. and Staessen, J., Eds., Elsevier, Amsterdam, 1989, 143.
63. **DeLeeuw, P. W. and Birkenhager, W. H.,** The impact of senescence on the renal vasculature and renin secretion, in *Handbook of Hypertension*, Vol. 12, *Hypertension in the Elderly*, Amery, A. and Staessen, J., Eds., Elsevier, Amsterdam, 1989, 99.
64. **Hallock, P. and Benson, I. C.,** Studies on the elastic properties of human isolated aorta, *J. Clin. Invest.*, 16, 595, 1987.
65. **Cleroux, J. C., Giannattasio, C., Bolla, G., et al.,** Decreased cardiopulmonary reflexes with aging in normotensive humans, *Am. J. Physiol.*, H968, 257, 1989.
66. **Goldstein, I. B. and Shapiro, D.,** The beat-to-beat blood pressure response to postural change in young and elderly healthy adult males, *J. Behav. Med.*, 13, 5, 1990.
67. **Shiraki, K., Sagawa, S., Yousef, M. K., Konda, N., and Miki, K.,** Physiological responses of aged men to head-up tilt during heat exposure, *J. Appl. Physiol.*, 63, 576, 1987.
68. **Storm, D. S., Metzger, B. L., and Therrien, B.,** Effects of age on autonomic cardiovascular responsiveness in healthy men and women, *Nurs. Res.*, 38, 326, 1989.
69. **Shimada, K., Kitazumi, T., Sadakane, N., Agura, H., and Ozawa, T.,** Age-related changes of baroreflex function, plasma norepinephrine, and blood pressure, *Hypertension*, 7, 113, 1985.
70. **Rothbaum, D. A., Shaw, D. J., Angell, C. S., and Shock, N. W.,** Cardiac performance in unanesthetized senescent male rat, *J. Gerontol.*, 28, 287, 1973.
71. **Rothbaum, D. A., Shaw, D. J., Angell, C. S., and Shock, N. W.,** Age differences in the baroreceptor response of rats, *J. Gerontol.*, 29, 488, 1974.
72. **Rowe, J. W. and Troen, B. R.,** Sympathetic nervous system and aging in man, *Endocr. Rev.*, 1, 167, 1980.
73. **Phillips, P. A., Hodsman, G. P., and Johnston, C. I.,** Neuroendocrine mechanisms and cardiovascular homeostasis in the elderly, *Cardiovasc. Drugs Ther.*, 4, 1209, 1990.
74. **Sowers, J. R. and Mohanty, P. K.,** Norepinephrine and forearm vascular resistance responses to tilt and cold pressor test in essential hypertension: effects of aging, *J. Vasc. Dis.*, 874, 1989.
75. **Bertel, O., Buhler, F. R., Kiowski, W., et al.,** Decreased β-adrenoceptor responsiveness as related to age, blood pressure and plasma catecholamines in patients with essential hypertension, *Hypertension*, 2, 130, 1980.
76. **Franco-Morselli, R., Elghozi, J. L., Joly, E., et al.,** Increased plasma noradrenaline concentrations in benign essential hypertension, *Br. Med. J.*, 2, 1251, 1977.
77. **Buhler, F. R. and Bolli, P.,** Changes in cardiovascular responsiveness caused by age and high blood pressure: implications for therapy, *J. Cardiovasc. Pharmacol.*, 7 (Suppl. 3), S206, 1985.

78. **Lake, C. R., Ziegler, M. G., Coleman, M. D., and Kopin, I. J.,** Age-adjusted plasma norepinephrine levels are similar in normotensive and hypertensive subjects, *N. Engl. J. Med.,* 208, 296, 1977.
79. **Sever, P. S., Owikowska, B., Birch, M., et al.,** Plasma noradrenaline in essential hypertension, *Lancet,* 1, 1078, 1977.
80. **Messerli, F. H., Sundagaard-Riise, K., Ventura, H. O., Dunn, F. G., Glade, L. B., and Frohlich, E. D.,** Essential hypertension in the elderly: haemodynamics, intravascular volume, plasma resin activity, and circulating catecholamine levels, *Lancet,* 2, 983, 1983.
81. **Roberts, J. and Tumer, N.,** Age-related changes in autonomic function of catecholamines, *Biol. Aging,* 3, 257, 1987.
82. **Cutler, N. R. and Hodes, J. E.,** Assessing the noradrenergic system in normal aging: a review of methodology, *Exp. Aging Res.,* 9, 123, 1983.
83. **Young, J. B., Rowe, J. W., Pallotta, J. A., Sparrow, D., and Landsberg, L.,** Enhanced plasma norepinephrine response to upright posture and oral glucose administration in elderly human subjects, *Metabolism,* 29, 532, 1980.
84. **Palmer, G. J., Ziegler, M. G., and Lake, C. R.,** Response of norepinephrine and blood pressure to stress increases with age, *J. Gerontol.,* 33, 482, 1978.
85. **Fiorica, V.,** Plasma norepinephrine levels of elderly men on a controlled sodium intake diet, *J. Am. Geriatr. Soc.,* 32, 576, 1984.
86. **Eisenhofer, G., Snolich, J., Cox, H., and Esler, M. D.,** Neuronal reuptake of norepinephrine and production of dihydroxyphenylglycol by cardiac sympathetic nerves in the anesthetized dog, *Circulation,* 84, B54, 1991.
87. **Guyton, A. C.,** Nervous regulation of the circulation and of respiration, in *Basic Neuroscience,* Guyton, A. C., Ed., W. B. Saunders, Philadelphia, 1987, chap. 27.
88. **Nyakas, C., Prins, A. J. A., and Bohus, B.,** Age-related alterations in cardiac response to emotional stress: relations to behavioral reactivity in the rat, *Physiol. Behav.,* 47, 273, 1990.
89. **Nyakas, C., Buwalda, B., Luiten, P. G. M., and Bohus, B.,** Effect of low amphetamine doses on cardiac responses to emotional stress in aged rats, *Neurobiol. Aging,* 13, 123, 1991.
90. **Welsh, K. A. and Gold, P. E.,** Age-related changes in brain catecholamine responses to a single footshock, *Neurobiol. Aging,* 5, 55, 1984.
91. **McIntosh, H. H. and Westfall, T. C.,** Influence of aging on catecholamine levels, accumulation, and release in F-344 rats, *Neurobiol. Aging,* 8, 233, 1987.
92. **Ferrari, A. U., Daffonchio, A., Gerosa, S., and Mancia, G.,** Alterations in cardiac parasympathetic function in aged rats, *Am. J. Physiol.,* 260, H647, 1991.
93. **Young, J. B. and Landsberg, L.,** Suppression of the sympathetic nervous system activity during fasting, *Science,* 196, 1473, 1977.
94. **Young, J. B. and Landsberg, L.,** Stimulation of the sympathetic nervous system during sucrose feeding, *Nature (London),* 269, 615, 1977.
95. **Rappaport, E. B., Young, J. B., and Landsberg, L.,** Effects of 2-deoxy-D-glucose on the cardiac sympathetic nerves and the adrenal medulla in the rat: further evidence for a dissociation of sympathetic nervous system and adrenal medullary responses, *Endocrinology,* 110, 650, 1982.
96. **Young, J. B., Einhorn, D., and Landsberg, L.,** Decreased sympathetic (SNS) activity in interscapular brown adipose tissue (IBAT) of streptozotocin-treated rats, *Diabetes,* 32 (Suppl.), 26A, 1983.
97. **Young, J. B. and Landsberg, L.,** Impaired suppression of sympathetic activity during fasting in the gold thioglucose treated mouse, *J. Clin. Invest.,* 65, 1086, 1980.
98. **Landsberg, L.,** The sympathoadrenal system, obesity and hypertension: an overview, *J. Neurosci. Methods,* 34, 179, 1990.
99. **Rappaport, E. B., Young, J. B., and Landsberg, L.,** Impact of age on basal and diet-induced changes in sympathetic nervous system activity of Fischer rats, *J. Gerontol.,* 36, 152, 1981.

100. **Davidson, M. B.,** The effect of aging on carbohydrate metabolism: a review of the English literature and a practical approach to the diagnosis of diabetes mellitus in the elderly, *Metabolism,* 28, 688, 1979.
101. **Fink, R. I., Kolterman, O. G., Griffen, J., and Olefsky, J. M.,** Mechanisms of the insulin resistance of aging, *J. Clin. Invest.,* 71, 1523, 1983.
102. **Rowe, J. W., Minaker, K. L., Pallotta, J. A., and Flier, J. S.,** Characterization of the insulin resistance of aging, *J. Clin. Invest.,* 71, 1581, 1983.
103. **Troisi, R. J., Weiss, S. T., Parker, D. R., Sparrow, D., Young, J. B., and Landsberg, L.,** Relation of obesity and diet to sympathetic nervous system activity, *Hypertension,* 17, 669, 1991.
104. **Krieger, D. R. and Landsberg, L.,** Mechanisms in obesity-related hypertension: role of insulin and catecholamines, *Am. J. Hypertens.,* 1, 84, 1988.
105. **Tumer, N., Ryan, P. J., and Roberts, J.,** Action of potassium on neurochemical transmission at the cardiac adrenergic neuroeffector junction with aging, *Mech. Ageing Dev.,* 52, 87, 1990.
106. **Mazzeo, R. S. and Horvath, S.,** A decline in myocardial and hepatic norepinephrine turnover with age in Fischer 344 rats, *Am. J. Physiol.,* 252, E762, 1987.
107. **Martinez, J. L., Vasquez, B. J., Messing, R. B., Jensen, R. A., Liang, K. C., and McGaugh, J. L.,** Age related changes in the catecholamine content of peripheral organs in male and female F344 rats, *J. Gerontol.,* 36, 280, 1981.
108. **Kreider, M. S., Goldberg, P., and Roberts, J.,** The effect of age on the tyramine-sensitive intraneuronal pool of norepinephrine in rat heart, *J. Cardiovasc. Pharmacol.,* 8, 137, 1986.
109. **Duckles, S. P.,** Effect of norepinephrine uptake blockade on contractile responses to adrenergic nerve stimulation of isolated rat blood vessels: influence of age, *J. Pharmacol. Exp. Ther.,* 243, 521, 1987.
110. **Daly, R. N., Goldberg, P. B., and Roberts, J.,** The effect of age on presynaptic α_2 adrenoceptor autoregulation of norepinephrine release, *J. Gerontol.,* 44, B59, 1989.
111. **Mortimer, M. L., Tumer, N., Johnson, M. D., and Roberts, J.,** Effect of age on presynaptic β_2 receptor mediated responses in the rat heart, *Mech. Ageing Dev.,* 59, 17, 1991.
112. **Kreider, M. S., Goldberg, P. B., and Roberts, J.,** Effect of age on adrenergic neuronal uptake in rat heart, *J. Pharmacol. Exp. Ther.,* 231, 367, 1984.
113. **Snyder, D. L., Aloyo, V. J., McIlvain, H. B., Johnson, M. D., and Roberts, J.,** Effect of age on potassium- and tyramine-induced release of norepinephrine from cardiac synaptosomes in male F344 rats, *J. Gerontol.,* 47, B190, 1992.
114. **Gey, K. F., Burkard, W. P., and Pletscher, A.,** Variation of the norepinephrine metabolism of the rat heart with age, *J. Gerontol.,* 11, 1, 1965.
115. **Limas, C. J.,** Comparison of the handling of norepinephrine in the myocardium of adult and old rats, *Cardiovasc. Res.,* 9, 664, 1975.
116. **Hody, G., Jonec, V., Morton-Smith, W., and Finch, C. E.,** Norepinephrine uptake by the myocardium of the senescent mouse *in vitro, J. Gerontol.,* 30, 275, 1975.
117. **Doggrell, S. A. and Vincent, L.,** The effects of cocaine and lignocaine on the rat right ventricle *in vitro, J. Auton. Pharmacol.,* 2, 1, 1982.
118. **Tessel, R. E., Smitt, C. B., Russ, D. N., and Hough, L. B.,** The effects of cocaine HCl and a quaternary derivative of cocaine, cocaine methiodide, on isolated guinea pig and rat atria, *J. Pharmacol. Exp. Ther.,* 205, 369, 1978.
119. **Borton, M. and Docherty, J. R.,** The effects of ageing on neuronal uptake of noradrenaline in the rat, *Naunyn-Schmeideberg's Arch. Pharmacol.,* 340, 139, 1989.
120. **Docherty, J. R. and Hyland, L.,** Age-related alteration in the cardiovascular responsiveness of the rat, *Pharmacologist,* 27, 156, 1985.
121. **Handa, R. K. and Duckles, S. P.,** Age related changes in adrenergic vasoconstrictor responses of the rat blood hind-limb, *Am. J. Physiol.,* 253, H1566, 1987.

122. **Parikh, A., Johnson, M., Aloyo, V., McIlvain, H. B., Bhavsar, V., and Roberts, J.,** Age-related decrease in omega conotoxin binding to rat cardiac synaptosomes, *P.S.E.B.M.*, 199, 453, 1992..
123. **Tumer, N., Mortimer, M. L., and Roberts, J.,** Gender differences in the effect of age on adrenergic neurotransmission in the heart, *Exp. Gerontol.*, 27, 301, 1992.
124. **Meyer, E. M., Momol, A. E., and Baker, S. P.,** Age-related reductions in rat atrial high affinity choline uptake, Ach synthesis, and Ach release. A brief note. *Mech. Ageing,* 30, 221, 1985.
125. **Shepherd, A. M. M. and Elfellah, M. S.,** Effect of age on muscarinic responses in the rat heart, *Clin. Res.*, 32, 862A, 1984.
126. **Abrass, I. B., Davis, J. L., and Scarpace, P. J.,** Isoproterenol responsiveness and myocardial β-adrenergic receptors in young and aging rats, *J. Gerontol.*, 37, 156, 1982.
127. **Vestal, R. E., Wood, A. J. J., and Shand, D. G.,** Reduced β-adrenoceptor sensitivity in the elderly, *Clin. Pharmacol. Ther.*, 26, 181, 1979.
128. **Narayanan, N. and Derby, J. A.,** Alterations in the properties of β-adrenergic receptors of myocardial membranes in aging: impairments in agonist-receptor interactions and guanine nucleotide regulation accompany diminished catecholamine-responsiveness of adenylate cyclase, *Mech. Ageing Dev.*, 19, 127, 1982.
129. **Fan, T. H. M. and Banerjee, S. P.,** Age-related reduction of β-adrenoceptor sensitivity in rat heart occurs by multiple mechanisms, *Gerontology,* 31, 373, 1985.
130. **Tumer, N., Bender, J., and Roberts, J.,** Absence of age-related changes in the binding of the β-adrenergic antagonist ^{125}I-iodohydroxybenzylpindolol in rat hearts, *Proc. Soc. Exp. Biol. Med.*, 186, 118, 1987.
131. **Tumer, N., Houck, W. T., and Roberts, J.,** Upregulation of adrenergic β receptor subtypes in the senescent rat heart, *Mech. Ageing Dev.*, 49, 235, 1989.
132. **Tumer, N., Houck, W. T., and Roberts, J.,** Effect of age on upregulation of the cardiac adrenergic β receptors, *J. Gerontol.*, 45, B48, 1990.
133. **Scarpace, P. J. and Abrass, I. B.,** β-Adrenergic agonist-mediated desensitization in senescent rats, *Mech. Ageing Dev.*, 35, 255, 1986.
134. **Pitha, J., Hughes, B. A., Kusiak, J. W., Dax, E. M., and Baker, S. P.,** Regeneration of β-adrenergic receptors in senescent rats: a study using an irreversible binding antagonist, *Proc. Natl. Acad. Sci. U.S.A.*, 79, 4424, 1982.
135. **Klein, C., Gerber, J. G., Gal, J., and Nies, A. S.,** β-Adrenergic receptors in the elderly are not less sensitive to timolol, *Clin. Pharmacol. Ther.*, 40, 161, 1986.
136. **Scarpace, P. J.,** Forskolin activation of adenylate cyclase in rat myocardium with age: effects of guanine nucleotide analogs, *Mech. Ageing Dev.*, 52, 169, 1990.
137. **Bruckner, R., Mugge, A., and Scholz, H.,** Existence and functional role of α_1 adrenoceptors in the mammalian heart, *J. Mol. Cell. Cardiol.*, 17, 639, 1985.
138. **Tung, L., Rand, M. J., and Louis, W. J.,** Cardiac α-adrenoceptors, *J. Cardiovasc. Pharmacol.*, 7 (Suppl. 6), S121, 1985.
139. **Partilla, J. S., Hoopes, M. T., Ito, H., Dax, E. M., and Roth, G. S.,** Loss of rat ventricular α_1 adrenergic receptors during aging, *Life Sci.*, 31, 2507, 1982.
140. **Borst, S. E., Narang, N., Crews, F. T., and Scarpace, P. J.,** Reduced α_1 adrenergic receptor-mediated inositide hydrolysis in cardiac atria of senescent rats, *J. Cardiovasc. Pharmacol.*, 16, 444, 1990.
141. **Schaffer, W. and Williams, R. S.,** Age-dependent changes in expression of α_1 adrenergic receptors in rat myocardium, *Biochem. Biophys. Res. Commun.*, 138, 387, 1986.
142. **Miyamoto, A. and Ohshika, H.,** Age-related changes in [^{3}H]prazosin binding and phosphoinositide hydrolysis in rat ventricular myocardium, *Gen. Pharmacol.*, 20, 647, 1989.
143. **Elfellah, M. S., Johns, A., and Shepherd, A. M. M.,** Effect of age on responsiveness of isolated rat atria to carbachol and on binding characteristics of atrial muscarinic receptors, *J. Cardiovasc. Pharmacol.*, 8, 873, 1986.
144. **Kennedy, R. H. and Seifen, E.,** Aging: effects on chronotropic actions of muscarinic agonists in isolated rat atria, *Mech. Ageing Dev.*, 51, 81, 1990.

145. **Baker, S. P., Marchand, S., O'Neil, E., Nelson, C. A., and Posner, P.,** Age-related changes in cardiac muscarinic receptors: decreased ability of the receptor to form a high affinity agonist binding state, *J. Gerontol.*, 40, 141, 1985.
146. **Narayanan, N. and Tucker, L.,** Autonomic interactions in the aging heart: age-associated decrease in muscarinic cholinergic receptor mediated inhibition of β-adrenergic activation of adenylate cyclase, *Mech. Ageing Dev.*, 34, 249, 1986.
147. **Shepherd, G. M.,** Neurotransmitters and neuromodulators, in *Neurobiology,* 2nd ed., Shepherd, G. M., Ed., Oxford University Press, New York, 1988, chap. 8.

Chapter 8

INFLUENCE OF AGE ON NEUROEFFECTOR MECHANISMS OF THE VASCULATURE

Sue P. Duckles

TABLE OF CONTENTS

0-8493-6981-9/93/$0.00 + $.50

I. INTRODUCTION

Alterations in homeostatic and adaptive capacities, including control of vascular function, have been shown to occur with advancing age. For example, an age-related rise in systemic blood pressure and increased levels of plasma catecholamines have been seen in both humans and animal models, including the Fischer 344 rat.[1-5] Changes in activity at a variety of sites, including baroreceptor sensing, central control, ganglionic transmission, adrenergic nerve endings, vascular smooth muscle, and endothelial cells, could all potentially contribute to such alterations of overall homeostatic function. It is the intent of this review, however, to focus primarily on the vascular neuroeffector unit: the blood vessel, including smooth muscle and endothelial cells, and its associated innervation. When pertinent, a wider view has been taken, but the principal focus is at the organ level, on the contractile function of the blood vessel and nerve endings themselves.

In any consideration of studies of aging, it is important to critically assess several important issues. In this review, postmaturational aging has been chosen as the exclusive focus, and the very interesting and complex question of the impact on vascular reactivity of development to maturity has not been addressed. Although age itself represents a continuous spectrum, the effects of development, either pre- or postnatal, need to be distinguished from the impact of aging from maturity to senescence.[6] In this regard, it is important to consider when the study of old age should begin, and the question, ''How old is old?'', needs to be addressed.[6] For this chapter, maturity for most rat strains, e.g., the F344, is considered to occur between 6 and 12 months.[7] Since the mean life span of this strain may occur at 24 months, depending on diet and other factors,[8] ideally, studies of aging in rats should include several groups of animals spanning the range between 12 and 24 months.[7] However, this standard has rarely been met in studies of the impact of age on vascular reactivity. In this review, a special effort has been made to emphasize the ages of groups that have been compared, and, with some exceptions which are noted, only studies of postmaturational aging have been included. Other sources should be consulted for information on development and maturation.[9,10]

Another important issue which is difficult to address in the investigation of aging concerns the health of the animal population being studied. In the case of humans, defining normal, healthy aging is an almost insoluble problem, and trying to separate effects of disease from effects of advancing age itself must always be considered in the interpretation of findings. In the case of animals, strikingly different results have been obtained when animals are raised in specific pathogen-free, barrier-protected conditions compared to animals maintained in a less protected environment.[146] Early studies of the impact of age on skeletal muscle function demonstrated a variety of changes, but more recent studies, done with barrier-protected rats, showed no such age-related alterations.[11] These studies suggest that earlier findings depended on exposure to a series of diseases throughout life and do not necessarily represent the inherent consequences of aging per se.

Another important variable relating to animal health that has not been fully explored in relation to its impact on vascular function in the aging organism is diet. Dietary restriction has been shown to produce a substantial prolongation of mean life span in rats.[5] Interestingly, development of a variety of pathological changes was also delayed, including nephropathy and cardiomyopathy.[12] However, the time course of development with age of increased blood pressure was not altered in animals with restricted food intake, even though life span was substantially prolonged.[5] These findings suggest that the increase in systemic blood pressure is not a consequence of renal pathology and follows a course independent of changes in diet.

II. FUNCTION OF ADRENERGIC NERVES

A. ADRENERGIC NERVE DENSITY

Two major approaches have been used to assess the influence of age on adrenergic nerve density: measurement of neurotransmitter content and histological assessment of nerve density. The second technique primarily involves fluorescence histochemistry for catecholamines. The results that have been obtained do not appear to differ based on the method of evaluation, but a more fundamental question concerns the extrapolation of assessments of nerve density or neurotransmitter content to predicting consequences for contractile function. Thus, as discussed below, even though there may be age-related changes in noradrenaline content, there may not be any alteration in contractile responses to adrenergic nerve stimulation. These findings imply that total transmitter content may not necessarily represent the most functionally active component, and a possible disparity between total and functional transmitter pools needs to be kept in mind.

There is no single pattern of the influence of age on noradrenaline content or adrenergic nerve density common to all species or all blood vessels, and various authors have reported an increase with age, no change, or a decrease. No change in noradrenaline content with age has been reported for a variety of veins and for a few arteries from the F344 rat when animals 6, 12, 20, and 27 months of age were examined, but in the same study, most arteries showed an age-related decline in noradrenaline content.[13] No change with age has also been reported in the aorta of three strains of mice,[14] rat spinal cord blood vessels[15] and coronary vein,[16] and guinea pig mesenteric, carotid, and femoral arteries.[17] The density and fluorescence intensity of the adrenergic innervation of the coronary artery and pancreatic vasculature did not change from 4 to 24 months in Wistar rats.[18] In contrast, adrenergic nerve density has been shown to decrease with age in human gingival arteries,[19] the rat coronary artery,[16] rat carotid artery and abdominal aorta,[18] and the tail artery of the Ivanos rat.[20] Noradrenaline content was shown to decrease with age in the human posterior tibial artery.[21]

An increase with age in nerve density has been reported in rat splanchnic vessels[22] and in the vasa nervorum,[23] but these involved comparisons between 26- or 27-month-old animals and 3-month old animals, so it is unclear whether part of the difference could be attributed to maturational changes, since 3-month-old animals are not fully grown.[8]

The noradrenaline content of the mesenteric artery and renal artery did not change with age, but decreased in the mesenteric vein, renal vein, and portal vein.[24] Serotonin fluorescence and serotonin uptake into adrenergic nerves also decreased with age in rats when 24- to 30-month-old animals were compared to 3-month-old ones.[25] Serotonin fluorescence histochemistry was also shown to decrease with age in the rat circle of Willis from 8 to 24 months of age.[26]

B. NORADRENALINE UPTAKE

Neuronal noradrenaline uptake is an additional measure of nerve function that has been considered to be correlated with nerve density, although this relationship is far from straightforward and has not been well substantiated.[27] Uptake of [^{3}H]-noradrenaline has been measured in femoral arteries and veins of F344 rats 6, 12, 20, and 27 months of age.[28] Neither neuronal nor extraneuronal uptake showed significant age-related changes except for an increase with age in neuronal uptake in the femoral vein.

The effect of cocaine to increase contractile responses to adrenergic nerve stimulation can also be considered as a functional index of the effectiveness of neuronal uptake. In renal and femoral arteries and veins as well as in the tail artery, the effect of cocaine to potentiate contractile responses to adrenergic nerve stimulation did not change with age, except for a decrease in the effect of cocaine in the tail artery between 6 and 12 months.[29] The effect of blockade of both uptake 1 and 2 by cocaine and deoxycorticosterone on noradrenaline overflow increased with age in the rat tail artery.[30] However, when corrected for the amount of noradrenaline present in the synapse, no change with age in the fraction of noradrenaline accumulated was seen, suggesting that in each age group the amount of noradrenaline taken up remains proportional to the amount of noradrenaline released.

The effect of cocaine to potentiate pressor responses of the rat perfused hindlimb to either noradrenaline or lumbar nerve stimulation increased with age from 6 to 20 months.[31] In contrast, cocaine produced a greater increase of the pressor response to noradrenaline in 4- to 6-month-old pithed rats[32] as well as in young anesthetized rats[33] compared to 21- to 24-month-old animals. However, these *in vivo* studies are complicated by the presence of endogenous vasoconstrictors, and circulating levels of these substances may be influenced by age and/or anesthesia.

C. CONTROL OF NORADRENALINE RELEASE

Only a few studies have explored the influence of age on noradrenaline release and prejunctional modulatory mechanisms. In the rat tail artery, endogenous noradrenaline release increases as a function of age from 12 to 27 months when expressed as a fraction of noradrenaline content.[30] One possible explanation for this age-related change would be a loss of function of presynaptic feedback modulation through α_2-adrenergic receptors. Indeed, there is some evidence that this does occur. In the pithed rat, the potency of the α_2-adrenoceptor agonist xylazine to decrease blood pressure declined in old animals.[33] Furthermore, in the rat tail artery, sensitivity to the effect of the α_2-antagonists, yohimbine and idazoxan, to increase noradrenaline release declines with age from 6 to 20 months, and the sensitivity to an α_2-agonist, UK 14304, is also decreased with age.[30,34]

While it has been suggested that there is a general decline with age in the function of presynaptic α_2-adrenergic receptors, there may be an alternative explanation for the results described above. There is some evidence that the function of presynaptic α_2-adrenergic receptors does not change with age. For example, in the human saphenous vein, the potency of the α_2-adrenoceptor agonist xylazine to inhibit [^{3}H]-noradrenaline release showed no age-related relationship.[35] Furthermore, in the rat tail artery, although there was an age-related decline in sensitivity to both agonists and antagonists selective for the α_2-adrenergic receptor, there was no age-related difference in the maximal effect of idazoxan.[34] The observed difference in sensitivity to either agonists or antagonists may be a consequence of the increased noradrenaline release with age. Increased noradrenaline release from nerves in older animals would lead to higher biophase concentrations of noradrenaline, and higher concentrations of either antagonists or agonists would be required to compete with higher synaptic concentrations of noradrenaline.

Interestingly, there is no age-related alteration from 6 to 20 months in the function of inhibitory presynaptic D_2 dopamine receptors.[34] The ability of the D_2 agonist, N-0923, to inhibit contractile responses to adrenergic nerve stimulation does not vary when tail arteries from 6- and 20-month-old F344 rats are compared. These findings suggest that the age-related increase in noradrenaline release may be due to some fundamental change in neuronal mechanisms controlling neurotransmitter release, such as alterations in calcium influx, intracellular buffering, or sensitivity of the transmitter release mechanism.

D. FUNCTIONAL RESPONSE TO ADRENERGIC NERVE STIMULATION

In spite of an age-related decrease in noradrenaline content and perhaps as a consequence of increased noradrenaline release, contractile responses of isolated blood vessels to adrenergic nerve stimulation remain quite stable with advancing age. As shown in Figure 1, in femoral and renal arteries and veins from F344 rats aged 6, 12, 20, and 27 months, there were no age-related changes in maximum contractile response or sensitivity to adrenergic nerve stimulation.[28] The same conclusion was reached by a study of renal and femoral arteries and veins and the tail artery, except that sensitivity of the

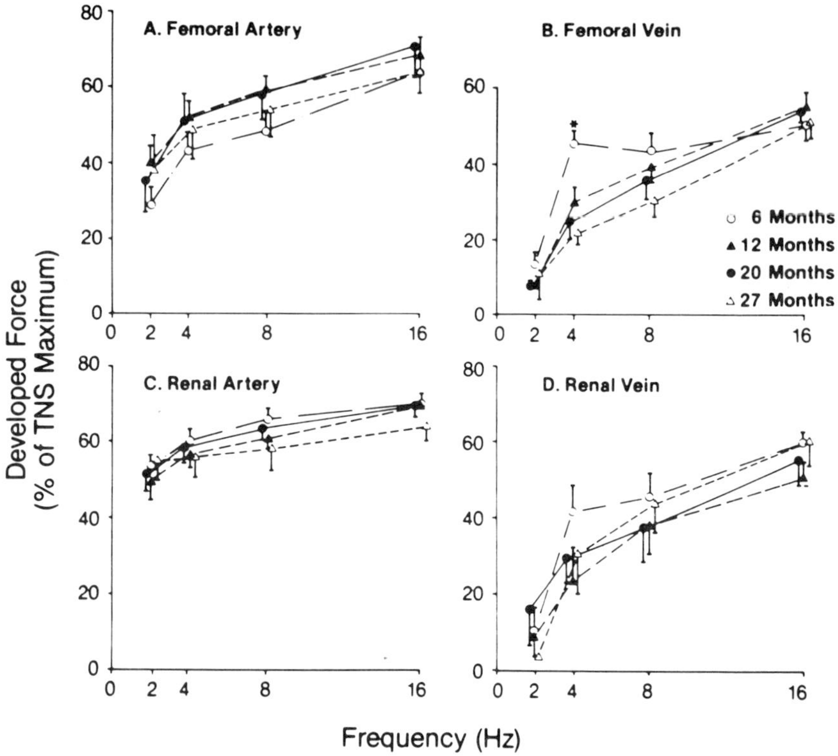

FIGURE 1. Contractile responses of isolated rat blood vessels to adrenergic nerve stimulation do not change as a function of age from 6 to 27 months. Responses to transmural nerve stimulation (200 pulses) are expressed as a percent of the maximum response and plotted as a function of stimulation frequency. (From Duckles, S. P., Carter, B. J., and Williams, C. L., *Circ. Res.*, 56, 109, 1985. With permission.)

tail artery was less at 6 months of age.[29] Contractile responses to adrenergic nerve stimulation of beagle mesenteric arterial strips also showed no change in sensitivity from 3 months to 12 years of age. In beagle coronary arteries, contractile responses to the indirectly acting sympathomimetic tyramine switch from relaxation to contraction as animals age from 2 to 12 years.[36] Furthermore, responses to transmural nerve stimulation are predominantly relaxations in coronary arteries from 2-year-old dogs, but contraction is seen from 2-year-old animals. This is thought to reflect a decline with age in relaxation responses to β-adrenergic activation with a concurrent increase in response to α-adrenergic receptors (see below). In contrast, in the perfused tail artery of Ivanos rats, sensitivity to nerve stimulation was greater at 6 months, with no further change between 16 and 30 months of age.[20] Thus, although there may be some alterations in adrenergic nerve function associated with maturation, overall contractile responsiveness to adrenergic nerve stimulation remains remarkably stable from maturity to senescence.

E. ROLE OF COTRANSMITTERS

The role of cotransmitters in adrenergic neurotransmission has become well established, and there is strong evidence that both ATP and neuropeptide Y are colocalized and coreleased along with noradrenaline.[37-40] Possible age-related changes in the synthesis, storage, release, or function of these co-transmitters could have marked effects on adrenergic nerve function. However, little is known about the influence of age on these substances or their actions. The content of neuropeptide Y did not change in cerebrovascular nerves of the rat from 4 to 27 months of age,[26] although levels of neuropeptide Y have been reported to decrease with age in human pial arteries.[41] Further investigations are required to determine whether aging alters the function of these adrenergic cotransmitters.

III. SENSORY INNERVATION

The presence of capsaicin-sensitive sensory nerves innervating blood vessels in a wide variety of organs has been shown in a number of species.[42-45] These vasodilator nerves contain both substance P and calcitonin gene-related peptide (CGRP); however, the major vasodilator substance appears to be CGRP.[46] The function of these nerves, which may serve to convey sensory information to the central nervous system as well as to release vasodilator transmitters locally, still remains unclear, but there is some evidence that age has an important influence on their function. Levels of substance P have been shown to decrease with age in human pial arteries,[41] but to markedly increase with age in rat mesenteric arteries and veins.[47] In contrast, fluorescence histochemistry of vessels of the circle of Willis in rats 4, 8, and 24 months old shows an increase in CGRP and no change in substance P.[26] Innervation density for both substance P and CGRP showed a gradual decrease in guinea pig mesenteric, carotid, femoral, and renal arteries from 4 months to 2 years.[17] Substance P content and transport were unchanged or slightly increased in the sensory ganglia of both the vagus and sciatic nerves of F344 rats from 12 to 25 months of age.[48]

The content of CGRP in the femoral, renal, and mesenteric arteries of the rat declined with age from 6 to 20 and 27 months of age, and vasodilator responses to sensory nerve stimulation of the perfused mesentery also decreased with increasing age.[49] In contrast, vasodilator sensitivity to exogenous CGRP increased between 20 and 27 months. Much remains to be learned about the impact of age on the function of this widespread vasodilator innervation of the vasculature.

IV. VASCULAR SMOOTH MUSCLE RESPONSIVENESS

A. MECHANICAL PROPERTIES

Investigation of mechanical properties of the vascular wall in species such as the rat, which are resistant to the development of atherosclerosis, indicate that aging may have relatively little impact. In rat carotid arteries from 12- and 24-month-old animals, characteristic impedance and incremental elastic modulus at specific pressure values did not substantially change as a function of age.[50] There was an increase with age in water content as well as collagen/elastin ratio, but the maximum active stress response to noradrenaline did not change with increasing age. In investigations of the effect of resting stretch on active responses of a variety of rat blood vessels *in vitro*, little effect of age from 12 to 27 months was seen.[28]

B. α-ADRENERGIC EFFECTS

As discussed above, vascular responsiveness to adrenergic nerve stimulation does not change with age, even though noradrenaline content may decline. The increase in fractional noradrenaline release with age that has been described in the rat tail artery[30,34] may partially compensate for the decline in noradrenaline content in some tissues. However, an important issue to consider is whether vascular smooth muscle responsiveness to the neurotransmitter noradrenaline is affected by advancing age.

Most of the early work on this question was done using isolated ring segments of the thoracic aorta. However, the aorta seems to be unique in showing significant age-related changes in reactivity, and it has been suggested that this tissue does not represent a good model for the study of vascular aging.[51] A number of investigators have described a decreased response to noradrenaline with increasing age in the rat aorta.[51-54] However, age-related changes in this tissue have been best described by Wanstall and O'Donnell.[55] Comparing animals 2 months old with those 19 to 21 months of age, the authors reported a decrease in potency to noradrenaline, no change in affinity, no change in absolute maximum response, and a decrease in α-adrenergic receptor reserve. Contractile responses to the nonspecific stimulus of depolarization produced by increased extracellular calcium also decline with age in the rat aorta.[56] Interestingly, this nonspecific decline in contractile response is not seen in animals restricted to a diet which is 60% of the consumption of animals fed ad libitum, a dietary regimen known to produce a substantial prolongation of the life span.[5]

In contrast, a variety of studies of other arteries show marked stability with increasing age in vascular responsiveness to noradrenaline or α_1-adrenergic receptor activation. The carotid and femoral arteries from 6-, 12-, 20-, and 27-month-old rats show no age-related changes in responsiveness to noradrenaline,[28] and maximum responses to noradrenaline in the renal and femoral arteries and veins and the tail artery do not change with age.[29] In the perfused rat tail artery, there is a decline in noradrenaline responsiveness from 6 to 16 months, but no further change to 30 months,[20] and no effect of age was seen in the rat pulmonary artery[55] or rat mesenteric artery.[57] In perfused rat mesenteric arteries, no age-related changes in response to noradrenaline are seen from 4 to 27 months.[58,59] In isolated beagle coronary arteries, contractile responses to noradrenaline in the presence of propranolol increase with age from 2 to 12 years.[36]

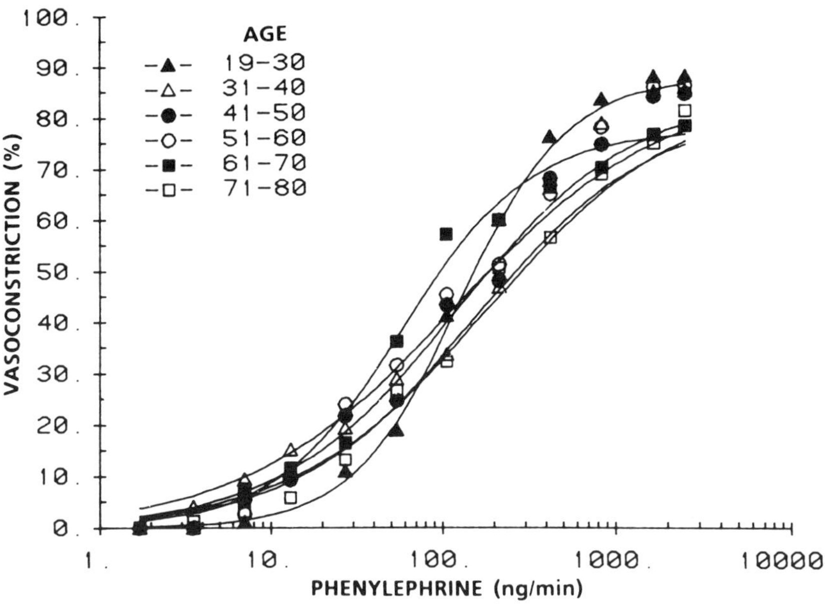

FIGURE 2. Vasoconstrictor response to phenylephrine infusion in human dorsal hand vein as a function of age. (From Pan, H. Y.-M., Hoffman, B. B., Persche, R. A., and Blaschke, T. F., *J. Pharmacol. Exp. Ther.*, 239, 802, 1986. With permission.)

In pithed rats, there is no age-related change in the pressor response to the α_1-agonist amidephrine and no change in response to noradrenaline in anesthetized animals.[33] Similarly, in isolated mesenteric arterial strips of beagles, responses to noradrenaline and sensitivity to the effects of the α_1-adrenergic receptor antagonist prazosin or the α_2-antagonist yohimbine did not change with age.[60] This study suggests that the ratio of functional α-adrenergic receptors is not influenced by age.

Similar age-related stability in vascular responsiveness to noradrenaline has been demonstrated in isolated human tissues, including saphenous veins,[35] splanchnic vessels,[61] digital arteries and metacarpal veins,[62] and erectile tissue of the male corpus cavernosum.[63] Contractile responses to phenylephrine of the human basilar artery are not correlated with age.[64] A few studies have also looked at the influence of age on vascular reactivity to α_1-adrenergic receptor stimulation in humans *in vivo*. As demonstrated in Figure 2, no age-related differences in vascular responsiveness to the α-adrenergic agonist phenylephrine were found using the dorsal hand vein technique to measure vasoconstriction.[65-67]

In terms of sensitivity to α_2-adrenergic receptor stimulation, a decrease in potency of yohimbine has been reported in isolated human saphenous veins,[35] and in the pithed rat, the pressor effect of the α_2-agonist xylazine has been shown to be reduced with age. Similarly, in the anesthetized rat, the pressor effect of noradrenaline in the presence of the α_2-antagonist prazosin was reduced with increasing age.[33] However, it has subsequently been demonstrated that in the pithed rat there is no difference with age in the effect of xylazine in the presence of captopril.[51] These findings suggest that the age-related differences reported earlier depend on different circulating levels of angiotensin, which is known to potentiate responses to α_2-adrenergic receptor stimulation, rather than on differences in the function of postjunctional α_2-adrenergic receptors per se.

A few investigations have focused on the impact of age on subcellular components involved in mediating vascular responses to adrenergic receptor activation. Noradrenaline-stimulated inositol phosphate accumulation, an index of inositol 1,4,5P_3 formation, increased in the rat aorta between 6 and 24 months.[68] A decrease in phosphokinase C activity was seen in the rat aorta, renal and iliac arteries, and vena cava between 6 and 24 months, and there was also a decreased effect of the phosphokinase C activator, PMA.[69] No change in basal ^{86}Rb efflux was seen in thoracic aorta, tail artery, or mesenteric artery branches when 6-, 12-, 24-, and 30-month-old rats were compared.[70] However, the initial efflux response to noradrenaline declined with age, although there was no age-related difference in the steady-state response to noradrenaline.

C. DEVELOPMENT OF DENERVATION SUPERSENSITIVITY

The phenomenon of denervation supersensitivity, increased postjunctional responsiveness after interruption of sympathetic activity, has been well described in a variety of adrenergically innervated organs.[71] In older humans after chronic suppression of sympathetic activity with guanadrel, there was no increase in responsiveness of the dorsal hand vein to α_1-adrenergic receptor stimulation, although such a response was seen in younger individuals.[67] In contrast, an increased venoconstriction to the α_2-adrenergic receptor agonist clonidine was seen in the older subjects after treatment with guanadrel. In the tail artery from rats treated with reserpine to deplete noradrenaline from adrenergic nerves, concentration response curves to phenylephrine or serotonin shift to the left, reflecting the nonspecific increase in vascular smooth muscle sensitivity characteristic of denervation supersensitivity.[72] Interestingly, the degree of denervation supersensitivity seen at 6, 12, 20, and 24 months of age did not differ in this model. In rats with unilateral lesions of the locus ceruleus to deplete cerebral noradrenaline, aging did not prevent the development of denervation supersensitivity in cerebral microvessels as measured by increased binding for the β-adrenergic ligand, [^{125}I]-pindolol.[73]

D. β-ADRENERGIC RESPONSIVENESS

One of the best described age-related changes in vascular reactivity concerns the loss of sensitivity to β-receptor stimulation described in arteries, but not in veins.[74] However, this change occurs predominantly during maturation. For example, relaxation to isoproterenol of the rat aorta or pulmonary

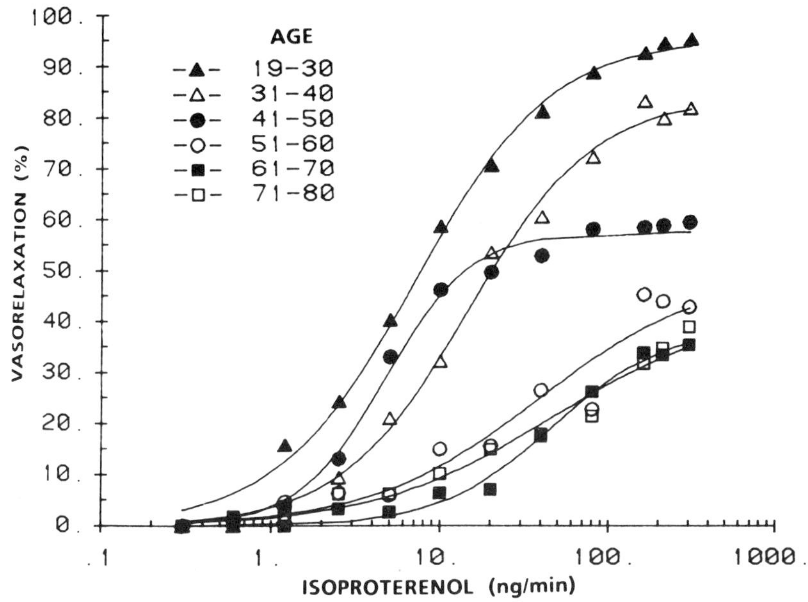

FIGURE 3. Vasodilator response to isoproterenol infusion in preconstricted human dorsal hand vein as a function of age. Veins were precontracted to approximately 20% of baseline value with phenylephrine. (From Pan, H. Y.-M., Hoffman, B. B., Persche, R. A., and Blaschke, T. F., *J. Pharmacol. Exp. Ther.*, 239, 802, 1986. With permission.)

artery declines after 1 or 2 months of age and almost completely disappears by 6 months.[54,75,76] In contrast, no age-related alterations in relaxation response to isoproterenol were seen in the rat jugular vein from 3 to 27 months.[75] No changes in vasodilator responses to noradrenaline or adenosine were seen in isolated rat basilar arteries from 12 to 30 months.[77]

There is some evidence that maturation affects responses to β_2-stimulation to a greater degree than β_1-effects.[78] A decline in responses to β-receptor activation with age from 2 to 12 years was also seen in beagle coronary arteries,[36] and an age-related loss in vasodilator response to both isoproterenol and adenosine was found in beagle mesenteric arteries.[79]

Using the human dorsal hand vein technique to measure vascular responsiveness *in vivo,* a decline with age in ED_{50} for isoproterenol as well as maximum relaxation response was seen (Figure 3).[80] This decline was progressive between ages 19 to 50, but no further change from 50 to 80 years was seen. In the oldest subjects who showed blunted responses to β-adrenergic receptor stimulation, full relaxation responses to nitroglycerin could still be achieved, suggesting that a general decline in vascular relaxation does not occur.

This loss in relaxation to β-adrenergic stimulation also occurs for adenosine, which acts through adenylate cyclase.[81,82] The number and affinity of β-adrenergic receptors in mesenteric arteries is unaffected by aging, and relaxation to forskolin or dibutyryl cAMP is unchanged.[81,83-85] Stability with aging of β-adrenergic receptor binding has also been demonstrated in rat cerebral microvessels.[73] Stimulation of cAMP accumulation and activation of cAMP-dependent protein kinase by isoproterenol declines during maturation.[81,83] Furthermore, an age-related deficit in isoproterenol-stimulated cAMP binding to the regulatory subunit of cAMP-dependent protein kinase has also been demonstrated.[86] These findings suggest that maturation influences a major rate-limiting step between the receptor and activation of adenylate cyclase, such as coupling of the receptor to G_s or the ability of G_s to activate adenylate cyclase.[85,86]

Continuous infusion of the β-adrenergic antagonist timolol for 1 week restored vascular relaxation to stimulation of β-adrenergic or adenosine receptors, suggesting that the decline in responsiveness may be due to an age-dependent increase in circulating catecholamines.[81] Furthermore, treatment with thyroid hormone has been shown to restore responsiveness to β-adrenergic stimulation in mature animals.[84,87,88] In mesenteric arteries, this restoration of β-adrenergic responsiveness after thyroid treatment occurred in the absence of any change in the number or affinity of β-adrenergic receptors.[84]

E. NONADRENERGIC VASOCONSTRICTORS

The effect of age on the overall contractile capability of vascular smooth muscle has not been systematically studied, but the lack of effect of age on α-adrenergic contractile responses, as discussed above, suggests that there is no general age-related change in the contractile machinery. Nevertheless, some nonreceptor-specific, age-related changes in contractile responses have been described. Some of these have been discussed above in the section on α-adrenergic responsiveness, including a decrease in phosphokinase C activity, decreased contractile effect of an inhibitor of phosphokinase C activity, and decreases in the initial ^{86}Rb efflux response to activation of α-adrenergic receptors.

In the rat aorta, and to a lesser extent in the pulmonary artery, there is an age-related reduction in the potency of potassium depolarization, responsiveness to calcium in potassium-depolarized tissues, or the effect of the calcium channel activator, Bay K 8644.[89] Blood vessels from young rats contract to Bay K 8644, but in aged rats it was necessary to depolarize the tissue before any contractile response to Bay K 8644 was seen. In contrast, no age-related changes in contractile responses to caffeine or the calcium ionophore, A23187, were found. It has been suggested that the resting membrane potential may be more negative in vascular smooth muscle from aged rats.

Contractile responsiveness to serotonin has been shown to increase with age, decrease, or remain unchanged. In dog coronary arteries, the magnitude of the maximum contractile response to serotonin increased from 2 to 12 years, with no change in ED_{50}.[90] In contrast, in rat coronary resistance vessels, there was both an increase in maximum contraction and sensitivity to serotonin in animals aged 2 years compared to 3-month-old rats.[91] This was shown to be due to an increase in 5-HT_2 receptor agonist affinity as well as efficiency of the receptor-effector coupling process.[92] In the rat aorta, there was no change in maximum contractile response to serotonin from 6 to 24 months of age, but a decrease in sensitivity was found.[51] In the rat basilar artery, maximum contractile response and sensitivity to serotonin did not change with age from 12 to 30 months.[77] Responsiveness to serotonin of forearm blood flow in the human was not age related,[93] and there was no correlation between serotonin potency and age in the isolated human saphenous vein.[94]

Contractile responses to angiotensin II showed no age-related differences in the dog coronary artery[90] or rat aortic strips,[95] and this was true regardless of the presence or absence of a functional endothelium. A decreased sensitivity to endothelin has been reported in rat mesenteric arteries, and the ability of endothelin to potentiate contractile responses to noradrenaline was shown to be greater in vessels from older rats.[58] Contractile responses to the thromboxane A_2 analog, U 46619, showed no age-related change in rat thoracic aorta[96] or in rat cerebral arterioles *in situ*.[97] Contractile responses of the rat basilar artery to high potassium, uridine triphosphate, or prostaglandin $F_{2\alpha}$ did not change with age from 12 to 30 months.[77]

F. NONADRENERGIC VASODILATORS

Since the discovery of the critical role of the endothelium in vascular responses,[98] no discussion of vasodilator effects is complete without a distinction between those substances which act through the endothelium and those that act directly on vascular smooth muscle. Accordingly, the impact of age on vasodilators thought to act directly on vascular smooth muscle is discussed first, followed by a discussion of the impact of age on endothelial function.

An important consideration in assessing the effect of age on vasodilator responses concerns equality of developed contractile tone, as it is well known that the degree of baseline tone can affect the magnitude of vasodilator response as well as the ability to observe age-related changes.[76] For example, in the rat aorta where the magnitude of response to α-adrenergic receptor activation declines with age (see above), vasodilator responses to the potassium channel activator, pinacidil,[99] or the calcium channel blocker, diltiazem,[55] are significantly greater in older animals. This age-related difference in vasodilator potency is thought to reflect reduced α-adrenergic receptor reserve in the aorta from old rats, and is not seen in the pulmonary artery, where age has no effect on α-adrenergic responsiveness.[55]

1. Endothelium-Independent Vasodilators

As discussed above, one of the most well-described age-related alterations in vascular responsiveness is the decline in response to the endothelium-independent vasodilation mediated by β-adrenergic receptors. However, this can best be described as a maturational change, as most investigators have found this to occur early in life. Most studies have shown that the inherent ability of vascular smooth muscle to relax does not change with age, as evidenced by no loss of vasodilator response to papaverine or the nitrovasodilators. For example, in rat mesenteric arteries there is no change in the relaxant response to papaverine from 14 to 25 months of age.[57] Responses to nitrovasodilators have been shown to be maintained with age in the human basilar artery[64] and in rat cerebral arterioles.[97] However, an age-related decline in relaxation to sodium nitroprusside has been demonstrated in rat aorta.[100]

Vasodilator responses to prostaglandin I_2 have been shown to be maintained with age in beagle mesenteric and coronary arteries.[79,90] Similarly, vasodilation to atrial natriuretic peptide is stable with age in the rat aorta, carotid, mesenteric, and renal arteries, and monkey renal arteries.[101-103] In the rat basilar artery, vasodilator responses to somatostatin did not differ from 12 to 30 months of age.[77]

In vivo, nitroprusside was shown to be more effective in 21- to 24- than 3- to 7-month-old conscious rats, but no difference was found in anesthetized animals; however, involvement of baroreflexes complicates the interpretation of this study.[104] In contrast, vasodilator effects of sodium nitroprusside decreased with age in pentobarbital-anesthetized rats, with no age-related differences in the vasodilator effect of nifedipine.[105] In the latter case, there was a lower resting blood pressure in the older animals which could have contributed to the age-related difference.

2. Endothelium-Dependent Vasodilators

Acetylcholine is the prototypical endothelium-dependent vasodilator, producing vascular smooth muscle relaxation through release of endothelium-derived relaxing factor, EDRF, thought to be nitric oxide.[106,107] *In vivo,* the ability of methacholine to decrease blood pressure in rats does not change with age from 1 to 2 years of age.[108] *In vitro,* vasodilator responses to acetylcholine have been shown to be stable with advancing age in beagle mesenteric[79] and coronary[90] arteries and in the rat basilar artery.[77] In human posterior ciliary arteries, sensitivity to acetylcholine was not correlated with age.[109] In the human basilar artery, vasodilator responses to bradykinin or the calcium ionophore, A-23187, did not change with age.[64] No change in sensitivity or maximum vasodilator response to cholinergic receptor activation was seen in the rat perfused caudal artery from 12 to 27 months of age.[59] Concentration response curves for acetylcholine to relax rat mesenteric resistance arteries precontracted with endothelin did not change with age from 9 to 27 months, and vascular effects of the nitric oxide donor, 3-morpholinosydonimine, did not alter with age.[58]

In contrast, in rat cerebral arterioles studied *in situ,* vasodilator responses to acetylcholine or bradykinin, as well as dilator responses to ADP or serotonin, decreased with age from 6 to 8 to 22 to 24 months.[97] Endothelium-dependent vasodilator effects of histamine have been reported to decline with age in rat mesenteric arteries and rat aorta,[57,100] but to be unchanged with age in beagle coronary and mesenteric arteries[79,90] and rat basilar arteries.[77]

Contractile responses of perfused rat mesenteric arteries to endothelin are augmented after removal of the endothelium, possibly by removal of basally released EDRF.[110] However, the degree of such augmentation was found to be greater in 9-month-old rats than in 27-month-old animals.[58] Blockade of the effects of EDRF with methylene blue also produced a greater augmentation of contractile responses of the isolated tail artery to adrenergic nerve stimulation in young rats (6, 12, or 20 months) than in older animals (27 months).[147]

V. *IN VIVO* HEMODYNAMICS

A high incidence of postural hypotension with advancing age has been well documented in humans, and this has led to speculation concerning the nature of functional changes that lead to alteration of control of vascular function. Several studies have found no evidence of autonomic neuropathy in older individuals who experience postural hypotension, and it has been suggested that the change of structure of the vascular tree and increased stiffness due to atherosclerosis may be more important as a determinant of postural hypotension than any change in autonomic function per se.[111-113] Indeed, the vascular alterations due to development of atherosclerosis, which are not seen in many animal models, may account for a substantial portion of the age-related alterations in vascular function seen in humans. A significant decline in renal blood flow with advancing age and trends toward increased systemic vascular resistance and decreased limb and hepatic blood flow have been seen in human subjects.[114]

In studies of skin blood flow in the human forearm using laser Doppler techniques, no age-related differences in red blood cell velocity or volume of red cells flowing through the vessels were seen at 30°C.[115] With increasing temperature to 35 or 40°C, significant increases in red blood cell velocity were found in all age groups, but the response in middle-aged (40 to 59 years) and older (60 to 79 years) subjects was less than that seen in the young group (20 to 39 years). These findings suggest that there was a reduced flow response to heating of individual microvessels in middle-aged and older men and, in addition, that vessel recruitment is decreased in old age.

Adaptive changes with age in hemodynamics may also occur as a consequence of the well-documented alterations in cardiac function. Lesser increases in heart rate and diastolic pressure were seen in older men with graded increments of tilt, suggesting a decrease in baroreflex responsiveness.[116] In animals that do not develop atherosclerosis, the influence of advancing age can be studied in the absence of this disease process. In young and old racing greyhounds, an animal that does not develop substantial atherosclerosis, characteristics of the carotid sinus baroreceptor control of arterial pressure-flow relations were well preserved in older animals.[117] In contrast, comparison of 1- and 7- to 8-year-old beagles showed a significant attenuation of arterial baroreflex control of heart rate, renal sympathetic activity, and arterial pressure, and the function of the cardiopulmonary baroreflex also appeared to be impaired in older animals.[118,119]

In conscious rats, no significant differences were seen from 12 to 24 months of age in central hemodynamics or regional blood flows, with the exception of a decrease with age in splenic blood flow. Interestingly, anesthesia with pentobarbital had a different effect on hemodynamics in aged compared to young animals.[120] Anesthesia produced a greater decrease in cardiac index and brain blood flow in older animals, and heart and kidney blood flow were reduced by anesthesia in old, but not young, animals. In F344 rats, there were no differences between 12- and 20-month-old animals in resting coronary blood flow or its distribution between endo- and epicardium either before or after maximal coronary vasodilation.[121]

Chronic exercise training affects cardiovascular homeostasis differently in adult and aged rats. The temporary cardiovascular adaptive response to exercise predominantly involves cardioinhibition in adult animals, but a vasodepressant component is more prominent in senescent animals.[122] In beagles, no significant age-related differences in resting vascular resistance were seen in any of the 13 regions examined.[123] With exercise, skeletal muscle blood flow was maintained in senescent beagles through a compensatory reduction in splanchnic blood flow, in spite of a significant reduction in cardiac output in older exercising animals.

In an intriguing study, chronic treatment with the calcium channel antagonist nimodipine produced a profound improvement in the walking pattern, nerve conduction velocity, and nerve fiber density in rats treated between 24 and 28 months of age.[124] It remains unclear, however, whether the improvements in neuromuscular function seen in these animals were due to vascular effects of the drug treatment or actions on neuronal function per se.

VI. CEREBRAL CIRCULATION

The influence of advancing age on brain circulation is a topic of special importance, and the ability to measure cerebral blood flow in humans with a series of noninvasive techniques has also facilitated investigation of the effect of age on this circulatory bed. Several investigations have demonstrated no significant differences with advancing age in global cerebral blood flow or cerebral metabolic rate for oxygen using a variety of different techniques.[125-128] A significant age-related depression in mean cerebral metabolic rate for glucose has been seen,[125] as well as an age-related progression of cerebral atrophy.[126,127] There is a preferential decline of regional cerebral blood flow in the frontal cortex with advancing age and increasing side-to-side asymmetry of cortical regional cerebral blood flow; however, white matter cerebral blood flow and cerebral metabolic rate for oxygen remain remarkably stable with advancing age.[127,129,130] A more specific definition of areas showing decreased regional blood flow in aging demonstrates that the cingulate gyri, parahippocampal gyri, superior temporal gyri, medial frontal gyri, and parietal cortex are impacted.[128] These studies have not clarified whether the primary change occurring with age and leading to cerebral atrophy and loss of function is vascular or neuronal in origin. However, the matched reduction in cerebral blood flow and cerebral metabolic rate for oxygen suggests that the regulatory mechanism coupling metabolic demand to oxygen supply is unaffected by age. Furthermore, the lack of a global change in cerebral blood flow suggests that age-related changes are a consequence of reduced metabolic needs. In this respect, neuronal cell depopulation and/or diminished activity stand out as the most likely hypothesis to explain age-related regional decreases in cerebral blood flow.

In young rats exposed to a complex environment, there is an increase in visual cortex thickness accompanied by angiogenesis such that intercapillary distance and capillary diameter remain constant.[131] While this response to the stimulation of a complex environment occurs in 14-month-old rats, it was not seen in 24-month-old animals. However, this study did not resolve the question of whether the old animals failed to make new neuronal synapses because they lacked the ability to generate the necessary vascular support or whether no new blood vessels were formed because synaptogenesis did not occur.

In studies of capillary morphometry in the rat brain, no age-related differences in total or perfused capillary volume or perfused capillary diameter were seen.[132] Measurements in conscious rats have shown no age-related alterations in global or regional cerebral blood flow or average oxygen supply-to-consumption ratio.[133] After deactivation of smooth muscle with EDTA, the diameter of pial arterioles was less in 9- to 12-month-old rats than 24- to 27-month-old animals, and vascular distensibility was also reduced in aging animals.[134] The ratio of nondistensible to distensible components was found to be greater in arterioles from aged compared to adult animals.

Activity in rat cerebral microvessels of five metabolic enzymes — myosin ATPase, lactate dehydrogenase, succinate dehydrogenase, cytochrome oxidase, and glucose-6-phosphate dehydrogenase — showed only minor differences as the animals aged from 10 to 12 months to 31 to 33 months.[135] The one exception was increased activity of myosin ATPase, which was particularly prominent in ganglionic vessels. Histochemical examination of a variety of structural parameters in the rat cerebral vasculature showed an increase from 9 to 30 months of age in basal lamina thickening, especially in capillaries, and an age-related increase in smooth muscle cell mitochondria in the hippocampus.[136,137] Advancing age did not affect the frequency of neurovascular appositions in the dentate fascia or in the hippocampus.[136,137]

The specialized functions of the blood-brain barrier reside in the arterioles and capillaries of the cerebral circulation. Since alterations in this barrier could have profound effects on brain function, possible age-related changes in the blood-brain barrier have been studied.[138] There is little evidence of functional abnormalities in the blood-brain barrier with age, and it has been pointed out that even if abnormalities were shown, these would have to be demonstrated to be causally related to aging and not merely secondary to other pathological conditions.[139]

The susceptibility of the blood-brain barrier to disruption during acute hypertension has been monitored by measurement of the leakage of fluorescent-labeled albumin. In these studies, the degree of albumin leakage was shown to be similar in 6- to 8- and 24- to 26-month-old rats, suggesting that there is no breakdown with aging of the integrity of the blood-brain barrier under control conditions or in the face of acute hypertension.[140]

VII. CONCLUSIONS

There is still much to be learned concerning the impact of age on vascular function, and in many studies, inadequate numbers of age groups have been used and the health of the animals has not been well documented. Nevertheless, some general conclusions concerning the impact of age on vascular function can be drawn. While a variety of conflicting effects of age on adrenergic nerve density have been reported, overall vascular contractile responses to adrenergic nerve activation appears to be well preserved with advancing age. However, this occurs in the face of a series of alterations of function of some of the component parts of this overall response. Stimulation-evoked noradrenaline release increases with age, and there is a loss of β-adrenergic relaxation, although this occurs primarily during maturation. Contractile responses to α-adrenergic receptor activation appear to be well preserved, but there has been little exploration to date of the contractile function of α-adrenergic receptor subtypes. The vasodilator function of the vascular endothelium may be well preserved in some respects, as vasodilator responses to acetylcholine or bradykinin have been demonstrated to be unchanged with age in most studies. However, there is some evidence that spontaneous release of EDRF may be reduced with age.

In vivo, some of these observations on isolated tissues have been corroborated, including maintenance of contractile response to α-adrenergic stimulation and loss of β-adrenergic response. In several species, no substantial alterations with advancing age in resting regional blood flow have been seen, with the exception of a decline in renal blood flow. Adaptive changes may be different with advancing age, including responses to exercise and changes in ambient temperature. Despite the high incidence of postural hypotension[141] as well as hypertension[142,143] in the elderly, several studies have found no evidence of autonomic neuropathy in older individuals, suggesting that postural hypotension and hypertension may be primarily consequences of the development of atherosclerosis.

In the cerebral circulation, some changes in regional blood flow have been associated with advancing age; however, the lack of a generalized change in cerebral blood flow suggests that regional changes occur as a consequence of diminished neuronal activity and do not reflect any primary abnormality of vascular function. Similarly, the function of the blood-brain barrier, although not extensively studied, may be well preserved with age.

As discussed in more detail in Chapter 7, profound declines in cardiac function with age have been described, including a decrease in adrenergic innervation and adrenergic nerve function and loss of contractile capacity. It is intriguing that the impact of age on the heart should be so different from its impact on the vasculature, and the reasons for this are not well understood. In this regard, recent studies exploring the impact of chronic exercise on the aged myocardium are most interesting. Chronic exercise of old rats was shown to reverse the age-related prolongation of isometric contraction and associated intracellular calcium transient, and also resulted in a significant increase in cardiac noradrenaline content.[144] Exercise ameliorated the slower oxalate-supported, ATP-dependent calcium transport seen in the older hearts, suggesting that increased calcium transport by the sarcoplasmic reticulum may be a potential mechanism which underlies the improvement of myocardial performance in exercised aged animals.[145] These findings are quite provocative and could be interpreted to suggest that some of the profound alterations in cardiac function seen in older animals are a consequence of their inactivity. To date, blood vessel function in exercised old animals has not been extensively explored, but it is intriguing to consider the possibility that peripheral adrenergic nerve function is generally well maintained with age except in an organ such as the heart where exercise and overall activity of the organism can have such a profound influence.

ACKNOWLEDGMENTS

Preparation of this manuscript was supported in part by a grant from the National Institute of Aging, AG06912. The competent assistance of Victor Barrios and Cindy Wyman is greatly appreciated.

REFERENCES

1. **Palmer, G. J., Ziegler, M. G., and Lake, C. R.,** Response of norepinephrine and blood pressure to stress increases with age, *J. Gerontol.*, 33, 482, 1978.
2. **Chuieh, C. C., Nespor, S. M., and Rapoport, S. I.,** Cardiovascular, sympathetic and adrenal cortical responsiveness of aged Fischer-344 rats to stress, *Neurobiol. Aging*, 1, 157, 1980.
3. **Esler, M., Skews, H., Lionard, P., Jackman, G., Bobik, A., and Korner, P.,** Age-dependence of noradrenaline kinetics in normal subjects, *Clin. Sci.*, 60, 217, 1981.
4. **Barnes, R. F., Rashkind, M., Gumbrecht, G., and Halter, J. B.,** The effects of age on the plasma catecholamine response to mental stress in man, *J. Clin. Endocrinol. Metab.*, 54, 64, 1982.

5. **Yu, B. P., Masoro, E. J., and McMahan, C. A.,** Nutritional influences on aging of Fischer 344 rats. I. Physical, metabolic, and longevity characteristics, *J. Gerontol.*, 40, 657, 1985.
6. **Coleman, P. D.,** Editorial. How old is old?, *Neurobiol. Aging,* 10, 115, 1989.
7. **Coleman, P. D., Finch, C., and Joseph, J.,** Editorial. The need for multiple time points in aging studies, *Neurobiol. Aging,* 11, 1, 1990.
8. **Masoro, E. J.,** Mortality and growth characteristics of rat strains commonly used in aging research, *Exp. Aging Res.*, 6, 219, 1980.
9. **Duckles, S. P. and Banner, W., Jr.,** Changes in vascular smooth muscle reactivity during development, *Annu. Rev. Pharmacol. Toxicol.*, 24, 65, 1984.
10. **Docherty, J. R.,** Cardiovascular responses in ageing: a review, *Pharmacol. Rev.*, 42, 103, 1990.
11. **Florini, J. R.,** Limitations of interpretation of age-related changes in hormone levels: illustration by effects of thyroid hormones on cardiac and skeletal muscle, *J. Gerontol.*, 44, B107, 1989.
12. **Maeda, H., Gleiser, C. A., Masoro, E. J., Murata, I., McMahan, C. A., and Yu, B. P.,** Nutritional influences on aging of Fischer 344 rats. II. Pathology, *J. Gerontol.*, 40, 671, 1985.
13. **Handa, R. K. and Duckles, S. P.,** Influence of age on norepinephrine content in arteries and veins of Fischer 344 rats, *Neurobiol. Aging,* 8, 511, 1987.
14. **Brennan, M. J., Pasagian, A., and Friedman, E.,** Genotypic variation in age-related changes in peripheral catecholamines, *Exp. Gerontol.*, 20, 285, 1985.
15. **Amenta, F., Bronzetti, E., Ferrante, F., and Ricci, A.,** The noradrenergic innervation of spinal cord blood vessels in old rats, *Neurobiol. Aging,* 11, 47, 1990.
16. **Amenta, F. and Mione, M. C.,** Age-related changes in the noradrenergic innervation of the coronary arteries in old rats: a fluorescent histochemical study, *J. Auton. Nerv. Syst.*, 22, 247, 1988.
17. **Dhall, U., Cowen, T., Haven, A. J., and Burnstock, G.,** Perivascular noradrenergic and peptide-containing nerves show different patterns of changes during development and ageing in the guinea-pig, *J. Auton. Nerv. Syst.*, 16, 109, 1986.
18. **Santer, R. M.,** Fluorescence histochemical observations on the adrenergic innervation of the cardiovascular system in the aged rat, *Brain Res. Bull.*, 9, 667, 1982.
19. **Waterson, J. G., Frewin, D. B., and Soltys, J. S.,** Age-related differences in catecholamine fluorescence of human vascular tissue, *Blood Vessels,* 11, 79, 1974.
20. **Fouda, A.-K. and Atkinson, J.,** Sensitivity to noradrenaline and electrical stimulation decreases with age in the rat tail artery, *Arch. Pharmacol.*, 334, 37, 1986.
21. **Neubauer, B. and Christensen, N. J.,** The decrease in noradrenaline concentration in the posterior tibial artery with age, *Gerontology,* 24, 299, 1978.
22. **Mione, M. C., Erdö, S. L., Kiss, B., Ricci, A., and Amenta, F.,** Age-related changes of noradrenergic innervation of rat splanchnic blood vessels: a histofluorescence and neurochemical study, *J. Auton. Nerv. Syst.*, 25, 27, 1988.
23. **Mione, M. C., Cavallotti, C., Collier, W. L., and Amenta, F.,** The noradrenergic innervation of the vasa nervorum in old rats: a fluorescence histochemical study, *J. Auton. Nerv. Syst.*, 18, 177, 1987.
24. **Amenta, F., Bronzetti, E., Felici, L., Ferrante, F., and Ricci, A.,** Age-related changes in adrenaline content of rat splanchnic blood vessels, *Mech. Ageing Dev.*, 51, 277, 1990.
25. **Cowen, T. and Thrasivoulou, C.,** Cerebrovascular nerves in old rats show reduced accumulation of 5-hydroxytryptamine and loss of nerve fibres, *Brain Res.*, 513, 237, 1990.
26. **Mione, M. C., Dhital, K. K., Amenta, F., and Burnstock, G.,** An increase in the expression of neuropeptidergic vasodilator, but not vasoconstrictor, cerebrovascular nerves in aging rats, *Brain Res.*, 460, 103, 1988.
27. **Duckles, S. P.,** Comparison of ^{3}H-norepinephrine and ^{3}H-metaraminol accumulation as indices of adrenergic nerve density in rabbit blood vessels, *Eur. J. Pharmacol.*, 67, 355, 1980.

28. **Duckles, S. P., Carter, B. J., and Williams, C. L.,** Vascular adrenergic neuroeffector function does not decline in aged rats, *Circ. Res.*, 56, 109, 1985.
29. **Duckles, S. P.,** Effect of norepinephrine uptake blockade on contractile responses to adrenergic nerve stimulation of isolated rat blood vessels: influence of age, *J. Pharmacol. Exp. Ther.*, 243, 521, 1987.
30. **Buchholz, J. N. and Duckles, S. P.,** The effect of age on prejunctional modulation of norepinephrine release, *J. Pharmacol. Exp. Ther.*, 252, 159, 1990.
31. **Handa, R. K. and Duckles, S. P.,** Age-related changes in adrenergic vasoconstrictor responses of the rat blood-perfused hindlimb, *Am. J. Physiol.*, 253, H1566, 1987.
32. **Borton, M. and Docherty, J. R.,** The effects of ageing on neuronal uptake of noradrenaline in the rat, *Naunyn-Schmiedeberg's Arch. Pharmacol.*, 340, 139, 1989.
33. **Docherty, J. R. and Hyland, L.,** α-Adrenoceptor responsiveness in the aged rat, *Eur. J. Pharmacol.*, 126, 75, 1986.
34. **Buchholz, J., Tsai, H., Friedman, D. J., and Duckles, S. P.,** Influence of age on control of norepinephrine release from the rat tail artery, *J. Pharmacol. Exp. Ther.*, 260, 722, 1992.
35. **Hyland, L. and Docherty, J. R.,** An investigation of age-related changes in pre- and postjunctional α-adrenoceptors in human saphenous vein, *Eur. J. Pharmacol.*, 114, 361, 1985.
36. **Toda, N. and Miyazaki, M.,** Senescent beagle coronary arteries in response to catecholamines and adrenergic nerve stimulation, *J. Gerontol.*, 42(2), 210, 1987.
37. **Burnstock, G.,** The changing face of autonomic neurotransmission, *Acta Physiol. Scand.*, 126, 67, 1986.
38. **Kennedy, C., Saville, V. L., and Burnstock, G.,** The contributions of noradrenaline and ATP to the responses of the rabbit central ear artery to sympathetic nerve stimulation depend on the parameters of stimulation, *Eur. J. Pharmacol.*, 122, 291, 1986.
39. **Machaly, M., Dalziel, H. H., and Sneddon, P.,** Evidence for ATP as a cotransmitter in dog mesenteric artery, *Eur. J. Pharmacol.*, 147, 83, 1988.
40. **Walker, P., Grouzmann, E., Burnier, M., and Waeber, B.,** The role of neuropeptide Y in cardiovascular regulation, *Trends Pharmacol. Sci.*, 12, 111, 1991.
41. **Edvinsson, L., Elsman, R., and Ottosson, A.,** Demonstration of perivascular peptides and changes in concentration with age in man, *Gerontology,* 32 (Suppl. 1), 50, 1986.
42. **Mayberg, M., Langer, R. S., Zervas, N. T., and Moskowitz, M. A.,** Perivascular meningial projections from cat trigeminal ganglia: possible neuroanatomical pathway for vascular headache in man, *Science,* 231, 228, 1981.
43. **Lundblad, L., Anggard, A., and Lundberg, J. M.,** Effects of antidromic trigeminal nerve stimulation in relation to parasympathetic vasodilation in cat nasal mucosa, *Acta Physiol. Scand.*, 119, 7, 1983.
44. **Lundberg, J. M., Brodin, E., Hua, Y., and Saria, A.,** Vascular permeability changes and smooth muscle contraction in relation to capsaicin-sensitive substance P afferents in the guinea pig, *Acta Physiol. Scand.*, 120, 217, 1984.
45. **Duckles, S. P.,** Effects of capsaicin on vascular smooth muscle, *Naunyn-Schmiedeberg's Arch. Pharmacol.*, 333, 59, 1986.
46. **Kawasaki, H., Takasaki, K., Saito, A., and Goto, K.,** Calcitonin gene-related peptide acts as a novel vasodilator neurotransmitter in mesenteric resistance vessels of the rat, *Nature (London),* 335, 164, 1988.
47. **Duckles, S. P.,** Effect of age on substance P levels in the rat mesenteric artery and vein, *Neurobiol. Aging,* 6, 237, 1985.
48. **MacLean, D. B., Eldridge, J. C., and Brodish, A.,** Substance P and somatostatin content and transport in the vagus and sciatic nerves of the aging Fischer 344 rat, *Neurobiol. Aging,* 9, 273, 1988.
49. **Li, Y. and Duckles, S. P.,** Effect of age on content of calcitonin gene related peptide and vasodilator nerve activity in blood vessels of the rat, submitted.
50. **Cox, R. H.,** Effects of age on the mechanical properties of rat carotid artery, *Am. J. Physiol.*, 233, H256, 1977.

51. **Docherty, J. R.**, The effects of ageing on vascular α-adrenoceptors in pithed rat and rat aorta, *Eur. J. Pharmacol.*, 146, 1, 1988.
52. **Tuttle, R. S.**, Age-related changes in the sensitivity of rat aortic strips to norepinephrine and associated chemical and structural alterations, *J. Gerontol.*, 21, 510, 1966.
53. **Simpkins, J. W., Field, F. P., and Ress, R. J.**, Age-related decline in adrenergic responsiveness of the kidney, heart and aorta of male rats, *Neurobiol. Aging*, 4, 233, 1983.
54. **Hyland, W., Warnock, P., and Docherty, J. R.**, Age-related alterations in α_1- and β-adrenoceptor mediated responsiveness of rat aorta, *Naunyn-Schmiedeberg's Arch. Pharmacol.*, 335, 50, 1987.
55. **Wanstall, J. C. and O'Donnell, S. R.**, Inhibition of norepinephrine contractions by diltiazem on aorta and pulmonary artery from young and aged rats: influence of *alpha*-adrenoceptor reserve, *J. Pharmacol. Exp. Ther.*, 245, 1016, 1988.
56. **Herlihy, J. T. and Yu, B. P.**, Dietary manipulation of age-related decline in vascular smooth muscle function, *Am. J. Physiol.*, 238, H652, 1980.
57. **Moritoki, H., Hosoki, E. and Ishida, Y.**, Age-related decrease in endothelium-dependent dilator response to histamine in rat mesenteric artery, *Eur. J. Pharmacol.*, 126, 61, 1986.
58. **Dohi, Y. and Lüscher, T. F.**, Aging differentially affects direct and indirect actions of endothelin-1 in perfused mesenteric arteries of the rat, *Br. J. Pharmacol.*, 100, 889, 1990.
59. **Hynes, M. R. and Duckles, S. P.**, Effect of increasing age on the endothelium-mediated relaxation of rat blood vessels *in vitro*, *J. Pharmacol. Exp. Ther.*, 241, 387, 1987.
60. **Toda, N. and Shimizu, I.**, Neuroeffector function in mesenteric arteries isolated from beagles of different ages, *J. Pharmacol. Exp. Ther.*, 240, 223, 1987.
61. **Scott, P. J. W. and Reid, J. L.**, The effect of age on the responses of human isolated arteries to noradrenaline, *Br. J. Clin. Pharmacol.*, 13, 237, 1982.
62. **Stevens, M. J., Lipe, S., and Moulds, R. F. W.**, The effects of age on the responses of human isolated arteries and veins to noradrenaline, *Br. J. Clin. Pharmacol.*, 14, 750, 1982.
63. **Christ, G. J., Maayani, S., Valcic, M., and Melman, A.**, Pharmacological studies of human erectile tissue: characteristics of spontaneous contractions and alterations in α-adrenoceptor responsiveness with age and disease in isolated tissues, *Br. J. Pharmacol.*, 101, 375, 1990.
64. **Hatake, K., Kakishita, E., Wakabayashi, I., Sakiyama, N., and Hishida, S.**, Effect of aging on endothelium-dependent vascular relaxation of isolated human basilar artery to thrombin and bradykinin, *Stroke*, 21, 1039, 1990.
65. **Pan, H. Y.-M., Hoffman, B. B., Pershe, R. A., and Blaschke, T. F.**, Decline in *beta* adrenergic receptor-mediated vascular relaxation with aging in man, *J. Pharmacol. Exp. Ther.*, 239, 802, 1986.
66. **Martin, S. A., Alexieva, S., and Carruthers, S. G.**, The influence of age on dorsal hand vein responsiveness to norepinephrine, *Clin. Pharmacol. Ther.*, 40, 257, 1986.
67. **Supiano, M. A., Hogikyan, R. V., Stoltz, A. M., Orstan, N., and Halter, J. B.**, Regulation of venous α-adrenergic responses in older humans, *Am. J. Physiol.*, 260, E599, 1991.
68. **Johnson, M. D. and Wray, A.**, Alpha 1 adrenergic receptor function in senescent Fischer 344 rat aorta, *Life Sci.*, 46, 359, 1990.
69. **Johnson, M. D., Wang, H.-Y., and Friedman, E.**, Protein kinase activity and contractile responsiveness in senescent blood vessels, *Eur. J. Pharmacol.*, 189, 405, 1990.
70. **Cox, R. H. and Tulenko, T. N.**, Effects of aging on agonist-activated ^{86}Rb efflux in arteries of Fischer 344 rats, *Am. J. Physiol.*, 257, H494, 1989.
71. **Fleming, W. W.**, Variable sensitivity of excitable cells: possible mechanisms and biological significance, *Rev. Neurosci.*, 2, 43, 1976.

72. **Duckles, S. P.,** Reserpine-induced supersensitivity in rat caudal artery: influence of age, *J. Pharmacol. Exp. Ther.,* 256, 513, 1991.
73. **Harik, S. I., Sromek, S. M., and Kalaria, R. N.,** Alpha- and beta-adrenergic receptors of the rat cerebral cortex and cerebral microvessels in aging, and their response to denervation, *Neurobiol. Aging,* 12, 567, 1991.
74. **Fleisch, J. H.,** Age-related changes in the sensitivity of blood vessels to drugs, *Pharmacol. Ther.,* 8, 477, 1980.
75. **Duckles, S. P. and Hurlbert, J. S.,** Effect of age on *beta* adrenergic relaxation of the rat jugular vein, *J. Pharmacol. Exp. Ther.,* 236, 71, 1986.
76. **Sawyer, R. and Docherty, J. R.,** Reduction with age in the relaxation to β-adrenoceptor agonists and other vasodilators in rat aorta, *Naunyn-Schmiedeberg's Arch. Pharmacol.,* 336, 60, 1987.
77. **Hamel, E., Assumel-Lurdin, C., Bouloy, M., and MacKenzie, E. T.,** Selective age-related changes in neuronal markers and smooth muscle reactivity in cerebrovascular beds of Fischer 344 rats, *Neurobiol. Aging,* 11, 631, 1990.
78. **O'Donnell, S. R. and Wanstall, J. C.,** *Beta*-1 and *beta*-2 adrenoceptor-mediated responses in preparations of pulmonary artery and aorta from young and aged rats, *J. Pharmacol. Exp. Ther.,* 228, 733, 1984.
79. **Shimizu, I. and Toda, N.,** Alterations with age of the response to vasodilator agents in isolated mesenteric arteries of the beagle, *Br. J. Pharmacol.,* 89, 769, 1986.
80. **Pan, H. Y.-M., Hoffman, B. B., Pershe, R. A., and Blaschke, T. F.,** Decline in *beta* adrenergic receptor-mediated vascular relaxation with aging in man, *J. Pharmacol. Exp. Ther.,* 239, 802, 1986.
81. **Deisher, T. A., Mankani, S., and Hoffman, B. B.,** Role of cyclic AMP-dependent protein kinase in the diminished *beta* adrenergic responsiveness of vascular smooth muscle with increasing age, *J. Pharmacol. Exp. Ther.,* 249, 812, 1989.
82. **Moritoki, H., Matsugi, T., Takase, H., Ueda, H., and Tanioka, A.,** Evidence for the involvement of cyclic GMP in adenosine-induced, age-dependent vasodilatation, *Br. J. Pharmacol.,* 100, 569, 1990.
83. **Tsujimoto, G., Lee, C.-H., and Hoffman, B. B.,** Age-related decrease in *beta* adrenergic receptor-mediated vascular smooth muscle relaxation, *J. Pharmacol. Exp. Ther.,* 239, 411, 1986.
84. **Tsujimoto, G., Hashimoto, K., and Hoffman, B. B.,** Effects of thyroid hormone on β-adrenergic responsiveness of aging cardiovascular systems, *Am. J. Physiol.,* 252, H513, 1987.
85. **Kazanietz, M. G. and Enero, M. A.,** Decreased β-adrenoceptor-mediated vasodilation in aorta from aged rats: possible involvement of a stimulatory GTP-binding protein, *Eur. J. Pharmacol.,* 198, 177, 1991.
86. **Chin, J. H. and Hoffman, B. B.,** Age-related deficit in beta receptor stimulation of cAMP binding in blood vessels, *Mech. Ageing. Dev.,* 53, 111, 1990.
87. **Parker, R. J., Berkowitz, B. A., Lee, C.-H., and Denckla, W. D.,** Vascular relaxation, aging and thyroid hormones, *Mech. Ageing. Dev.,* 8, 397, 1978.
88. **O'Donnell, S. R. and Wanstall, J. C.,** Thyroxine treatment of aged or young rats demonstrates that vascular responses mediated by β-adrenoceptor subtypes can be differentially regulated, *Br. J. Pharmacol.,* 88, 41, 1986.
89. **Wanstall, J. C. and O'Donnell, S. R.,** Age influences responses of rat isolated aorta and pulmonary artery to the calcium channel agonist, Bay K 8664, and to potassium and calcium, *J. Cardiovasc. Pharmacol.,* 13, 709, 1989.
90. **Toda, N., Bian, K., and Inoue, S.,** Age-related changes in the response to vasoconstrictor and dilator agents in isolated beagle coronary arteries, *Naunyn-Schmiedeberg's Arch. Pharmacol.,* 336, 359, 1987.
91. **Nyborg, N. C. B. and Mikkelsen, E. O.,** Serotonin response increases with age in rat coronary resistance arteries, *Cardiovasc. Res.,* 22, 131, 1988.

92. **Nyborg, N. C. B.,** Ageing is associated with increased 5-HT_2-receptor affinity and decreased receptor reserve in rat isolated coronary arteries, *Br. J. Pharmacol.,* 102, 282, 1991.
93. **Blauw, G. J., Van Brummelen, P., Chang, P. C., Vermeij, P., and Van Zwieten, P. A.,** Arterial dilation and venous constriction induced by serotonin in the elderly, *Drugs,* 36, 74, 1988.
94. **Docherty, J. R. and Hyland, L.,** 5-Hydroxytryptamine receptors in the human saphenous vein, *Br. J. Pharmacol.,* 86, 735P, 1985.
95. **Wakabayashi, I., Sakamoto, K., Hatake, K., Yoshimoto, S., and Kurahashi, M.,** Effect of age on contractile response to angiotensin II in rat aorta, *Life Sci.,* 47, 771, 1990.
96. **Shams, G., Wallace, L. J., Miller, D. D., and Feller, D. R.,** Effects of thromboxane A_2 on thoracic aorta of young and old rats: use of selective thromboxane receptor antagonists, *Pharmacology,* 40, 27, 1990.
97. **Mayhan, W. G., Faraci, F. M., Baumbach, G. L., and Heistad, D. D.,** Effects of aging on responses of cerebral arterioles, *Am. J. Physiol.,* 258, H1138, 1990.
98. **Furchgott, R. F.,** The role of endothelium in the responses of vascular smooth muscle to drugs, *Annu. Rev. Pharmacol. Toxicol.,* 24, 175, 1984.
99. **Wanstall, J. C., O'Donnell, S. R., and Zheng, X.-P.,** Age-dependence of the effects of pinacidil on rat aorta, *J. Pharm. Pharmacol.,* 41, 641, 1989.
100. **Moritoki, H., Tanioka, A., Maeshiba, Y., Iwamoto, T., Ishida, Y., and Araki, H.,** Age-associated decrease in histamine-induced vasodilatation may be due to reduction of cyclic GMP formation, *Br. J. Pharmacol.,* 95, 1015, 1988.
101. **Emmick, J. T. and Cohen, M. L.,** Aging and vasodilation to atrial peptides, *Clin. Exp. Theory Pract.,* A8, 75, 1986.
102. **Duckles, S. P.,** Influence of age on relaxation of rat blood vessels to atriopeptin II, *J. Pharmacol. Exp. Ther.,* 240, 697, 1987.
103. **Kawai, Y. and Ohhashi, T.,** Age-related changes in relaxant response of vascular smooth muscles to atrial natriuretic peptide, *J. Pharmacol. Exp. Ther.,* 252, 1234, 1990.
104. **Docherty, J. R., Fitzgerald, D., and O'Malley, K.,** Age-related reduction of baroreflex tachycardia without loss of β-adrenoceptor-mediated tachycardia in Sprague-Dawley rats, *J. Cardiovasc. Pharmacol.,* 8, 376, 1986.
105. **Docherty, J. R.,** Age-related differences in depressor responses to sodium nitroprusside and nifedipine in anaesthetized rats, *J. Auton. Pharmacol.,* 10, 49, 1990.
106. **Palmer, R. M. J., Ferrige, A. G., and Moncada, S.,** Nitric oxide release accounts for the biological activity of endothelium-derived relaxing factor, *Nature (London),* 327, 524, 1987.
107. **Ignarro, L. J., Buga, G. M., Wood, K. S., Byrns, R. E., and Chaudhuri, G.,** Endothelium-derived relaxing factor produced and released from artery and vein is nitric oxide, *Proc. Natl. Acad. Sci. U.S.A.,* 84, 9265, 1987.
108. **Kelliher, G. J. and Conahan, S. T.,** Changes in vagal activity and response to muscarinic receptor agonists with age, *J. Gerontol.,* 35, 842, 1980.
109. **Nyborg, N. C. B. and Nielsen, P. J.,** The level of spontaneous myogenic tone in isolated human posterior ciliary arteries decreases with age, *Exp. Eye Res.,* 51, 711, 1990.
110. **Martin, W., Furchgott, R. F., Villani, G. M., and Jothianandan, D.,** Depression of contractile responses in rat aorta by spontaneously released endothelium-derived relaxing factor, *J. Pharmacol. Exp. Ther.,* 198, 529, 1986.
111. **MacLennan, W. J., Hall, M. R. P., and Timothy, J. I.,** Postural hypotension in old age: is it a disorder of the nervous system or of blood vessels?, *Age Ageing,* 9, 25, 1980.
112. **Smith, S. A. and Fasler, J. J.,** Age-related changes in autonomic function: relationship with postural hypotension, *Age Ageing,* 12, 206, 1983.
113. **Lipsitz, L. A.,** Altered blood pressure homeostasis in advanced age: clinical and research implications, *J. Gerontol.,* 44, M179, 1989.

114. **Leithe, M. E., Hermiller, J. B., Magorien, R. D., Unverferth, D. V., and Leier, C. V.,** The effect of age on central and regional hemodynamics, *Gerontology,* 30, 240, 1984.
115. **Richardson, D.,** Effects of age on cutaneous circulatory response to direct heat on the forearm, *J. Gerontol.,* 44, M189, 1989.
116. **Smith, J. J., Hughes, C. V., Ptacin, M. J., Barney, J. A., Tristani, F. E., and Ebert, T. J.,** The effect of age on hemodynamic response to graded postural stress in normal man, *J. Gerontol.,* 42, 406, 1987.
117. **Cox, R. H., Bagshaw, R. J., and Detweiler, D. K.,** Arterial baroreceptor reflexes in young and old racing greyhounds, *Am. J. Physiol.,* 240, H383, 1981.
118. **Hajduczok, G., Chapleau, M. W., Johnson, S. L., and Abboud, F. M.,** Increase in sympathetic activity with age. I. Role of impairment of arterial baroreflexes, *Am. J. Physiol.,* 260, H1113, 1991.
119. **Hajduczok, G., Chapleau, M. W., and Abboud, F. M.,** Increase in sympathetic activity with age. II. Role of impairment of cardiopulmonary baroreflexes, *Am. J. Physiol.,* 260, H1121, 1991.
120. **Tuma, R. F., Irion, G. L., Vasthare, U. S., and Heinel, L. A.,** Age-related changes in regional blood flow in the rat, *Am. J. Physiol.,* 249, H485, 1985.
121. **Hachamovitich, R., Wicker, P., Capasso, J. M., and Anversa, P.,** Alterations in coronary blood flow and reserve with aging in Fischer 344 rats, *Am. J. Physiol.,* 256, H66, 1989.
122. **Wei, J. Y., Li, Y.-X., and Ragland, J.,** Effect of exercise training on resting blood pressure and heart rate in adult and aged rats, *J. Gerontol.,* 42, 11, 1987.
123. **Haidet, G. C. and Parsons, D.,** Reduced exercise capacity in senescent beagles: an evaluation of the periphery, *Am. J. Physiol.,* 260, H173, 1991.
124. **Van der Zee, C. E. E. M., Schuurman, T., Gerritsen van der Hoop, R., Traber, J., and Gispen, W. H.,** Beneficial effect of nimodipine on peripheral nerve function in aged rats, *Neurobiol. Aging,* 11, 451, 1990.
125. **Dastur, D. K.,** Cerebral blood flow and metabolism in normal human aging, pathological aging, and senile dementia, *J. Cereb. Blood Flow Metab.,* 5, 1, 1985.
126. **Itoh, M., Hatazawa, J., Miyazawa, H., Matsui, H., Meguro, K., Yanai, K., Kubota, K., Watanuki, S., Ido, T., and Matsuzawa, T.,** Stability of cerebral blood flow and oxygen metabolism during normal aging, *Gerontology,* 36, 43, 1990.
127. **Waldemar, G., Hasselbalch, S. G., Andersen, A. R., Delecluse, F., Petersen, P., Johnsen, A., and Paulson, O. B.,** ^{99m}Tc-*d,l*-HMPAO and SPECT of the brain in normal aging, *J. Cereb. Blood Flow Metab.,* 11, 508, 1991.
128. **Martin, A. J., Friston, K. J., Colebatch, J. G., and Frackowiak, R. S. J.,** Decreases in regional cerebral blood flow with normal aging, *J. Cereb. Blood Flow Metab.,* 11, 684, 1991.
129. **Pantano, P., Baron, J.-C., Lebrun-Grandie, P., Duquesnoy, N., Bousser, M.-G., and Comar, D.,** Regional cerebral blood flow and oxygen consumption in human aging, *Stroke,* 15, 635, 1984.
130. **Leenders, K. L., Perani, D., Lammerstma, A. A., Heather, J. D., Buckingham, P., Healy, M. J. R., Gibbs, J. M., Wise, R. J. S., Hatazawa, J., Herold, S., Beaney, R. P., Brooks, D. J., Spinks, T., Rhodes, C., Frackowiak, R. S. J., and Jones, T.,** Cerebral blood flow, blood volume and oxygen utilization-normal values and effect of age, *Brain,* 113, 27, 1990.
131. **Black, J. E., Polinsky, M., and Greenough, W. T.,** Progressive failure of cerebral angiogenesis supporting neural plasticity in aging rats, *Neurobiol. Aging,* 10, 353, 1989.
132. **Buchweitz-Milton, E. and Weiss, H. R.,** Perfused capillary morphometry in the senescent brain, *Neurobiol. Aging,* 8, 271, 1987.
133. **Buchweitz-Milton, E. and Weiss, H. R.,** Cerebral oxygen consumption and blood flow in Fischer-344 rats of different ages, *Neurobiol. Aging,* 8, 55, 1987.

134. **Hajdu, M. A., Heistad, D. D., Siems, J. E., and Baumbach, G. L.,** Effects of aging on mechanics and composition of cerebral arterioles in rats, *Circ. Res.*, 66, 1747, 1990.
135. **Rieke, G. K. and Cannon, M. S.,** A histochemical study of cerebral cortical vessels and ganglionic vessels of the caudatoputamen in aging normotensive rats, *Stroke,* 16, 285, 1985.
136. **Topple, A., Fifkova, E., Baumgardner, D., and Cullen-Dockstader, K.,** Effect of age on blood vessels and neurovascular appositions in the CA1 region of the rat hippocampus, *Neurobiol. Aging,* 12, 211, 1991.
137. **Topple, A., Fifkova, E., and Cullen-Dockstader, K.,** Effect of age on blood vessels and neurovascular appositions in the rat dentate fascia, *Neurobiol. Aging,* 11, 371, 1990.
138. **Mooradian, A. D.,** Effect of aging on the blood-brain barrier, *Neurobiol. Aging,* 9, 31, 1988.
139. **Harik, S. I.,** Some comments on the ''Effect of aging on the blood-brain barrier'', *Neurobiol. Aging,* 9, 50, 1988.
140. **Mayhan, W. G.,** Disruption of blood-brain barrier during acute hypertension in adult and aged rats, *Am. J. Physiol.*, 258, H1735, 1990.
141. **Lipsitz, L. A.,** Orthostatic hypotension in the elderly, *N. Engl. J. Med.*, 321, 952, 1989.
142. **Niarchos, A. P. and Laragh, J. H.,** Hypertension in the elderly. I. Pathophysiology, *Mod. Conc. Cardiovasc. Dis.*, 49, 43, 1980.
143. **Niarchos, A. P. and Laragh, J. H.,** Hypertension in the elderly. II. Diagnosis and treatment, *Mod. Conc. Cardiovasc. Dis.*, 49, 43, 1980.
144. **Gwathmey, J. K., Slawsky, M. T., Perreault, C. L., Biggs, G. M., Morgan, J. P., and Wei, J. Y.,** Effect of exercise conditioning on excitation-contraction coupling in aged rats, *J. Appl. Physiol.*, 69, 1366, 1990.
145. **Tate, C. A., Taffet, G. E., Hudson, E. K., Blaylock, S. L., McBride, R. P., and Michael, L. H.,** Enhanced calcium uptake of cardiac sarcoplasmic reticulum in exercise-trained old rats, *Am. J. Physiol.*, 258, H431, 1990.
146. **Duckles, S. P.,** unpublished observations.
147. **Gonzalez, R. and Duckles, S. P.,** unpublished results.

Chapter 9

THE RESPIRATORY SYSTEM

Charles Brink

TABLE OF CONTENTS

0-8493-6981-9/93/$0.00 + $.50

I. INTRODUCTION

The adult lung is composed of a series of tissue networks (nerves, airways, blood vessels, and lymphatics) and a considerable number of cell populations (fixed and transitory). These components act in an integrated fashion and permit the lung to satisfy the O_2/CO_2 metabolic demands of the organism by extracting O_2 from ambient air and eliminating the CO_2 produced. In addition to this gas-exchange activity, the lung is also considered to be a metabolically active and immunologically responsive organ. How these three pulmonary functions are controlled and altered during development and aging are essentially unknown.

The aim of this chapter is to (1) describe the autonomic inputs to the lung, (2) present evidence which supports their role in pulmonary function, and (3) highlight some of the changes which occur in this organ during development and aging.

II. DEVELOPMENTAL ASPECTS

The adult human lung is derived from two embryonic sources: (1) an enterodermal airway tube, originally composed of columnar epithelial cells, and (2) a mass of mesenchymal cells associated with the splanchnopleura. The epithelial tube, which is present when the human fetus is approximately 3 mm in length, develops continuously throughout the complex period of morphogenesis and leads to the complete differentiated alveolated organ (Figure 1). The mesenchymal cell mass evolves into two portions: (1) the continuous endothelial lining found in all blood vessels and (2) a group of interstitial cells which are associated with the fibrous connective tissue skeleton extending from the hilum to the pleural surfaces. In the final stages of development, a number of cells of extrapulmonary origin such as macrophages, mast cells, and plasma cells are also found in the lung. The initial morphometric findings have led to a general scheme[2] of human lung growth *in utero* (Figure 2). This classification suggests an early period in which budding and branching occurs, leading eventually to complete bronchial formation. As development progresses, the newly formed bronchi divide and the amount of connective tissue between the airways decreases. Vascularization occurs during this period, leading to the development of the airspaces. Even though the airway epithelium flattens out and there is an increased number of blood vessels in the lung, there remains sufficient tissue between the epithelium and the vessels to prevent adequate gas exchange. However, the primitive airways are open at all stages of lung development and at no point is the tissue solid.[3] In the last week of gestation, when alveolar formation begins, the elastic fiber network becomes extensive and quite dense.[4] From the period just prior to birth to the age of 7 in humans,[2,5-9] there is an increase in alveoli. This phenomenon was thought to be associated with the disappearance of conducting airways;[10] however, recent evidence suggests that this may not

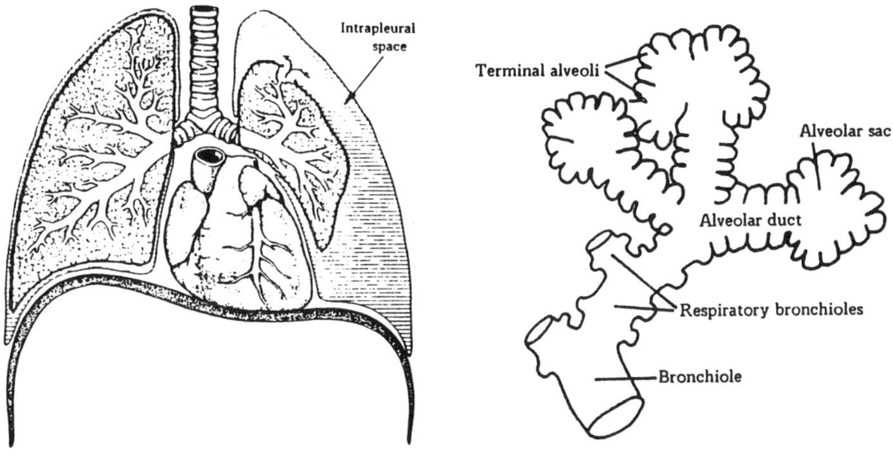

FIGURE 1. The adult differentiated alveolated lung.

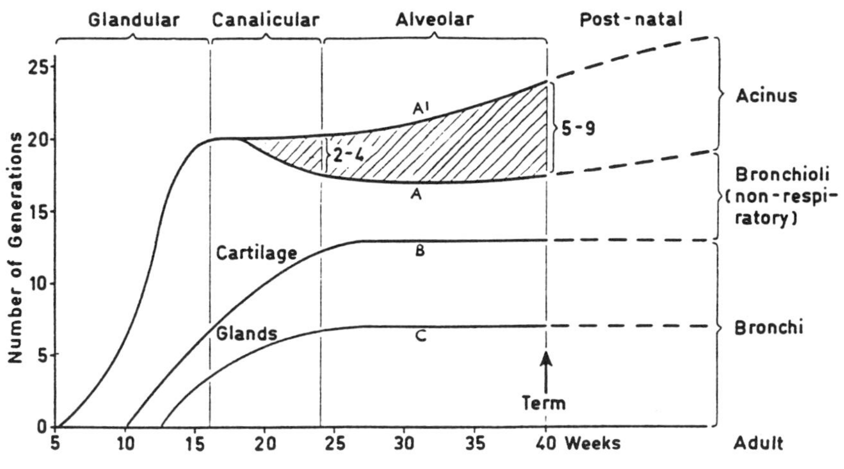

FIGURE 2. Summary of intrauterine development of the intrasegmental bronchial tree. (From Butcher, U. and Reid, L., *Thorax,* 16, 207, 1961. With permission.)

be the case in humans.[11] Langston and co-workers[11] have reclassified the major phases of human intrauterine lung development. According to their data (Table 1), the major components in lung development suggest that normal infants are more mature at birth than has previously been suggested.[10] While this classification is useful for understanding the stages of development, these stages are not precisely defined and considerable overlap in tissue development may occur within the lung.

While there are a considerable number of studies concerning the generation of various structures which develop in the lung, few reports have been

TABLE 1
Phases of Intrauterine Lung Development

Phase	Gestation period	Major event
Embryonic	26 d–6 weeks	Development of major airways
Pseudoglandular	6–16 weeks	Development of airways to terminal bronchioles
Canalicular	16–28 weeks	Development of acinus and its vascularization
Saccular	28–36 weeks	Subdivision of saccules by secondary crests
Alveolar	36 weeks–term	Alveolar acquisition

From Langston, C., Kida, K., Reed, M., and Thurlbeck, W. M., *Am. Rev. Respir. Dis.*, 129, 607, 1984. With permission.

published which describe the neuronal pathways in this organ. Generally, lung innervation in mammals occurs at an early embryonic stage. In the human fetus of approximately 35 mm, bundles of nerve fibers are initially observed around the trachea. By the time the fetus has tripled in size, these nerve fibers have penetrated the tracheal wall to form an intrachondral plexus.[12] The nerves are also associated with other structures in newborns, such as mucous glands.[13] Histochemical studies in tissues from the adult human lung have clearly demonstrated that a similar pattern exists, since nerve fibers are also found in the trachea and close to the glands.[14-20] These data have been confirmed using immunohistochemical techniques for the localization of nerves in the human lung during aging.[21] Taylor and Smith[12] initially reported that nerve fibers to pulmonary arteries and veins could be observed in the 65-mm-long fetus. Other investigators have shown that adult human pulmonary vasculature is innervated by noradrenergic and cholinergic nerves.[22] However, these nerves contain several other putative neurotransmitters, including peptides, which may have a regulatory role in the pulmonary circulation.[23,24] Hislop and co-workers[25] demonstrated a considerable difference in the relative number of immunoreactive nerves present in the human lung during development. Using both general and specific neural markers, they observed that between birth and 3 years of age there was no change in the patterns. However, in the adult lungs, the number and density of neuropeptide-containing nerves decreased compared with the results observed in the neonatal lungs. This reduction of neuropeptide content in human lung tissue had previously been reported by Laitinen and co-workers.[20] While these data demonstrated that innervation of the airways appears to decrease with age, the innervation of the arteries in the lung is the opposite. While one third of the arteries associated with the alveolar ducts were not innervated just after birth, all were innervated by the age of 3 to 4 years, and in the adult lung these distal airways were well innervated.[19] Allen and co-workers[23] have shown that the human pulmonary veins were less densely innervated than the pulmonary artery. Thus, innervation of the human pulmonary vasculature generally increased during aging.

However, the increase was more pronounced in the arteries than in the veins. While the immunocytochemical techniques have allowed an identification of the peptide-containing nerves in the lung, their function during development and aging has been less frequently explored.

III. AUTONOMIC NERVES IN THE LUNG

The development of pulmonary innervation in the human lung is still not completely characterized and understood. However, the pattern of innervation in the adult human lung shows that the autonomic nerve supply to this organ is from the sympathetic trunk and the vagus (parasympathetic) nerve. The upper portion of the respiratory tract (trachea) receives nerve branches from the recurrent and superior laryngeal nerves. The nerve trunks which enter the lung divide into the periarterial and peribronchial plexuses. The latter plexus branches into an extrabronchial plexus which is external to the cartilage, and a subcondrial plexus lying between the cartilage and the epithelium. The preganglionic fibers end at the peribronchial ganglia in both large and small airways, where these terminals synapse with the cell bodies of the postganglionic parasympathetic nerves. The axons of these postganglionic nerves travel a short distance and end at or in close association with a variety of effector cells such as airway smooth muscle cells, mucous glands, epithelial cells, and mast cells. The autonomic nerves have been classified according to the neurotransmitter released at the effector organ (cholinergic and adrenergic).

A. CHOLINERGIC TRANSMISSION

Both electron microscopy and histochemical studies[26] have shown that the postganglionic terminals, which are highly concentrated in the proximal airways, release acetylcholine. Stimulation of the vagal parasympathetic cholinergic system provokes bronchoconstriction,[27,28] mucous secretion,[29,30] and ciliary activity,[31] effects which are abolished by atropine.[32] In addition, atropine and other anticholinergic drugs are known to induce a bronchodilator response in adults,[33,34] suggesting that there is vagal tone in normal airways. This tone is due to the release of acetylcholine, since inhalation of an acetylcholinesterase inhibitor produces bronchoconstriction.[35] In human isolated airway muscle preparations, the postganglionic cholinergic innervation has been examined by electrical field stimulation.[36,37] The contraction induced by electrical field stimulation is blocked by atropine and tetrodotoxin (nerve blocker), suggesting the presence of cholinergic nerves in the isolated bronchial tissues.[38] However, the contraction induced by electrical field stimulation is less in bronchioles than in bronchi.[39] These results corroborate histochemical evidence suggesting that the distal airways are less innervated than the proximal bronchial airways.[40] In addition, recent evidence using autoradiographic mapping[41] suggests that the density of muscarinic receptors in the peripheral human lung is also less than that observed in the central large airways. Similarly, *in vivo,* flow-volume curves demonstrated that anticholinergic drugs induce a greater bronchodilation in larger than in small airways.[42]

The vagal nerve not only regulates airway tone, but stimulation also increases mucous secretion from the submucosal glands both *in vivo* and *in vitro*.[43,44] In fact, electrical field stimulation of the human bronchial mucosa using the Ussing chamber technique has demonstrated that mucous production under these conditions can be blocked by atropine and tetrodotoxin.[45] In addition, cholinergic agonists are potent stimulators of submucosal gland secretion[46] as well as ion transport across airway epithelium in animals.[47] Similar effects have also been obtained in human airways *in vitro*.[30,48,49] Data derived from animal studies with cholinergic agonists indicate that these agents stimulate both mucous and serous cells of submucosal glands,[50] leading to an increase in the volume of mucous secreted without affecting viscosity.[51] However, atropine, which has a small effect on sputum volume, has little effect on viscosity[52] and only a minor inhibitory effect on mucociliary transport in normal subjects.[53]

Innervation of the adult human lung pivots around the cholinergic system, which controls airway caliber by directly altering airway muscle tone or by narrowing the airways via mucosal edema and/or excessive mucosal production. However, in the embryonic and/or fetal lung, little is known about the cholinergic input at the airway smooth muscle or at the level of the mucosal glands during development.

B. ADRENERGIC TRANSMISSION

In the adult human lung, the sympathetic nerve supply originates from the upper six thoracic segments of the spinal cord, and postganglionic fibers synapse in the middle and inferior cervical ganglia and the upper four thoracic ganglia. Postganglionic fibers from these ganglia enter the lung at the hilum and run with the cholinergic nerves to form a dense plexus around airways and vessels.[14,18,26] β-Adrenergic nerves in the adult lung, while present, are sparse compared with cholinergic fibers. In the adult human lung, adrenergic innervation has been demonstrated by fluorescence histochemistry. These fibers have been found in close association with submucosal glands[18,54] and bronchial arteries,[18,55] but have not been demonstrated in smooth muscle of intrapulmonary airways.[18,55,56] These data have been confirmed[57] using an antibody to dopamine-β-hydroxylase (highly selective for adrenergic nerves). Adrenergic nerve terminals end on the ganglionic cells of the peribronchial plexus, suggesting that the β-adrenergic system may influence airway tone via modulation of parasympathetic activity at the level of the postganglionic cells.[58] Data from *in vivo* studies with β-blockers indirectly suggest that an endogenous control of cholinergic input may exist. These results are derived from the effects of β-blockers in asthmatics, where pindolol or timolol induce bronchoconstriction in asthmatics and patients with COPD (chronic obstructive pulmonary disease).[59-61] This effect is not observed in normal subjects, since β-antagonists have no effect on basal tone in healthy individuals.[62-64] These data suggest an increased adrenergic input to the airways in asthmatics,[64,65] an effect which may be provided by the circulating catecholamine adrenaline, since there is an absence of functional sympathetic nerves at the airway level. Indeed, results derived from a number of studies have shown that electrical stimulation of sympathetic fibers produced a slow bronchodilation,[55,66,67] an effect that depended on the level of vagal tone. Sympathetic stimulation induced marked bronchodilation when vagal tone was present and little or no relaxant effect when vagal activity was blocked.[69] In addition, β-adrenergic blockade had little bronchoconstrictor effect when vagal activity was normal[70] or had been abolished by atropine.[71] In contrast, β-blockers produced a significant bronchoconstriction only under conditions where vagal activity was greater than normal.[72] These findings suggest that the sympathetic system works via the parasympathetic system, i.e., sympathetic bronchodilation of normal airways depends on the level of intrinsic parasympathetic tone.

While little is known about the function and connection of the vagal nuclei in the brainstem, the assumption has been made that they receive inputs from higher brain centers. The vagal nuclei also receive inputs from the pulmonary vagal afferents which form the afferent limb of a reflex whose motor limb is the vagal efferent supply.

Vagal afferents are found throughout the entire human respiratory tract. Reflex bronchospasm can result from stimuli originating anywhere in the nasopharynx,[73] airways,[74] and esophagus.[75] At least three types of receptors of vagal afferent nerves have been described. The first type is the slowly adapting receptor (stretch) found mainly in smooth muscle. These receptors are generally stimulated by changes in tension and are responsible for the Hering-Breuer inflation reflex, which inhibits sustained inspiratory activity. Their tonic discharge at high lung volumes prolongs expiration.[76] Stimulation of these receptors also leads to a reflex bronchodilation by inhibiting vagal tone. Stretch-adapting receptors may also be responsible for the bronchodilator response to lung inflation in humans, particularly after induced bronchoconstriction.[77,78] The second type of vagal afferent receptor is the rapidly adapting or "irritant" receptor, which is located beneath the tight junction between epithelial cells.[17,79] While the extrapulmonary irritant receptors are sensitive to mechanical stimuli (particulates, cigarette smoke), the intrapulmonary irritant receptors are triggered by chemicals (gases, mediators). The third type of afferent receptor is the nonmyelinated C-fiber, which is also located in the epithelial layer.[17,80] The C-fibers play a role in reflex bronchoconstriction. The reflex bronchospasm via the vagal afferent and efferent pathway has been supported by results obtained using SO_2 delivered to the isolated upper airway, which results in constriction of the lower airways,[81] or allergen when delivered to only one lung, which produced a bilateral bronchoconstriction.[82]

In addition, histamine-induced bronchoconstriction, which is largely prevented by vagotomy,[70] atropine, or the ganglionic blocking agent hexamethonium, provides further support for the existence of C-fibers in the lung. However, the effect of the muscarinic receptor agonist methacholine is blocked by atropine but not hexamethonium.[83]

C. NONADRENERGIC, NONCHOLINERGIC TRANSMISSION

The preganglionic fibers of the nonadrenergic, noncholinergic nerves follow those of the vagal parasympathetic branches, and the postganglionic fibers terminate on smooth muscle cells, where their activation results in long-lasting relaxation. Since this effect is not blocked by adrenergic or cholinergic antagonists, the system has been referred to as NANC.[26,56,84,85] NANC nerves also stimulate mucous secretion, at least in the ferret,[86] mediate relaxation of pulmonary vascular cells,[87,88] and may regulate mediator release from mast cells.[89] Considerable evidence suggests that in this nervous system, the neurotransmitters for the nonadrenergic inhibitory nerves, which relax airway smooth muscle, are vasointestinal peptide (VIP)[90] and peptide histidine methionine (PHM).[91] Substance P (SP) and calcitonin gene-related peptide (CGRP) are the principal neurotransmitters in the nonadrenergic excitatory nerves. Stimulation of the latter system induces a bronchoconstrictory response which is not blocked by atropine. In human airways *in vitro,* a contractile response to SP can be demonstrated,[92] but *in vivo* SP infusions have profound effects on the cardiovascular system at doses which have little or no effect on airway function.[93] The peptides mentioned above and other peptides (neuropeptide Y, galanin, gastrin-releasing peptide), have been demonstrated in the adult lung. However, there is a paucity of information as to whether they are found in the embryonic or fetal lung. Many of these peptides have potent effects on a variety of tissue and/or cellular components in the adult lung; however, their actions on fetal and neonatal lungs have been largely neglected.

IV. AUTONOMIC RECEPTORS IN THE LUNG

For a neurotransmitter such as acetylcholine to activate a target tissue such as airway smooth muscle, the agent must first be bound in a highly specific manner by specific recognition sites. When the receptor complex is formed, the binding proteins must also possess the ability to initiate a series of membrane and intracellular signals which eventually lead to the biological response.

A. CHOLINERGIC RECEPTORS

Activation of vagal efferent neurons by central discharges from the vagal nuclei in the brain cause the release of acetylcholine from preganglionic fibers located in the airway ganglia. At the ganglionic neurons, acetylcholine stimulates nicotinic cholinergic receptors.[94] A recent report[95] has shown that nicotinic agonists such as dimethylphenylpiperazinium caused contraction of airway smooth muscle if the ganglia were intact, a response which was inhibited by hexamethonium, the nicotinic antagonist.[94] Stimulation of these receptors generates a series of fast excitatory postsynaptic electrical potentials leading to an activation of the short postganglionic neurons. However, there is also a series of slow inhibitory postsynaptic potentials which determine the firing rate of the postganglionic neurons, and these potentials are inhibited by atropine and, more selectively, by pirenzepine, suggesting the presence of a muscarinic site on these ganglia. The presence of muscarinic sites in airway ganglia has received some support from experiments employing pirenzepine (M_1 antagonist), which blocks the bronchoconstriction provoked by vagal nerve stimulation[95-98] at lower concentrations than those required to inhibit the vagally induced decrease in heart rate. Since airway muscles have a low affinity for this antagonist,[96] the suggestion has been made that M_1 receptors found on airway ganglia may be responsible for controlling airway muscle tone. Results obtained in the rabbit provide support for this notion, since in this species, pirenzepine has been shown to be a potent antagonist of preganglionic vagal stimulation when compared with postganglionic vagal stimulation.[95] In humans, pirenzepine has been reported to be a bronchodilator when injected intravenously,[99] albeit at doses which are probably not M_1 selective. However, pirenzepine given by aerosol inhibited the reflex bronchoconstriction provoked by sulfur dioxide without attenuating the bronchoconstriction provoked by methacholine.[97] However, the presence and role of M_1 receptors remains to be established, since only a limited number of investigations have been performed to examine the role of these receptors in airway diseases.[100]

Conduction through the airway ganglia leads to activation of the postganglionic fibers and release of acetylcholine at the level of the airway muscle. The binding of acetylcholine to muscarinic receptors on effector cells is well documented. Autoradiographic studies of adult human lung have demonstrated high densities of muscarinic receptors on central airway smooth muscles, submucosal glands, parasympathetic ganglia,[41] and in the alveolar region of the human lung.[101] In pharmacological studies on isolated tissue preparations using various antagonists, prejunctional and postjunctional muscarinic sites have been identified.[102,103] In fact, data from the airways of various species have shown inhibition of the release of acetylcholine from prejunctional sites at the level of the airway effector cells, since blockade of these receptors by galamine (M_2 antagonist) or AF-DX 116[96] potentiated the contraction provoked by electrical field stimulation, and in human isolated airways, the selective M_2 agonist pilocarpine inhibited the contractions induced by electrical field stimulation, although this effect was quite variable and some preparations were not at all affected by this agonist.[98] In human subjects, Minette and Barnes[98] have demonstrated that pilocarpine blocked the bronchoconstriction induced by sulfur dioxide, while in asthmatics no such effect was observed.

While these data indicate that cholinergic innervation of the respiratory tract in the adult human lung is regulated by the action of the neurotransmitter on muscarinic receptors (M_1, M_2, and M_3), little information is available concerning the presence of these sites in fetal or neonatal lungs. Recently, Wills and Douglas[104] have shown that the response of bovine airway muscle preparations is significantly reduced during aging, i.e., a decrease in sensitivity of these tissues to muscarinic agonists was observed. In addition, other reports[105-107] demonstrated a significant reduction in binding sites and loss of acetylcholinesterase activity in membrane fractions from bovine lung. These results suggest major alterations in the receptors and those enzymes associated with the breakdown of this neurotransmitter. Such modifications suggest that there are marked changes in neuronal inputs and regulation in the lung during aging. In fact, the therapeutic value of anticholinergic agents such as ipratropium bromide in young infants is at best controversial.[108-110] Indeed, ipratropium bromide was ineffective in children suffering from bronchiolitis.[108,109] These clinical results are rather surprising, since anticholinergic agents affect central airways in adults,[33] which coincides with the maximum density of the muscarinic sites in the adult lung.[41] In contrast, Prendiville and co-workers[110] have recently shown that the effects of this anticholinergic agent were similar to those observed in adult patients, i.e., there was a significant reduction in specific airway resistance after ipratropium bromide nebulization in wheezing infants. However, in this study, most infants failed to show any improvement in both peak expiratory flow rate and V_{max}FRC, suggesting the poor bronchodilatory activity of anticholinergics in wheezing infants. Further studies are needed to clarify these differences which have been reported for anticholinergics in infants.

In man, a major part of the cholinergic innervation is regulated by enzymes which are responsible for degradation of the neurotransmitter acetylcholine. Acetylcholinesterase and pseudocholinesterase (butrylcholinesterase) hydrolyze the cholinergic neurotransmitter acetylcholine. Pseudocholinesterase activity has been determined in normal adults and neonates, and the levels in the neonate may be as low as 40% of the normal adult. However, the physiological or functional significance of these enzyme differences in airway responses has only recently been explored in the adult lung.[111,112]

In neonates, the relative absence of muscle tone can easily be demonstrated by artificial ventilation of the lungs under light anesthesia.[113] In small babies, inflation of the lungs can be achieved by surprisingly low pressures, in marked contrast to the resistance to inflation which the adult patient exhibits under similar conditions. In this respect, the behavior of the lightly anesthetized infant bears a striking resemblance to the curarized and anesthetized adult. The use of suxamethonium as a muscle relaxant in neonates as an anesthetic agent has demonstrated that newborns are resistant to this agent.[114] Stead[114] initially reported the extreme sensitivity of the neonate to D-tubocurarine, demonstrating that infants exhibited a myasthenic type of reaction subsequent to the administration of anesthesia. In fact, another report[115] had shown that reversal of the action of D-tubocurarine by neostigmine was doubtful in the neonate. Thus, in neonates, there appears to be a reciprocal relationship between competitive blocking and depolarizing agents. These neonatal data suggest that the cholinergic system may be altered during aging, since these effects are not observed in adult patients.

B. ADRENERGIC RECEPTORS

There is considerable evidence for a role of the adrenergic system in the human lung not only in the adult, but also in fetal lungs prior to and during birth. Clinical evidence suggests that infants delivered by cesarean section without labor have a higher incidence of breathing difficulties,[116] less mature biochemical indices of lung maturity,[117,118] and lower tidal volumes and lung compliance[119] than do other infants. In contrast, infants born to mothers who received β-adrenergic agonists immediately before delivery are reported to have a lower incidence of hyaline membrane disease.[120] Data derived from a variety of animal species indicate that β-adrenergic responses are involved in fetal lung maturation.[121] Davis and co-workers[122] demonstrated that β-adrenergic receptors are present and functional in human fetal lungs as early as the second trimester of gestation and that their concentration increases with gestational age. However, the receptor concentration throughout the second trimester remains significantly lower than that reported for the adult human lung.[123] In contrast, Falkay and co-workers[124] failed to show an increase in the concentration of β-adrenergic receptors with advancing gestational age. These differences may be due to the membrane fractions obtained for the radioligand binding techniques which were used. However, radioligand binding studies of membrane fractions from the adult human lung have consistently demonstrated a high density of β-receptors in this organ, and these data are supported by results obtained using autoradiographic techniques.[125-127] These results suggest that β-adrenoceptors increase during development and with age in the human lung. These data are similar to what has been reported in the guinea pig[128] and rat[129] during aging. Duncan and co-workers[130] also demonstrated that receptor density is increased from the proximal to the distal portions of the guinea pig lung. This distribution profile is exactly opposite that of the cholinoceptors (see above). Thus, for β-adrenoceptors, the highest densities are found in the periphery of the lung and the lowest in tracheal muscle. These receptor density data, at least in the guinea pig,[130] are at odds with the functional results, i.e., increased receptor density was never associated with increased tissue reactivity. Tracheal tissues taken from animals of any age were always more sensitive to isoproterenol than bronchial preparations, but the latter tissues had a fivefold greater density of β-adrenoceptors. In addition, reduced sensitivity to isoproterenol is not necessarily associated with increased receptor density. For example, bronchial tissues from old guinea pigs were less sensitive to isoproterenol than tissues derived from middle-aged animals, yet the receptor densities were identical in these preparations. Therefore, additional factors contribute to the reduced potency of isoproterenol which occurs in bronchial tissues during development. Since the stimulation of β-adrenoceptors involves the receptor molecule, with an agonist binding site on the outside of the cell membrane, linked by a GTP-binding protein to a catalytic unit (adenylate cyclase), the physiological response observed during aging may involve one or a combination of these three structures. While some workers suggested that the transduction events (receptor-G-protein coupling) may be altered during aging,[131,132] Vanscheeuwijck and co-workers[129] have shown that in the rat lung there is an increase in the number of receptors with age (data similar to that reported for the guinea pig), but that there was a reduced percentage of the high-affinity binding sites without alteration in cAMP production. Such data would imply that in the lung, aging is not accompanied by a decreased homeostasis including a reduction in the adrenoceptor-mediated responses. What remains to be determined are those mechanisms which are responsible for increasing receptor density. Previous studies *in vivo*[133] had shown that hydrocortisone reduced airway reactivity in guinea pigs by a mechanism involving β-adrenoceptors. Since radioligand binding studies have systematically demonstrated that receptor numbers increase during aging, the reasonable assumption was that hormonal levels may be responsible for the increased expression in β-receptors. The actions of hormonal treatment on fetal lung development of β-adrenoceptors are well known;[134] in fact, hormonal therapy is known to accelerate general lung maturation.[135,136] While β-adrenoceptors are present throughout the process of development and aging, what is surprising is the therapeutic inefficacy of salbutamol during infancy,[137,138] since in children and adults, β-agonists remain the principal therapy for patients with asthma.

V. GENERAL COMMENTS

While a considerable amount of information demonstrates that neuronal inputs in the respiratory tract play a role during development, maturation, and aging some data suggest that the role may be quite limited. Mortola and co-workers[139] demonstrated in rats and rabbits which were subjected to cervical vagotomy and sympathectomy after weaning that neither proximal nor distal airways were altered. Thus, the lungs on the denervated side showed no mechanical changes (static and dynamic lung compliance or lung resistance) or modifications in breathing patterns. These data are further supported by the results obtained in patients who have undergone heart-lung transplantation, since the airways are denervated due to severance of the cholinergic, adrenergic, and sensory nerves. Stretton and co-workers[140] have shown that the denervated airways from these patients have normally functioning postganglionic nerves and no change in the cholinergic responsiveness of receptors. Thus, the clinical denervation hypersensitivity which has been reported in heart-lung transplant patients to methacholine-induced provocation may be due to other factors such as pulmonary infection and/or processes associated with rejection of the lung tissues subsequent to surgical intervention.

However, previous studies in the adult animal had demonstrated that vagotomy produced modifications in type II alveolar cells and pulmonary surfactant.[141] In fact, vagal afferent activity is known to regulate the depth and frequency of breathing, while vagal efferent activity is reported to have direct trophic actions on lung tissue, at least in the fetal animal. The suggestion that neurotransmitters may act as trophic factors during development has received some support from several lines of investigation in rats.[142] The postulation that neurons provide signals which can regulate different phases of cell development[143] has received some support from the experiments performed by Slotkin and co-workers.[144,145] When 6-OHDA was injected systemically, the central lesions had little or no effect on receptor binding, but destroyed the peripheral nerve terminals immediately after birth, leading to impaired receptor acquisition. These data provided support for the notion that the presence of nerve terminals, and not neural activity, appeared critical in the receptor ontogeny which occurred in the rat lung. A second abnormality in lung characterized by sympathetic hyperreactivity was also reported by these same authors[146] subsequent to the 6-OHDA treatment. This appeared near the end of the first postnatal week and lasted for about 1 month. Thus, these investigators demonstrated that neural activity and the presence of the nerve terminals provide trophic inputs which influence development of sympathetic target organs. The ablation of nerve terminals (systemic 6-OHDA) or reductions in impulse flow as a result of central catecholaminergic lesions (intracisternal 6-OHDA) compromise developmental patterns of postsynaptic adrenergic receptors. These data suggest a role for the neuronal inputs at the level of the fetal respiratory tract and a diminished role after birth.

There are significant alterations in the autonomic nervous system within the respiratory tract during aging. These modifications result in dramatic changes in how the lung responds to neuronal inputs. While considerable effort has been expended to understand the adult respiratory tract, the developmental and neonatal periods have received less attention. Further physiological and pharmacological investigations at the neonatal level, at which period there are marked changes in pulmonary function, may help to elucidate the mechanisms involved in aging in the lung.

REFERENCES

1. **Weibel, E. R.,** Morphological basis of alveolar-capillary gas exchange, *Physiol. Rev.*, 53, 419, 1973.
2. **Butcher, U. and Reid, L.,** Development of the intrasegmental bronchial tree: the pattern of branching and development of cartilage at various stages of intra-uterine life, *Thorax*, 16, 207, 1961.
3. **Emery, J. L. and Mithal, A.,** The number of alveoli in the terminal respiratory unit of man during the late intrauterine life and childhood, *Arch. Dis. Child.*, 35, 544, 1960.

4. **Boyden, E. A.,** The pattern of the terminal airspaces in a premature infant of 30–32 weeks that lived nineteen and a quarter hours, *Am. J. Anat.*, 126, 31, 1969.
5. **Boyden, E. A.,** The terminal airsacs and their blood supply in a 37-day infant lung, *Am. J. Anat.*, 116, 413, 1965.
6. **Boyden, E. A.,** Notes on the development of the lung in infancy and early childhood, *Am. J. Anat.*, 121, 749, 1967.
7. **Boyden, E. A.,** The structure of the pulmonary acinus in a child of six years and eight months, *Am. J. Anat.*, 132, 275, 1971.
8. **Dunnill, M. S.,** Postnatal growth of the lungs, *Thorax,* 17, 329, 1962.
9. **Staub, N. C.,** The interdependence of pulmonary structure and function, *Anesthesiology,* 24, 831, 1963.
10. **Burri, P. H.,** Development and growth of the human lung, in *Handbook of Physiology, Section 3. The Respiratory System,* Vol. 1, *Circulation and Nonrespiratory Function,* Fishman, A. P. and Fisher, A. B., Eds., American Physiological Society, Bethesda, 1985, 1.
11. **Langston, C., Kida, K., Reed, M., and Thurlbeck, W. M.,** Human lung growth in late gestation and in the neonate, *Am. Rev. Respir. Dis.*, 129, 607, 1984.
12. **Taylor, I. M. and Smith, R. B.,** Intrinsic innervation of the human foetal lung between 35 and 140 mm crown-rump length stages, *Biol. Neonate,* 18, 193, 1971.
13. **Pessacq, T. P.,** The innervation of the lung of newborn children, *Acta Anat.*, 79, 93, 1971.
14. **Larsell, G. and Dow, L. S.,** The innervation of the human lung, *Am. J. Anat.*, 52, 125, 1933.
15. **Gaylor, J. B.,** The intrinsic nervous mechanism of the human lung, *Brain,* 57, 143, 1934.
16. **Spencer, H. and Leof, D.,** The innervation of the human lung, *J. Anat.*, 98, 599, 1964.
17. **Laitinen, A., Laitinen, L. A., Hervonen, A., and Partanen, M.,** Innervation of the human lower airway tract, *Eur. J. Respir. Dis.*, 63 (Suppl. 124), 83, 1982.
18. **Partanen, M., Laitinen, A., Hervonen, A., Toivanen, M., and Laitinen, L. A.,** Catecholamine and acetylcholinesterase-containing nerves in human lower respiratory tract, *Histochemistry,* 76, 175, 1982.
19. **Laitinen, A., Partanen, M., Hervonen, A., and Laitinen, L. A.,** Electron microscopic study of the innervation of the human lower respiratory tract, *Eur. J. Respir. Dis.*, 67, 209, 1985.
20. **Laitinen, L. A., Laitinen, A., Panula, P. A., Partanen, M., Tervo, K., and Tervo, T.,** Immunohistochemical demonstration of substance P in the lower respiratory tract of the rabbit and not in man, *Thorax,* 38, 531, 1983.
21. **Daniel, E. E., Kannan, M., Davis, C., and Posey-Daniel, V.,** Ultrastructural studies on the neuromuscular control of human tracheal and bronchial muscle, *Respir. Physiol.*, 63, 109, 1986.
22. **Amenta, F., Cavalloti, C., Ferrante, F., and Tonelli, F.,** Cholinergic innervation of the human pulmonary circulation, *Acta Anat.*, 117, 58, 1983.
23. **Allen, K. M., Wharton, J., Polak, J. M., and Haworth, S. G.,** A study of nerves containing peptides in the pulmonary vasculature of healthy infants and children and of those with pulmonary hypertension, *Br. Heart J.*, 62, 353, 1989.
24. **Sheppard, M. N., Marangos, P. J., Bloom, S. R., and Polak, J. M.,** Neuron specific enolase: a marker for the early development of nerves and endocrine cells in the human lung, *Life Sci.*, 34, 265, 1984.
25. **Hislop, A. A., Wharton, J., Allen, K. M., Polak, J. M., and Haworth, S. G.,** Immunohistochemical localization of peptide-containing nerves in human airways: age related changes, *Am. J. Respir. Cell. Mol. Biol.*, 3, 191, 1990.
26. **Richardson, J. B.,** Nerve supply to the lung, *Am. Rev. Respir. Dis.*, 119, 785, 1979.
27. **Colebatch, H. J. H. and Halmagyi, D. F. J.,** Effects of vagotomy and vagal stimulation on lung mechanics and circulation, *J. Appl. Physiol.*, 18, 881, 1963.

28. **Olsen, C. R., Colebatch, H. J. H., Mebel, P. E., Nadel, J. A., and Staub, N. C.,** Motor control of pulmonary airways studied by nerve stimulation, *J. Appl. Physiol.*, 20, 202, 1965.
29. **Pavia, D. and Thomson, M. L.,** Inhibition of mucocillary clearance from the human lung by hyoscine, *Lancet,* 1, 449, 1971.
30. **Sturgess, J. and Reid, L.,** An organ culture study of the effects of drugs on the secretory activity of the human bronchial submucosal gland, *Clin. Sci.*, 43, 553, 1972.
31. **Corssen, G. and Allen, C. R.,** Acetylcholine: its significance in controlling ciliary activity of human respiratory epithelium in vitro, *J. Appl. Physiol.*, 14, 401, 1959.
32. **Nadel, J. A. and Widdicombe, J. G.,** Reflex control of airway size, *Ann. N.Y. Acad. Sci.*, 108, 712, 1963.
33. **Ingram, R. H., Wellman, J. J., McFadden, E. R., and Mead, J.,** Relative contributions of large and small airways to flow limitation in normal subjects before and after atropine and isoproterenol, *J. Clin. Invest.*, 59, 696, 1977.
34. **DeTroyer, A., Yernault, J. C., and Rodenstein, D.,** Effects of vagal blockade on lung mechanics in normal man, *J. Appl. Physiol.*, 46, 217, 1979.
35. **Quigley, C., Fuller, R. W., Dixon, C. M. S., and Barnes, P. J.,** Inhaled acetylcholinesterase inhibitors cause dose related bronchoconstriction in asthmatics and normal subjects, *Am. Rev. Respir. Dis.*, 131, 283, 1985.
36. **Davis, C., Kannan, M. S., Jones, T. R., and Daniel, E. E.,** Control of human airway smooth muscle in vitro studies, *J. Appl. Physiol.*, 53, 1080, 1982.
37. **Taylor, S. M., Pare, P. D., and Schellenberg, R.,** Cholinergic and nonadrenergic mechanisms in human and guinea pig airways, *J. Appl. Physiol.*, 56, 958, 1984.
38. **DeJongste, J. C., Mons, H., Bonta, I. L., and Kerrebijn, K. F.,** Nonneural components in the response of fresh human airways to electric field stimulation, *J. Appl. Physiol.*, 63, 1558, 1987.
39. **Palmer, J. B., Cuss, F. M. C., and Barnes, P. J.,** VIP and PHM and their role in nonadrenergic inhibitory responses in isolated human airways, *J. Appl. Physiol.*, 61, 1322, 1986.
40. **Nadel, J. A., Cabezas, G. A., and Austin, J. H. M.,** In vivo roentgenographic examination of parasympathetic innervation of small airways: use of powdered tantalum and a fine focal spot X-ray, *Invest. Radiol.*, 6, 9, 1971.
41. **Mak, J. C. and Barnes, P. J.,** Autoradiographic visualization of muscarinic receptor subtypes in human and guinea-pig lung, *Am. Rev. Respir. Dis.*, 141, 1559, 1990.
42. **Hensley, M. J., O'Cain, C. F., McFadden, E. R., and Ingram, R. H.,** Distribution of bronchodilatation in normal subjects: beta agonists versus atropine, *J. Appl. Physiol.*, 45, 778, 1978.
43. **Davis, B., Martin, M., Fisher, S., Graf, P., Widdicombe, J., and Nadel, J. A.,** New method for study of canine mucous gland secretion in vivo: cholinergic regulation, *Am. Rev. Respir. Dis.*, 113, 257, 1976.
44. **Borson, D. B., Charlin, M., Gold, B. D., and Nadel, J. A.,** Neural regulation of $^{35}SO_4$-macromolecule secretion from tracheal glands of ferrets, *J. Appl. Physiol.*, 57, 457, 1984.
45. **Baker, B., Peatfield, A. G., and Richardson, P. S.,** The nervous control of mucus secretion from human bronchus, *J. Physiol.*, 365, 297, 1985.
46. **Ueki, I., German, V. F., and Nadel, J. A.,** Micropipette measurements of airway submucosal gland secretion, *Am. Rev. Respir. Dis.*, 121, 351, 1980.
47. **Marin, M. G., Davis, B., and Nadel, J. A.,** Effects of acetylcholine on Cl^- and Na^+ fluxes across dog tracheal epithelium in vitro, *Am. J. Physiol.*, 231, 1546, 1980.
48. **Boat, T. F. and Kleinerman, J. L.,** Human respiratory tract secretions. II. Effects of cholinergic and adrenergic agents on in vitro release of protein and mucus glycoprotein, *Chest,* 67, 32, 1975.
49. **Shelhamer, J., Maroni, Z., and Kaliner, M.,** Immunologic and neuropharmacologic stimulation of mucous glycoprotein release from human airways in vitro, *J. Clin. Invest.*, 66, 1400, 1980.

50. **Basbaum, C. B., Ueki, I., Berzina, L., and Nadel, J. A.,** Tracheal submucosal gland serous cells stimulated in vitro with adrenergic and cholinergic agonists, *Cell Tissue Res.*, 220, 481, 1981.
51. **Leikhauf, G. D., Ueki, I., and Nadel, J. A.,** Selective autonomic regulation of the viscoelastic properties of submucosal gland secretions from cat trachea, *J. Appl. Physiol.*, 56, 426, 1984.
52. **Lopez-Vidreiro, M. T., Costello, J., Clark, T. J. H., Das, I., Keal, E. E., and Reid, L.,** Effects of atropine on sputum production, *Thorax*, 30, 543, 1975.
53. **Yeates, D. B., Aspin, N., Levison, H., Jones, M. T., and Bryan, A. C.,** Mucociliary tracheal transport rates in man, *J. Appl. Physiol.*, 39, 487, 1975.
54. **Pack, R. J. and Richardson, P. S.,** The adrenergic innervation of the human bronchus: a light and electron microscopic study, *J. Anat.*, 138, 493, 1984.
55. **Doidge, J. M. and Satchell, D. G.,** Adrenergic and nonadrenergic inhibitory nerves in the mammalian lung, *J. Auton. Nerv. Syst.*, 5, 83, 1982.
56. **Richardson, J. B. and Béland, J.,** Nonadrenergic inhibitory nervous system in human airways, *J. Appl. Physiol.*, 41, 764, 1976.
57. **Sheppard, M. N., Kurian, S. S., Henzen-Logmans, S. C., Michetti, F., Cocchia, D., Cole, P., Rush, R. A., Marangos, P. J., Bloom, S. R., and Polak, J. M.,** Neuronal-specific enolase and S-100. New markers for delineating the innervation of the respiratory tract in man and other animals, *Thorax*, 38, 333, 1983.
58. **Richardson, J. B. and Ferguson, C. C.,** Neuromuscular structure and function in the airways, *Fed. Proc.*, 38, 202, 1979.
59. **Ruffin, R. E., Frith, M. B., Anderson, R. C., Kumana, C. R., Newhouse, M. T., and Hargreave, F. E.,** Selectivity of beta-adrenoceptor antagonist drugs assessed by histamine bronchial provocation, *Clin. Pharmacol. Ther.*, 25, 536, 1979.
60. **McNeill, R. S.,** Effect of beta-adrenergic blocking agent propranolol on asthmatics, *Lancet*, 2, 1101, 1964.
61. **Van Herwaarden, C. L. A.,** β-Adrenergic blockade and pulmonary function in patients suffering from chronic obstructive lung disease, *J. Cardiovasc. Pharmacol.*, 5, 546, 1983.
62. **Zaid, G. and Beall, G. N.,** Bronchial response to beta-adrenergic blockade, *N. Engl. J. Med.*, 275, 580, 1966.
63. **Tattersfield, A. E., Leaver, D. G., and Pride, N. B.,** Effect of beta-adrenergic blockade and stimulation on normal human airways, *J. Appl. Physiol.*, 35, 613, 1973.
64. **Richardson, P. S. and Sterling, G. M.,** Effect of beta-adrenergic blockade on airway conductance and lung volume in normal and asthmatic subjects, *Br. Med. J.*, 3, 143, 1969.
65. **McNeill, R. S. and Ingram, C. G.,** Effect of propranolol on ventilatory function, *Am. J. Cardiol.*, 18, 473, 1966.
66. **Grundstrom, N., Andersson, R. G. G., and Wilkberg, J. E. S.,** Pharmacological characterization of the autonomous innervation of guinea-pig tracheobronchial smooth muscle, *Acta Pharmacol. Toxicol.*, 49, 150, 1981.
67. **Dixon, W. E. and Ranson, F.,** Bronchodilator nerves, *J. Physiol. (London)*, 45, 413, 1911.
68. **Cabezas, G. A., Graf, P. D., and Nadel, J. A.,** Sympathetic versus parasympathetic nervous regulation of airways in dogs, *J. Appl. Physiol.*, 31, 651, 1971.
69. **Zuskin, E., Mitchell, C. A., and Bouhuys, A.,** Interactions between the effects of beta-blockade and cigarette smoke on airways, *J. Appl. Physiol.*, 36, 449, 1974.
70. **DeKock, M. A., Nadel, J. A., Zwi, S., Colebatch, H. J. H., and Olsen, C. R.,** New method for perfusing bronchial arteries: histamine bronchoconstriction and apnea, *J. Appl. Physiol.*, 21, 185, 1966.
71. **MacDonald, A. G., Ingram, C. G., and McNeill, R. S.,** The effect of propranolol on airway resistance, *Br. J. Anaesth.*, 39, 919, 1967.

72. **Langer, I.,** The bronchoconstrictor action of propranolol aerosol in asthmatic subjects, *J. Physiol.,* 190, 41, 1967.
73. **Nadel, J. A. and Widdicombe, J. G.,** Reflex effects of upper airway irritation on total lung resistance and blood pressure, *J. Appl. Physiol.,* 17, 861, 1962.
74. **Widdicombe, J. G.,** Receptors on the trachea and bronchi of the cat, *J. Physiol. (London),* 123, 71, 1954.
75. **Mansfield, L. E., Hameister, H. H., Spaulding, H. S., Smith, N. J., and Glab, N.,** The role of the vagus nerve in airway narrowing caused by intraesophageal hydrochloric acid provocation and esophageal distention, *Ann. Allergy,* 47, 431, 1981.
76. **Bartoli, A., Bystryzycka, E., Guz, A., Jain, S. K., Noble, M. J. M., and Trenchard, D.,** Studies on the pulmonary vagal control of central respiratory movements in the absence of breathing movements, *J. Physiol. (London),* 230, 449, 1973.
77. **Nadel, J. A. and Tierney, D. F.,** Effect of a previous deep inspiration on airway resistance in man, *J. Appl. Physiol.,* 16, 717, 1961.
78. **Orehek, J., Gayrard, P., Grimaud, C., and Charpin, J.,** Effect of maximal respiratory maneouvres on bronchial sensitivity of asthmatic patients as compared to normal people, *Br. Med. J.,* 1, 123, 1975.
79. **Jeffrey, P. and Reid, L.,** Intraepithelial nerves in normal rat airways: a quantitative electron microscopic study, *J. Anat.,* 114, 35, 1973.
80. **Rhodin, J. A. G.,** The ciliated cell: ultrastructure and function of human tracheal mucosa, *Am. Rev. Respir. Dis.,* 93, 1, 1966.
81. **Nadel, J. A., Salem, H., Tamplin, B., and Yokina, Y.,** Mechanics of bronchoconstriction during inhalation of sulfur dioxide, *J. Appl. Physiol.,* 20, 164, 1965.
82. **Gold, W. M., Kessler, G. F., and Yu, D. Y. C.,** Role of the vagus nerves in experimental asthma in allergic dogs, *J. Appl. Physiol.,* 33, 719, 1972.
83. **Holtzman, M. J., Sheller, J. R., DiMeo, M., Nadel, J. A., and Boushey, H. A.,** Effect of ganglion blockade on bronchial reactivity in atopic subjects, *Am. Rev. Respir. Dis.,* 122, 17, 1980.
84. **Ito, Y. and Takeda, K.,** Non-adrenergic inhibitory nerves and putative transmitters in the smooth muscle of cat trachea, *J. Physiol. (London),* 330, 497, 1982.
85. **Barnes, P. J.,** The third nervous system in the lung: physiological and clinical perspectives, *Thorax,* 39, 561, 1984.
86. **Peatfield, A. C., Barnes, P. J., Bratcher, C., Nadel, J. A., and Davis, B.,** Vasoactive intestinal peptide stimulates tracheal submucosal gland secretion in ferret, *Am. Rev. Respir. Dis.,* 128, 89, 1983.
87. **Palmer, J. B. A. and Barnes, P. J.,** Neuropeptides and airway smooth muscle function, *Am. Rev. Respir. Dis.,* 136, 550, 1987.
88. **Greenberg, B., Rhoden, K., and Barnes, P. J.,** Vasoactive intestinal peptide causes non-endothelial cell dependent relaxation in human and bovine pulmonary arteries, *Blood Vessels,* 24, 45, 1987.
89. **Barnes, P. J.,** Non-adrenergic non-cholinergic neural control of human airways, *Arch. Int. Pharmacodyn. Ther.,* 280 (Suppl.), 208, 1986.
90. **Said, S. I.,** Vasoactive intestinal peptide (VIP): current status, *Peptides,* 5, 143, 1984.
91. **Lundberg, J. M., Fahrenkrug, J., Hokfelt, T., Martling, C. R., Larsson, O., Tatemoto, K., and Angaard, A.,** Coexistence of peptide HI (PHI) and VIP in nerves regulating blood flow and bronchial smooth muscle tone in various mammals including man, *Peptides,* 5, 593, 1984.
92. **Lundberg, J. M., Martling, C. R., and Saria, A.,** Substance P and capsaicin induced contraction of human bronchi, *Acta Physiol. Scand.,* 119, 49, 1983.
93. **Fuller, R. W., Maxwell, D. L., Dixon, C. M. S., McGregor, G. P., Barnes, V. F., Bloom, S. R., and Barnes, P. J.,** The effect of substance P on cardiovascular and respiratory function in human subjects, *J. Appl. Physiol.,* 62, 1473, 1987.
94. **Skoogh, B. E.,** Transmission through airway ganglia, *Eur. J. Respir. Dis.,* 64 (Suppl. 131), 159, 1983.

95. **Bloom, J. W., Baumgartener-Folkerts, C., Palmer, J. D., Yamamura, H. I., and Halonen, M.,** A muscarinic receptor subtype modulates vagally stimulated bronchial contractions, *J. Appl. Physiol.*, 65, 2144, 1988.
96. **Minette, P. A. and Barnes, P. J.,** Muscarinic receptor subtypes in the lung, *Am. Rev. Respir. Dis.*, 141, S162, 1990.
97. **Lammers, J. W., Minette, P. A., McCusker, M., and Barnes, P. J.,** The role of pirenzepine sensitive (M_1) muscarinic receptors in vagally mediated bronchoconstriction in humans, *Am. Rev. Respir. Dis.*, 139, 446, 1989.
98. **Minette, P. A. and Barnes, P. J.,** Prejunctional inhibitory muscarinic receptors on cholinergic nerves in human and guinea-pig airways, *J. Appl. Physiol.*, 64, 2532, 1988.
99. **Cazzola, M., Russo, F., De Santis, D., Principe, P., and Marmo, E.,** Respiratory responses to pirenzepine in healthy subjects, *Int. J. Clin. Pharmacol. Ther. Toxicol.*, 25, 105, 1987.
100. **Cazzola, M., D'Amato, G., Guidetti, E., Staudinger, H., Steinijans, V. W., and Kilian, U.,** An M_1-selective muscarinic receptor antagonist Telenzepine improves lung function in patients with chronic obstructive bronchitis, *Pulm. Pharmacol.*, 3, 185, 1991.
101. **Casale, T. B. and Eckland, P.,** Characterization of muscarinic receptor subtypes in human peripheral lung, *J. Appl. Physiol.*, 65, 594, 1988.
102. **Roffel, A. F., Elzinga, C. R. S., Van Amsterdam, R. G. M., De Zeeuw, R. A., and Zaagsma, J.,** Muscarinic M_2 receptors in bovine tracheal smooth muscle discrepancies between binding and function, *Eur. J. Pharmacol.*, 153, 73, 1988.
103. **Van Koppen, C. J., Rodriguez de Miranda, J. F., Beld, A. J., Hermanussen, M. W., Lanners, J. W., and Van Ginneken, C. A. M.,** Characterization of the muscarinic receptor in human tracheal smooth muscle, *Naunyn-Schmiedeberg's Arch. Pharmacol.*, 331, 247, 1985.
104. **Wills, M. and Douglas, J. S.,** Aging and cholinergic responses in bovine trachealis muscle, *Br. J. Pharmacol.*, 93, 918, 1988.
105. **Rothberg, K. G., Morris, P. L., and Douglas, J. S.,** Characterization of cholinergic muscarinic receptors in cow tracheal muscle membranes: effect of maturation, *Biochem. Pharmacol.*, 36, 1687, 1987.
106. **Mitchell, R. W., Murphy, T. M., Kelly, E., and Leff, A. R.,** Maturation of acetylcholinesterase expression in tracheal smooth muscle contraction, *Am. J. Physiol.*, 259, L130, 1990.
107. **Murphy, T. M., Mitchell, R. W., Phillips, I. J., and Leff, A. R.,** Ontogenic expression of acetylcholinesterase activity in trachealis of young swine, *Am. J. Physiol.*, 261, 322, 1991.
108. **Henry, R. L., Milner, A. D., and Stokes, G. M.,** Ineffectiveness of ipratropium bromide in acute bronchiolitis, *Arch. Dis. Child.*, 58, 925, 1983.
109. **Henry, R. L., Hiller, E. J., Milner, A. D., Hodges, I. C. G., and Stokes, G. M.,** Nebulized ipratropium bromide and sodium cromoglycatein the first two years of life, *Arch. Dis. Child.*, 59, 54, 1984.
110. **Prendiville, A., Green, S., and Silverman, M.,** Ipratropium bromide and airways function in wheezy infants, *Arch. Dis. Child.*, 62, 397, 1987.
111. **Adler, M., Reutter, S. A., Moore, D. H., and Filbert, M. G.,** Regulation of acetylcholine hydrolysis in canine tracheal smooth muscle, *Eur. J. Pharmacol.*, 205, 73, 1991.
112. **Norel, X., Angrisani, M., Labat, C., Gorenne, I., Dulmet, E., and Brink, C.,** Degradation of acetylcholine in human airways: role of butyrylcholinesterase, *Br. J. Pharmacol.*, in press.
113. **Jackson, G. and Rees, M. B.,** Anaesthesia in the newborn, *Br. Med. J.*, 2, 1419, 1950.
114. **Stead, A. L.,** The response of the newborn infant to muscle relaxants, *Br. J. Anaesth.*, 27, 124, 1955.
115. **Bush, G. H. and Stead, A. L.,** The use of d-turbocurarine in neonatal anaesthesia, *Br. J. Anaesth.*, 34, 721, 1962.

116. **Fedrick, J. and Butler, N. R.,** Hyaline membrane disease, *Lancet,* 2, 768, 1987.
117. **Whitle, M. J., Hill, C. M., and Harkes, A.,** Effects of labor on the lecithin/sphingomyelin ratio in serial samples of amniotic fluid, *Br. J. Obstet. Gynaecol.,* 84, 500, 1977.
118. **Callen, P., Goldsworthy, S., Graves, L., Harvey, D., Mellows, H., and Parkinson, C.,** Mode of delivery and the lecithin/sphingomyelin ratio, *Br. J. Obstet. Gynaecol.,* 86, 965, 1979.
119. **Faxelius, G., Hagnevik, K., Lagercrantz, H., Lundell, B., and Irestedt, L.,** Catecholamine surge and lung function after delivery, *Arch. Dis. Child.,* 58, 262, 1983.
120. **Bergman, B. and Hedner, T.,** Antipartum administration of terbutaline and the incidence of hyaline membrane disease in preterm infants, *Acta Obstet. Gynecol. Scand.,* 57, 217, 1978.
121. **Roberts, J. M. and Musci, T. J.,** Alveolar beta-adrenoceptors: modulation and role in perinatal adaptation, in *The Perinatal Development of the Respiratory System,* Gluckman, P. and Johnson, B., Eds., Perinatal Press, Ithaca, NY, 1986, 135.
122. **Davis, D. J., Dattel, B. J., Ballard, P. L., and Roberts, J. M.,** β-Adrenergic receptors and cyclic adenosine monophosphate generation in human fetal lung, *Pediatr. Res.,* 21, 142, 1987.
123. **Engel, G.,** Subclass of beta-adrenoceptors. A quantitative estimation of $beta_1$ and $beta_2$ adrenoceptors in guinea-pig and human lung, *Postgrad. Med. J.,* 57, 77, 1981.
124. **Falkay, G., Nemeth, G., and Kovacs, L.,** Binding properties of beta-adrenergic receptors in early human fetal lung, *Biochem. Biophys. Res. Commun.,* 135, 816, 1986.
125. **Barnes, P. J., Karliner, J. S., and Dollery, C. T.,** Human lung adrenoceptors studied by radioligand binding, *Clin. Sci.,* 58, 457, 1980.
126. **Carstairs, J. R., Nimmo, A. J., and Barnes, P. J.,** Autoradiographic localisation of beta-adrenoceptors in human lung, *Eur. J. Pharmacol.,* 103, 189, 1984.
127. **Carstairs, J. R., Nimmo, A. J., and Barnes, P. J.,** Autoradiographic visualization of beta-adrenoceptor subtypes in human lung, *Am. Rev. Respir. Dis.,* 132, 541, 1985.
128. **Brink, C., Douglas, J. S., and Duncan, P. G.,** Changes in the response of guinea pig airways in vivo and in vitro to cimetidine and propranolol during development, *Br. J. Pharmacol.,* 75, 531, 1982.
129. **Vanscheeuwijck, P., and Van de Velde, E., and Fraeyman, N.,** Effects of aging on properties and function of β-adrenoceptors in rat lung, *Eur. J. Pharmacol.,* 172, 373, 1989.
130. **Duncan, P. G., Brink, C., and Douglas, J. S.,** β-Receptors during aging in respiratory tissues, *Eur. J. Pharmacol.,* 78, 45, 1982.
131. **Scarpace, P. J. and Abrass, I. B.,** Decreased beta-adrenergic agonist affinity and adenylate cyclase activity in senescent rat lung, *J. Gerontol.,* 38, 143, 1983.
132. **Scarpace, P. J. and Yu, B. P.,** Diet restriction retards the age-related loss of beta-adrenergic receptors and adenylate cyclase in rat lung, *J. Gerontol.,* 42, 442, 1987.
133. **Brink, C., Ridgway, P. G., and Douglas, J. S.,** Modification of airway smooth muscle responses by hydrocortisone: in vitro and in vivo, *J. Pharmacol. Exp. Ther.,* 203, 1, 1977.
134. **Cheng, J. B., Goldfien, A., and Ballard, P. L.,** Glucocorticoids increase pulmonary β-adrenergic receptors in fetal rabbit, *Endocrinology,* 107, 1646, 1980.
135. **Rooney, S. A.,** The surfactant system and lung phospholipid biochemistry, *Am. Rev. Respir. Dis.,* 131, 439, 1985.
136. **Kresch, M. J. and Gross, I.,** The biochemistry of fetal lung development, *Clin. Perinatol.,* 14, 481, 1987.
137. **Lenney, W. and Milner, A. D.,** At what age do bronchodilators work?, *Arch. Dis. Child.,* 53, 532, 1978.
138. **Prendivile, A., Green, S., and Silverman, M.,** Paradoxical response to nebulised salbutamol in wheezy infants, assessed by partial expiratory flow-volume curves, *Thorax,* 42, 86, 1987.

139. **Mortola, J. P., Saetta, M., and Bartlett, D.,** Postnatal development of the lung following denervation, *Respir. Physiol.*, 67, 137, 1987.
140. **Stretton, C. D., Mak, J. C. W., Belvisi, M. G., Yacoub, M. H., and Barnes, P. J.,** Cholinergic control of human airways in vitro following extrinsic denervation of the human respiratory tract by heart-lung transplantation, *Am. Rev. Respir. Dis.*, 142, 1030, 1990.
141. **Klaus, M., Reiss, O. K., Tooley, W. H., Piel, C., and Clements, J. A.,** Alveolar epithelial cell mitochondria as source of the surface active lung lining, *Science,* 137, 750, 1962.
142. **Slotkin, T. A., Whitmore, W. L., Orband-Miller, L., Queen, K. L., and Haim, K.,** β-adrenergic control of macromolecule synthesis in neonatal rat heart, kidney and lung: relationship to sympathetic neuronal development, *J. Pharmacol. Exp. Ther.*, 243, 101, 1987.
143. **Burnstock, G.,** Purinergic nerves, *Pharmacol. Rev.*, 24, 509, 1972.
144. **Slotkin, T. A., Lau, C., Kavlock, R. J., Whitmore, W. L., Queen, K. L., Orband-Miller, L., Bartolome, M., Baker, F. E., Cameron, A. M., Antolick, L., and Bartolome, J. V.,** Trophic control of lung development by sympathetic neurons: effects of neonatal sympathectomy with 6-hydroxydopamine, *J. Dev. Physiol.*, 10, 577, 1988.
145. **Slotkin, T. A., Whitmore, W. L., Orband-Miller, L., Queen, K. L., and Haim, K.,** Beta adrenergic control of macromolecule synthesis in neonatal rat heart, kidney and lung: relationship to sympathetic neuronal development, *J. Pharmacol. Exp. Ther.*, 243, 101, 1987.
146. **Slotkin, T. A., Seidler, F. J., Haim, K., Cameron, A. M., Antolick, L., and Lau, C.,** Neonatal central catecholaminergic lesions with intracisternal 6-hydroxydopamine: effects on development of presynaptic and postsynaptic pathways and on the ornithine decarboxylase/polyamine system in heart, lung and kidney, *J. Pharmacol. Exp. Ther.*, 247, 975, 1988.

Chapter 10

ENTERIC SYSTEM

Robert M. Santer and Daryll M. Baker

TABLE OF CONTENTS

0-8493-6981-9/93/$0.00 + $.50

I. INTRODUCTION

Nervous regulation of the gastrointestinal tract is a highly complex mechanism that is effected by the enteric nervous system, located within the intestinal wall, in conjunction with extrinsic autonomic innervation provided by sympathetic and parasympathetic nerves. The enteric nervous system consists of enormous numbers of neurons and supporting cells which were originally described in the latter half of the nineteenth century, and is now known to involve neuronal circuitry more complex than any other hitherto investigated part of the peripheral nervous system. Our current state of knowledge of the neuronal circuitry, the neurotransmitters involved, and the structure and arrangement of the enteric nervous system has recently been described by Furness and Costa.[1] Much of the credit for our contemporary view of the enteric nervous system must go to these authors and their co-workers who, over the past 2 decades, have used many elegant experimental and histochemical approaches in furthering our understanding of this complex component of the peripheral nervous system.

II. GASTROINTESTINAL INNERVATION

A. ENTERIC NERVOUS SYSTEM

It has long been known that the principal nervous plexuses of the enteric system are the myenteric plexus of Auerbach, which lies between the outer longitudinal and inner circular layers of the muscularis externa, and the submucous plexus of Meissner within the submucosa. Nerve bundles are also found within the muscularis itself. The two main plexuses consist of ganglia containing nerve cell bodies interconnected by bundles of nerve axons to form a pattern which is distinctive in different regions of the intestine and also characteristically different between species. The myenteric plexus contains by far the largest number of neurons, and within it three parts can be discerned: a primary plexus made up of the ganglia and the internodal strands that interconnect them, a secondary plexus of finer nerve bundles oriented in relation to the circular muscle, and a tertiary plexus of very small nerve bundles ramifying within the spaces of the mesh formed by the primary plexus. The intrinsic neural circuitry for the propulsion of food and digesta along the gut is contained within the myenteric plexus. The submucous plexus has smaller ganglia and is less regularly arranged. In 1864, Auerbach[2] described nerve bundles connecting the myenteric and submucous plexuses. Neurons within these ganglionated plexuses project orally or anally for variable distances and make complex intraganglionic connections. They also send axons to the muscle layers and to the mucosae by which the respective functions of intestinal motility and secretion are controlled with the involvement of an array of amine and peptide neurotransmitter candidates. The gastrointestinal tract also has a sensory innervation with cell bodies in dorsal root or certain cranial sensory ganglia. The peripheral processes of these sensory nerves are also components of the enteric plexuses and may participate in peripheral circuitry involving prevertebral ganglia.

B. SYMPATHETIC INNERVATION

Sympathetic preganglionic control of the intestine is derived from cholinergic neurons in the thoracic and upper lumbar spinal cord projecting primarily via the splanchnic nerves to sympathetic ganglia. The postganglionic sympathetic axons passing to the intestine originate from neurons for the most part in the prevertebral sympathetic ganglia, but the esophagus and upper part of the abdominal intestine receive their sympathetic innervation from paravertebral ganglia of the upper parts of the sympathetic chain.[3] The postganglionic axons reach the gut as paravascular nerve bundles, but then ramify within the gut wall to supply the mucosa, the enteric vasculature, and the ganglia of the myenteric plexus. Only a small number of axons innervate the smooth muscle cells of the muscularis directly,[4] but the sympathetic innervation of the sphincteric musculature is more dense. The distribution of the sympathetic innervation has been elucidated over the past 3 decades by the detection of noradrenaline (NA) by fluorescence histochemical methods and by the immunohistochemical localization of tyrosine hydroxylase (TH), the rate-limiting enzyme in the biosynthesis of NA. The sympathetic innervation of the intestine is involved in reducing intestinal motility, regulating intestinal blood flow, and suppressing mucosal transport.

C. PARASYMPATHETIC INNERVATION

Parasympathetic innervation of the intestine is derived from preganglionic neurons in the dorsal motor nucleus of the vagus and in the sacral parasympathetic nucleus of the spinal cord (sacral levels 2 to 4) which project to the gut via vagal and pelvic nerve trunks. The postganglionic neurons are located predominantly in the ganglia of the submucous and myenteric plexuses, from where relatively short axons pass to their targets within the gut wall. The parasympathetic innervation of the intestine has an excitatory effect on gastrointestinal motility and promotes gastric acid and pepsin secretion in the stomach and water and electrolyte transport in the small and large intestine.

Although it is not normally considered together with the autonomic nervous system, the innervation of the pharynx by special visceral efferent neurons located in the nucleus ambiguus of the medulla oblongata must be included here.

III. GASTROINTESTINAL FUNCTION IN OLD AGE

Certain aspects of gastrointestinal function in the elderly are suggestive of autonomic dysfunction. Moreover, there is evidence of similar manifestations in the laboratory rat which promote it as a suitable animal model for studies on aging of the gastrointestinal tract and the enteric nervous system.

In man, there is an increased incidence with advancing age of problems associated with motility of the intestinal system,[5,6] including a decrease in the amplitude of esophageal peristalsis[7-9] during swallowing, which may be combined with the failure of the esophageogastric junction to relax, increased gastric emptying time,[10,11] and increased intestinal transit time[12,13] with associated fecal stasis. Constipation in patients who seem less aware of the fullness of the rectum might also be attributable to altered sensory innervation of the large bowel. In addition, secretory dysfunction may commonly occur, resulting in dryness of the mouth,[6] which may affect the initiation of swallowing and decreased gastric acid secretion[14] in the stomach. There is evidence also for a decline in the ability of the intestine to absorb calcium,[15,16] xylose,[17] and fat[5] with increasing age.

Comparable effects have been reported in the gastrointestinal system of aged rats. Two studies[18,19] have demonstrated increased transit time in the large intestine, and these results must be viewed in light of the fact that the overall length of the colon remains unchanged throughout adult life,[20] nor is the thickness of the muscularis altered.[18] The increased transit time can be attributed to changes in cholinergically mediated neuromuscular transmission.[18,21] With regard to absorbative function in the rat intestine, there is considerable evidence for reductions with age in the transport of water,[22,23] sugar,[23-25] and amino acids.[22] Thus, the rat can be considered to be a useful model for the study of changes in the gastrointestinal system with age.

IV. GASTROINTESTINAL INNERVATION IN OLD AGE

A. ENTERIC NERVOUS SYSTEM

It is possible to make accurate counts of the numbers of neurons in the myenteric plexus after performing the NADH-tetrazolium reductase histochemical reaction of Gabella[26,27] on portions of fresh intestine, from which laminar preparations of the muscularis externa are then prepared. Neuron counts of representative areas can then be multiplied by the length of parts of the intestine to give estimates of the total numbers in the myenteric plexus of the different parts.

Extensive studies on neuron number in the rat[28] and guinea pig[29] have shown large decreases in neuron number in both the small and large intestine of aged animals (Figure 1). In the aged guinea pig, small intestine myenteric neuron numbers fell to 40 to 60%[29] of the numbers in young adults, and in the rat, comparable values were 40, 64, and 42% in the small intestine, colon, and rectum, respectively.[28] Measurements of neuron size indicate that the age-associated neuronal fallout affected the population equally in the rat[28] and guinea pig.[29] The results suggested that structural reorganization of the myenteric ganglia may take place in old age. The number and appearance of myenteric neurons also changes with age in the human colon,[30] and this has been implicated in idiopathic constipation. However, in man, the thickness of the human muscularis externa makes neuron myenteric number assessment and arrangement problematical in contrast to the examination of intramural

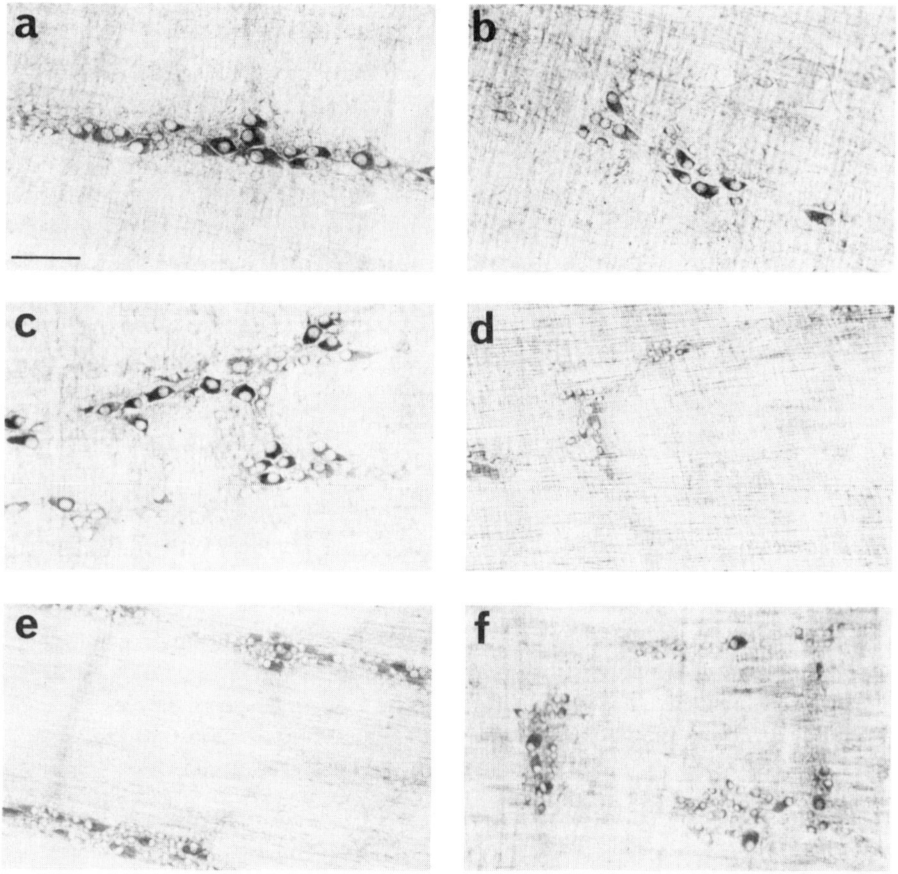

FIGURE 1. Laminar preparations of the muscularis externa of the jejunum (a, b), colon (c, d), and rectum (e, f) showing neurons of the myenteric plexus of Auerbach stained by the NADH-tetrazolium reductase method in 4-month-old (a, c, e) and 24-month-old (b, d, f) rat. There are marked reductions in the numbers of neurons in the old animals in all three regions of the intestine. (Bar for a–e = 100 μm.)

nerves, for which methodology has been developed.[31] The range of neurotransmitter phenotypes of myenteric neurons has yet to be investigated in aged animals. It will be of great interest to see whether a particular population is affected by age-associated neuronal fallout, as it will then be possible, because of our current understanding of enteric neuronal circuitry,[1] to predict functional perturbations in old age.

Histochemical and immunohistochemical investigations of nerve axons within the gut wall also indicate that reductions in nerve density occur with age. In a quantitative analysis of the density of sympathetic, noradrenergic innervation of the rat jejunum by Baker and Santer,[32] the total amount of fluorescent nerve axons was reduced in aged rats to approximately 50% of that seen in the young animal and, furthermore, the number of axon varicosities — the points of neurotransmitter release — were reduced by almost 75%. The orderly appearance of the noradrenergic innervation seen in young rats was greatly disrupted in the old animals and the tertiary plexus was virtually nonexistent. These facts, together with pharmacological evidence,[33] suggest a decline in sympathetic control of intestinal function in old rats. Degenerating axons were also observed in this study and, interestingly, were confined to just one or two sets of circularly arrayed myenteric ganglia, which is suggestive of a discrete projection of individual postganglionic neurons to the gut. The occurrence of TH- and neuropeptide Y (NPY)-immunopositive nerves in the villi and mucosa of the aged rat small intestine is also very sparse.[73] In contrast, however, the density and arrangement of the noradrenergic innervation in the human small intestine did not noticeably change with age in a series of patients which included octagenarians.[34,74]

Peptidergic nerves have been identified in all layers of the wall of the rat small intestine, and obvious decreases in the densities of substance P (SP)-, somatostatin (SOM)-, and vasoactive intestinal polypeptide (VIP)-containing nerves have been reported[35] in aged animals. In a report on the rat colon which included assays of VIP, SP, NPY, and calcitonin gene-related peptide (CGRP) over the first year of life,[36] decreases in the levels of these peptides were not recorded, which suggests that the observed decreases[35] may occur later in life and not as a gradual decline. In man, peptidergic nerves have been detected by immunohistochemistry in subjects up to almost 80 years of age,[37-40] but assays of gut peptide levels reveal an upward trend at least until the mid-60s.[41]

B. POSTGANGLIONIC SYMPATHETIC INNERVATION

There is now a considerable amount of evidence which indicates that the postganglionic sympathetic neurons of the coeliac-superior mesenteric ganglion (CSMG) complex are more susceptible to aging than their counterparts in the superior cervical ganglion (SCG). The aforementioned dramatic decrease in sympathetic innervation of the rat small intestine,[32] and also in the spleen,[42] is indicative of this. In aged rats, CSMG neurons exhibit reduced noradrenaline fluorescence,[43,45] reduced levels of noradrenaline synthesis,[44] increased amounts of age pigment and other morphological alterations,[45-47] and reduced levels of succinic dehydrogenase activity[48] (Figure 2), indicative of a general reduction in metabolic activity. Acid phosphatase activity, indicative of lysosomal action, is also increased in the CSMG in old rats.[75] Such changes are not as developed in the superior SCG.[48-51] On the other hand, the amount of the growth-associated peptide GAP-43 remains unaltered in aged rat CSMG.[52]

Such age-associated changes in prevertebral ganglia projecting to the enteric plexuses and other parts of the intestine have not been as extensively investigated in man, but they do show large aggregations of autofluorescent

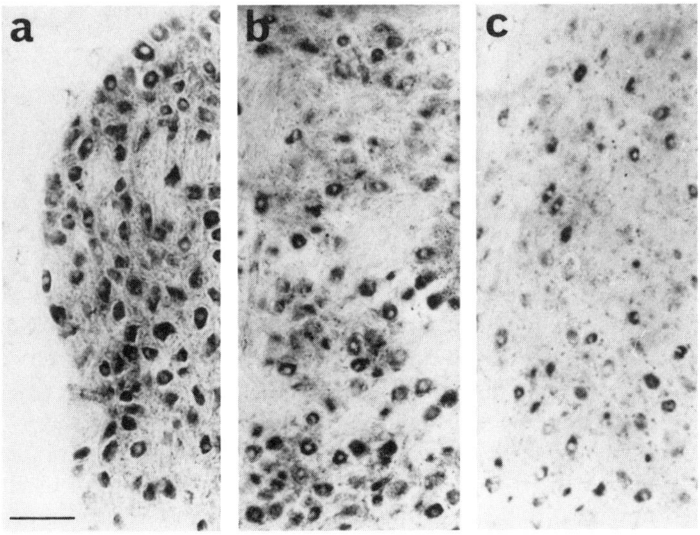

FIGURE 2. Succinate dehydrogenase activity in neurons of the rat coeliac-superior cervical ganglion complex in animals 6 (a), 12 (b), and 24 (c) months of age showing a decrease in the staining intensity of many neurons. (Bar for a–c = 100 μm.)

lipopigment[53] and Lewy bodies,[54] and are frequently in close contact with highly swollen, dystrophic axons.[55] Such axons, which stain positively for TH and NPY, are found much more frequently in prevertebral than paravertebral sympathetic ganglia, as they are in the aged rat.[56]

C. PREGANGLIONIC INNERVATION

The origin of the swollen, dystrophic axons that occur frequently in the prevertebral ganglia[55,56] is unknown, but it has been suggested, in the rat[56] at least, that they may be preganglionic in nature. In a series of investigations, Low et al.[57-59] studied the sympathetic preganglionic neurons of the intermediolateral cell column of the thoracic spinal cord projecting into the splanchnic nerves in human subjects up to 79 years of age. The authors concluded, as a result of neuron and myelinated axon counts, that there is a decline in the sympathetic preganglionic neuron number of about 8% per decade. Whether any compensatory development of collaterals takes place is unknown, but the implication is that a decline in sympathetic control of the intestine occurs with advancing age in man. Recently, evidence has emerged that neuron numbers in the dorsal motor nucleus of the vagus decline in the mouse over time,[60] such that by 31 months of age, only 62% of the neurons remain. Neuron counts in this vagal nucleus in a 63-year-old human did not reveal such a decline.[61] Neuronal fallout during life therefore may affect both sympathetic and parasympathetic preganglionic neurons involved in control of the enteric nervous system. Neuron numbers have also been shown to decline with age in the nucleus ambiguus,[60] which provides a motor innervation to the pharynx.

V. DISCUSSION AND CONCLUSIONS

This short review has revealed that investigations into the effects of age on the enteric plexuses are not numerous and, to date, principally confined to the myenteric plexus and its extrinsic autonomic innervation. Nevertheless, the available information from work carried out on small rodents points to a decrease in enteric neuron number[28,29] in both the small and large intestine and to a decrease in both peptide- and NA-containing nerves within the myenteric plexus and the muscularis externa.[32,35] It is clear that no comprehensive study of any transmitter-identified neuron population in any part of the intestine has been undertaken with respect to aging. Moreover, studies are needed to investigate any temporal sequences or progressions in the fallout of transmitter-identified enteric neurons during life. Because so much information is now available on the transmitter identity of enteric neurons, their connectivity and projections within the gut,[1] such studies can be contemplated.

From a functional point of view, there is some evidence[33] that the observed reduction in noradrenergic innervation[32] is reflected in decreased responsiveness of the muscularis externa to β-agonists, but sympathetic control of the enteric vasculature and mucosal functions have yet to be assessed in aged animals. Feher and Penzes[35] reported significant decreases with age in certain peptide-containing nerves in all layers of the rat small intestine by electron microscopic immunocytochemistry, but thus far no correlations have been made with the functions with which these peptides are thought to be involved. For example, is the decrease in SP-containing nerves reflected in changes in sensory innervation[62] of the gut or in its other roles within the enteric nervous system?[63] Is the decrease in VIP-containing nerves reflected in changes in gut relaxation[64] and associated transit times, and also in vasodilation[65] within the intestine? Is the decrease in intestinal SOM-containing nerves also associated with inhibitory[66,67] changes in gut motility with age? These functional questions, in addition to those relating to other amine and peptide neurotransmitter candidates so far uninvestigated in the enteric nervous system of aged animals, need to be addressed before any predictions relating to the human condition can be made.

The extrinsic autonomic innervation of the enteric plexuses shows distinct changes with age which together are suggestive of a decline in the level of autonomic control of intestinal function. The degenerative changes with age described above[32,42-48,53-60] in pre- and postganglionic neurons supplying the intestine are very consistent with the recently expounded neurotrophic theory of Purves,[68] in which the existence of target tissues is vital for the continued maintenance and survival of the neurons that innervate them. It is generally agreed that the expression of growth factors by the target tissues is responsible for this phenomenon. With regard to the peripheral nervous system, nerve growth factor (NGF) is considered to be the most potent factor discovered thus far, but it is likely that other growth factors are also involved. The influence of NGF on the survival and morphology of sympathetic neurons is apparently lifelong[69,70] and its withdrawal results in neuronal degeneration. In the intestine, the expression of the gene for NGF is among the lowest in any tissue investigated,[71] but information on the level of NGF gene expression in aged tissue is not available. The evidence presented in this review highlights declines in the preganglionic and postganglionic innervation of the intestine, and it is possible to suggest that these effects might be interdependent. Thus, it may be that the fallout of enteric neurons deprives the neurons of the prevertebral sympathetic ganglia of their targets and that subsequent degeneration of these neurons deprives the sympathetic preganglionic neurons of their targets, with resultant degeneration. A similar effect may take place in the dorsal motor nucleus of the vagus, the survival of whose neurons is known to be dependent on the presence of gastric tissue.[72] The factor or factors responsible for the fallout of the enteric neurons themselves, presumably produced by the intestine, remain unknown.

Studies on the aging of intestinal innervation and in particular of the enteric plexuses are in their infancy, but our detailed knowledge of the circuitry of the enteric nervous system now provides a means by which to examine intestinal innervation in old age and understand functional perturbations.

REFERENCES

1. **Furness, J. B. and Costa, M.,** *The Enteric Nervous System,* Churchill Livingstone, New York, 1987.
2. **Auerbach, L.,** Fernere vorlaufige Mitteilung über den Nervenapparat des Darmes, *Arch. Pathol. Anat. Physiol.,* 30, 457, 1864.
3. **Lundberg, J. M., Dahlström, A., Larsson, I., Petterson, G., Ahlman, K., and Kewenter, J.,** Efferent innervation of the small intestine by adrenergic neurons of the cervical sympathetic and stellate ganglia, studied by retrograde transport of peroxidase, *Acta Physiol. Scand.,* 104, 33, 1978.
4. **Furness, J. B. and Costa, M.,** The adrenergic innervation of the gastrointestinal tract, *Ergeb. Physiol.,* 69, 1, 1974.
5. **Brocklehurst, J. C.,** *Textbook of Geriatric Medicine and Gerontology,* Churchill Livingstone, New York, 1985, 526.
6. **Anderson, F. and Williams, B.,** *Practical Management of the Elderly,* 5th ed., Blackwell Scientific, Oxford, 1989, 49.
7. **Soergel, K. H., Zboralske, F. F., and Amberg, J. R.,** Presbyesophagus: esophageal motility in nonagenarians, *J. Clin. Invest.,* 43, 1472, 1964.
8. **Hollis, J. B. and Castell, D. O.,** Esophageal function in elderly men. A new look at 'presbyoesophagus', *Ann. Intern. Med.,* 80, 371, 1974.
9. **Khan, T. A., Shragge, B. W., Crispin, J. S., and Lind, J. F.,** Esophageal motility in the elderly, *Dig. Dis. Sci.,* 20, 1049, 1977.
10. **Moore, J. G., Tweedy, C., Christian, P. E., and Datz, F. L.,** Effect of age on gastric emptying of liquid-solid meals in man, *Dig. Dis. Sci.,* 28, 340, 1983.

11. **Horowitz, M., Maddern, G. J., Chatterton, B. E., Collins, P. J., Harding, P. E., and Shearman, D. J. C.,** Changes in gastric emptying rates with age, *Clin. Sci.*, 67, 213, 1984.
12. **Anuras, S. and Leoning-Baucke, V.,** Gastrointestinal motility in the elderly, *J. Am. Geriatr. Soc.*, 32, 386, 1984.
13. **Brocklehurst, J. D. and Khan, M. Y.,** A study of fecal stasis in old age and the use of 'Dorbanex' in its prevention, *Gerontol. Clin.*, 11, 293, 1969.
14. **Baron, J. H.,** Studies of basal and peak acid output with an augmented histamine test, *Gut*, 4, 136, 1963.
15. **Avioli, L. V., McDonald, J. E., and Lee, S. W.,** The influence of age on the intestinal absorption of ^{47}Ca in women and its relation to ^{47}Ca absorption in postmenopausal osteoporosis, *J. Clin. Invest.*, 44, 1960, 1965.
16. **Bullamore, J. R., Wilkinson, R., Gallagher, R. C., Nordin, B. E., and Marshall, D. H.,** Effect of age on calcium absorption, *Lancet*, 11, 535, 1970.
17. **Webster, S. G. P. and Leeming, J. T.,** Assessment of small bowel function in the elderly using a modified xylose tolerance test, *Gut*, 16, 109, 1975.
18. **McDougal, J. N., Miller, M. S., Burks, T. F., and Kreulen, D. L.,** Age-related changes in colonic function in rats, *Am. J. Physiol.*, 247, G542, 1984.
19. **Varga, F.,** Transit time changes with age in the gastrointestinal tract of the rat, *Digestion*, 14, 319, 1976.
20. **Goodlad, R. A. and Wright, N. A.,** Changes in intestinal cell proliferation, absorptive capacity and structure in young, adult and old rats, *J. Anat.*, 173, 109, 1990.
21. **Nowak, T. V., Harrington, B., and Kalbfleisch, J.,** Age-related changes in enteric neuromuscular transmission, *J. Pharmacol. Exp. Ther.*, 253, 683, 1990.
22. **Treves, C., Favilli, F., Stio, M., Iantomasi, T., and Vincenzini, T.,** Changes in metabolite transport by small intestine and kidney of young and old rats, *Mech. Ageing Dev.*, 52, 263, 1990.
23. **Esposito, G., Faelli, A., Tosco, M., Orsenigo, M. N., and Battistessa, R.,** Age-related changes in rat intestinal transport of D-glucose, sodium and water, *Am. J. Physiol.*, 248, R282, 1985.
24. **Thomson, A. B. R.,** Unstirred water layer and age-dependent changes in jejunal D-glucose transport, *Am. J. Physiol.*, 236, E685, 1979.
25. **Doubek, W. G. and Armbrecht, H. J.,** Changes in intestinal glucose transport over the lifespan of the rat, *Mech. Ageing Dev.*, 39, 91, 1987.
26. **Gabella, G.,** Detection of nerve cells by a histochemical technique, *Experientia*, 25, 218, 1969.
27. **Gabella, G.,** Neuron size and number in the myenteric plexus of the newborn and adult rat, *J. Anat.*, 109, 81, 1971.
28. **Santer, R. M. and Baker, D. M.,** Enteric neuron numbers and sizes in Auerbach's plexus in the small and large intestine of adult and aged rats, *J. Auton. Nerv. Syst.*, 25, 59, 1988.
29. **Gabella, F.,** Fall in the number of myenteric neurons in aging guinea pigs, *Gastroenterology*, 96, 1487, 1989.
30. **Krishnamurti, S., Schuffler, M. D., Rohrmann, C. A., and Pope, C. E.,** Severe idiopathic constipation is associated with a distinctive abnormality of the colonic myenteric plexus, *Gastroenterology*, 88, 26, 1985.
31. **Ferri, G.-L.,** Human gut neuroanatomy: methodology for a quantitative analysis of nerve elements and neurotransmitter diversity in the human 'enteric nervous system', *Basic Appl. Histochem.*, 32, 117, 1988.
32. **Baker, D. M. and Santer, R. M.,** A quantitative study of the effects of age on the noradrenergic innervation of Auerbach's plexus in the rat, *Mech. Ageing Dev.*, 42, 147, 1988.
33. **Baker, D. M., Watson, S. P., and Santer, R. M.,** Evidence for a decrease in sympathetic control of intestinal function in the aged rat, *Neurobiol. Aging*, 12, 363, 1991.

34. **Llewellyn-Smith, I. J., Furness, J. B., O'Brien, P. E., and Costa, M.,** Noradrenergic nerves in human small intestine, *Gastroenterology,* 87, 513, 1984.
35. **Feher, E. and Penzes, L.,** Density of substance P, vasoactive intestinal polypeptide and somatostatin-containing nerve fibers in the ageing small intestine of the rats, *Gerontology,* 33, 341, 1987.
36. **Belai, A., Lincoln, J., Milner, P., and Burnstock, G.,** Progressive changes in adrenergic, serotinergic and peptidergic nerves in proximal colon of streptozotocin-diabetic rats, *Gastroenterology,* 95, 1234, 1988.
37. **Polak, J. M., Bloom, S. R., Sullivan, S. N., Facer, P., and Pearse, A. G. E.,** Enkephalin-like immunoreactivity in the human gastrointestinal tract, *Lancet,* 1, 972, 1977.
38. **Ferri, G.-L., Botti, P., Biliotti, G., Rebecchi, L., Bloom, S. R., Tonelli, L., Labo, G., and Polak, J. M.,** VIP-, substance P- and met-enkephalin-immunoreactive innervation of the human gastroduodenal mucosa and Brunner's glands, *Gut,* 25, 948, 1984.
39. **Ferri, G.-L., Adrian, T. E., Allen, J. M., Soimero, L., Cancellieri, A., Yeats, J. C., Blank, M., Polak, J. M., and Bloom, S. R.,** Intramural distribution of regulatory peptides in the sigmoid-recto-anal region of the human gut, *Gut,* 29, 762, 1988.
40. **Llewellyn-Smith, I. J., Furness, J. B., Murphy, R., O'Brien, P. E., and Costa, M.,** Substance-P containing nerves in the human small intestine, *Gastroenterology,* 86, 421, 1984.
41. **Ferri, G.-L., Soimero, L., Morreale, R. A., and Rebecchi, L.,** A new approach to gut neuro-endocrinology in ageing: towards the integrated assessment of neuroendocrine structure and intramural distribution of regulatory peptides, in *Neuroendocrine System and Aging,* Vezzadini, P., Facchini, A., and Labo, G., Eds., EURAGE, Rijswijk, The Netherlands, 1986, 313.
42. **Felten, S. Y., Bellinger, D. L., Collier, T. J., Coleman, P. D., and Felten, D. L.,** Decreased sympathetic innervation of spleen in aged Fischer 344 rats, *Neurobiol. Aging,* 8, 159, 1987.
43. **Santer, R. M.,** Fluorescence histochemical evidence for decreased noradrenaline synthesis in sympathetic neurones of aged rats, *Neurosci. Lett.,* 15, 177, 1979.
44. **Partanen, M., Waller, S. B., London, E. D., and Hervonen, A.,** Indices of neurotransmitter synthesis and release in aging sympathetic nervous system, *Neurobiol. Aging,* 6, 227, 1985.
45. **Santer, R. M., Partanen, M., and Hervonen, A.,** Glyoxylic acid fluorescence and ultrastructural studies of neurones in coeliac-superior mesenteric ganglion of the aged rat, *Cell Tissue Res.,* 211, 475, 1980.
46. **Baker, D. M. and Santer, R. M.,** Morphometric studies on pre- and paravertebral neurons in the rat: changes with age, *Mech. Ageing Dev.,* 42, 139, 1988.
47. **Hervonen, A., Partanen, M., Helen, P., Koistinaho, J., Alho, H., Baker, D. M., Johnson, J. E., and Santer, R. M.,** The sympathetic neuron as a model of neuronal aging, in *Neurohistochemistry: Modern Methods and Applications,* Panula, P. et al., Eds., Alan R. Liss, New York, 1986, 569.
48. **Baker, D. M. and Santer, R. M.,** Development of a quantitative histochemical method for the determination of succinate dehydrogenase activity in autonomic neurons and its application to the study of aging in the autonomic nervous system, *J. Histochem. Cytochem.,* 38, 525, 1990.
49. **Santer, R. M.,** Morphological evidence for the maintenance of the cervical sympathetic system in aged rats, *Neurosci. Lett.,* 130, 248, 1991.
50. **Hervonen, A., Kimura, A., Sato, A., and Sato, Y.,** Responses of cortical cerebral blood flow produced by stimulation of cervical sympathetic trunks are well maintained in aged rats, *Neurosci. Lett.,* 120, 55, 1990.
51. **Reis, D. J., Ross, D. A., and Joh, T. H.,** Change in the activity and amounts of enzymes synthesising catecholamines and acetylcholine in the brain, adrenal medulla and sympathetic ganglia of aged rat and mouse, *Brain Res.,* 136, 465, 1977.

52. **Schmidt, R. E., Spencer, S. A., Coleman, W. D., and Roth, K. A.,** Immunohistochemical localisation of GAP-43 in the rat and human sympathetic nervous system — effect of aging and diabetes, *Brain Res.,* 552, 190, 1991.
53. **Koistinaho, J., Sorvaniemi, M., Alho, H., and Hervonen, A.,** Microspectrofluorometric quantitation of autofluroescent lipopigment in the human sympathetic ganglia, *Mech. Ageing Dev.,* 37, 79, 1986.
54. **Wakabayashi, K.,** Parkinson's disease: the distribution of Lewy bodies in the peripheral autonomic nervous system, *Brain Nerve,* 41, 965, 1989.
55. **Schmidt, R. E., Chae, H. Y., Parvin, C. A., and Roth, K. A.,** Neuroaxonal dystrophy in aging human sympathetic ganglia, *Am. J. Pathol.,* 136, 1327, 1990.
56. **Schmidt, R. E., Plurad, S. B., and Modert, C. W.,** Neuroaxonal dystrophy in the autonomic ganglia of aged rats, *J. Neuropathol. Exp. Neurol.,* 42, 376, 1983.
57. **Low, P. A., Okazaki, H., and Dyck, P. J.,** Splanchnic preganglionic neurons in man. I. Morphology of preganglionic cytons, *Acta Neuropathol.,* 40, 55, 1977.
58. **Low, P. A. and Dyck, P. J.,** Splanchnic preganglionic neurons in man. II. Morphometry of myelinated fibers of T7 ventral spinal root, *Acta Neuropathol.,* 40, 219, 1977.
59. **Low, P. A. and Dyck, P. J.,** Splanchnic preganglionic neurons in man. III. Morphometry of myelinated fibers of rami communicantes, *J. Neuropathol. Exp. Neurol.,* 37, 734, 1978.
60. **Sturrock, R. R.,** A comparison of age-related changes in neuron number in the dorsal motor nucleus of the vagus and the nucleus ambiguus of the mouse, *J. Anat.,* 173, 169, 1990.
61. **Nara, T., Goto, N., and Hamano, S.,** Development of the human dorsal nucleus of vagus nerve: a morphometric study, *J. Auton. Nerv. Syst.,* 33, 267, 1991.
62. **Matthews, M. R. and Cuello, A. C.,** Substance P-immunoreactive branches of sensory neurons innervate guinea-pig sympathetic neurons, *Proc. Natl. Acad. Sci. U.S.A.,* 79, 1668, 1982.
63. **Costa, M., Furness, J. B., Llewellyn-Smith, I. J., Murphy, R., Bornstein, J. C., and Keast, J. R.,** Functional roles for substance P-containing neurons in the gastrointestinal tract, in *Substance P Metabolism and Biological Actions,* Jordan, C. C. and Ohme, P., Eds., Taylor & Francis, London, 1985, 99.
64. **Fahrenkrug, J., Galbo, H., Holst, J. J., and Schaffalitzky de Muckadell, O. B.,** Influence of the autonomic nervous system on the release of vasoactive intestinal polypeptide from the porcine gastrointestinal tract, *J. Physiol.,* 280, 405, 1978.
65. **Said, S. I. and Mutt, V.,** Potent peripheral and splanchnic vasodilator peptide from normal gut, *Nature (London),* 225, 863, 1970.
66. **Guillemin, R.,** Somatostatin inhibits the release of acetylcholine induced electrically in the myenteric plexus, *Endocrinology,* 99, 1653, 1976.
67. **Furness, J. B. and Costa, M.,** Actions of somatostatin on excitatory and inhibitory nerves in the intestine, *Eur. J. Pharmacol.,* 56, 69, 1979.
68. **Purves, D.,** *Body and Brain. A Trophic Theory of Neural Connections,* Harvard University Press, Cambridge, 1988.
69. **Ruit, K. G., Osborne, P. A., Schmidt, R. E., Johnson, E. M., and Snider, W. D.,** Nerve growth factor regulates sympathetic ganglion cell morphology and survival in the adult mouse, *J. Neurosci.,* 10, 2412, 1990.
70. **Anand, P., Rudge, P., Matthias, C. J., Springall, D. A., Ghatei, M. A., Naher-Noe, M., Sharief, M., Misra, V. P., Polak, J. M., Bloom, S. R., and Thomas, P. K.,** New autonomic and sensory neuropathy with loss of adrenergic sympathetic function and sensory neuropeptides, *Lancet,* 337, 1253, 1991.
71. **Shelton, D. L. and Reichardt, L. F.,** Expression of the β-nerve growth factor gene correlates with the density of sympathetic innervation in effector organs, *Proc. Natl. Acad. Sci. U.S.A.,* 81, 7951, 1984.
72. **Rinaman, L. and Levitt, P.,** Access to gastric tissue promotes the survival of axotomised neurons in the dorsal motor nucleus of the vagus in neonatal rats, *J. Comp. Neurol.,* 313, 213, 1991.

73. **Santer, R. M. and Baker, D. M.,** personal observations.
74. **Llewellyn-Smith, I. J.,** personal communication.
75. **Santer, R. M.,** personal observations.

Chapter 11

RENAL AUTONOMIC NEUROEFFECTOR MECHANISMS IN AGING

Alberto Ricci

TABLE OF CONTENTS

0-8493-6981-9/93/$0.00 + $.50

I. RENAL NEUROANATOMY

The intrinsic innervation of the kidney has been studied for over 100 years. In the last few years, a combination of electromicroscopic, histochemical, and immunohistochemical techniques has given a more precise definition to the localization and distribution of nerves and receptors in the kidney. Although many physiological studies have been done, much controversy has arisen regarding the methods used.

This chapter summarizes and demonstrates the current views concerning the function of renal nerves.

A. GROSS AND INTRINSIC INNERVATION

Around 1840, a detailed description of the innervation of the kidney was attempted.[1] Early observations showed that nerve fibers traveling with the renal artery enter the kidney.

The renal nerves derive from the coeliac plexus, mesenteric plexus, intermesenteric plexus, superior hypogastric plexus, and splanchnic nerves originating from thoracic and lumbar branches.[2] These plexuses derive from the T^5 to L^3 spinal cord segments.[3] Combined, these fibers generate the renal plexus which lies close to the aorticorenal ganglion. This is the largest ganglion associated with the renal plexus, positioned close to the junction between the aorta and the renal artery.[4] The nerve fibers that arise from the renal plexus run along the renal artery and vein, and enter the hilus.

1. Adrenergic Nerves

The combined use of fluorescence, histochemical, and autoradiographic techniques to demonstrate the localization of endogenous noradrenaline in adrenergic nerves, as well as the thocholine histochemical method for acetylcholinesterase-containing nerves, has shown that in some species studied, the majority of adrenergic innervation supplies arterioles in the renal cortex. This is predominantly in the space between the afferent and efferent glomerular arterioles, and close to the tubules often filling the space between it and the arterioles. The tubules receive a distinct tubular innervation following the contour of both proximal and distal tubules. These branches derive from periarteriolar nerves.[5]

The authors observed species-specific differences in the nerve distribution. The nerves, examined by electron microscopy, are composed of axons surrounded by Schwann cells. Also observed were varicosities, which are neuroeffector junctions between a neuron and arteriolar walls and tubular epithelial cells. These varicosities contain mitochondria and vesicles (small, dense core vesicles) similar to those contained in autonomic nerves.

Using light and electron microscopic autoradiography and exogenous tritiated noradrenaline, silver grains were precipitated onto those areas which contained [^{3}H]-noradrenaline. This demonstrated accumulation within axonal varicosities, more specifically packed with dense core vesicles. The varicosities are positioned near juxtaglomerular granular cells and proximal and distal tubules.[6] These findings were observed in several mammalian and nonmammalian species, and in 13- to 16-week-old human fetuses.[6-8]

2. Cholinergic Nerves

Acetylcholinesterase-containing nerves are densely distributed close to the proximal and distal convoluted tubules, afferent and efferent glomerular arterioles, and in the juxtaglomerular region.

The histochemical studies which demonstrated the presence acetylcholinesterase in the kidney indicated the existence of a distinct cholinergic innervation.[9,10] Past studies suggested that acetylcholinesterase-containing nerves might be adrenergic. This hypothesis was tested by using 6-hydroxydopamine (6-OHDA), a drug able to destroy adrenergic fiber and/or nerve terminals. The use of 6-OHDA resulted in the disappearance of both acetylcholinesterase- and catecholamine-containing nerves in the rat kidney.[11] The results of this study, in accordance with the hypothesis of Burn and Rand,[12] support the existence of an adrenergic innervation of the glomerular arterioles and renal tubules. Rat kidney contains myelinated and unmyelinated axons that innervate the corticomedullary region, the parietal loose connective tissue and subepithelial connective tissue of the calyxes.[6,11] The unmyelinated fibers, with an average diameter of 0.5 μm, appear to contain noradrenaline and to be adrenergic. Parallel physiological studies showed that myelinated fibers are afferent and that unmyelinated fibers are efferent and exclusively adrenergic.[13]

In the juxtaglomerular apparatus, a dense catecholaminergic plexus associated with the glomerular afferent and efferent arterioles and adjacent tubules has been described.[9-12] The authors observed that in the juxtaglomerular region, a large portion of the vascular area is not innervated. Moreover, the nerve terminals are in direct contact with both vascular and tubular cells. This relationship suggests a possible role of nerves in the control of juxtaglomerular apparatus activity, particularly in renin secretion from granular cells, and in changing the diameter of arterioles by acting on vascular smooth muscle. Recent data suggest that nerves which supply the juxtaglomerular apparatus are mainly adrenergic and rich in acetylcholinesterase. Their varicosities are packed with vesicles of the small, dense core variety. These studies are able to differentiate between sympathetic and sensory innervation in the kidney.

Myelinated fibers, mostly observed around intrarenal vessels in the corticomedullary area and close to the renal tubules and glomeruli, have often been cited as evidence of a sensory renal innervation. Moreover, anatomical studies carried out in recent years have also identified nerve fibers containing neuropeptides, such as substance P (SP) and calcitonin gene-related peptide (CGRP), that are frequently found in primary sensory neurons and their peripheral processes.[14-16]

Some authors have shown that kidney denervation produces a supersensitivity to acetylcholine that modifies the responses of the kidney in the excretion of Na and water. These effects were blocked by atropine and muscarinic blockers.[17] According to different authors, it could be suggested that Na^+ transport depends on the balance of cholinergic and adrenergic activities. The finding of choline acetyltransferase (CAT) in renal tissue, unaffected by sympathetic denervation, seems to confirm the presence of cholinergic nerve terminals. Moreover, CAT activity, the conversion of choline into acetylcholine, and high-affinity choline uptake are suggestive of renal cholinergic innervation; the presence of muscarinic receptors supports the existence of a cholinergic system in the kidney, although they may be present in noninnervated areas.

Recent data[18] demonstrated complex acetylcholine-bradykinin interactions, able to increase cholinergic vascular responses through stimulation of muscarinic receptors in rat brain. The same phenomenon was observed in the kidney for the control of renal hemodynamics and Na reabsorption.

B. NEUROTRANSMITTER RECEPTORS

1. Adrenoceptors

Many studies have been performed on renal catecholamine receptors. Insel and Snavely,[19] using fluorescence techniques, demonstrated β-adrenoceptors in the epithelium of proximal and distal convoluted tubules, the collecting ducts, the afferent arterioles of the glomerular vascular pole, and the juxtaglomerular apparatus.

The α_2-adrenoceptors are mainly located within proximal tubules in guinea pig kidney.[20]

Radioligand binding techniques have been used to identify renal adrenoceptors. Many discrepancies exist among the different species examined; α-adrenoceptors are found in renal tubules in the rat, but not in the dog.[21]

Radioligand binding and autoradiographic techniques have also been used to identify β-adrenoceptors in the kidney.[19,22,23] β-Adrenoceptors have been found in rat and dog kidneys;[24] β_1- and β_2-adrenoceptors coexist in the kidney and are involved in β-adrenoceptor-mediated responses.

2. Dopamine Receptors

It is well known that in the kidney dopamine (DA) exerts many effects by interacting with the two subtypes of peripheral DA receptors, DA-1 (D-1) and DA-2 (D-2).[25,26] Its functional effects are characterized by vasodilation, decreased renal resistance, natriuresis, and enhancement of urine outflow.[25,26]

The natriuretic effect of DA is probably mediated by direct action on the proximal tubules. DA-1 binding sites have been shown, using autoradiographical techniques, to be located within proximal cortical tubules.[27] This may suggest an important role of DA in sodium excretion activity mediated through DA-1 receptors.[28-30]

Many findings confirm, from a pharmacological, biochemical, and anatomical point of view, the existence of DA-2 receptors in the kidney[25,26] and their localization within the renal cortex, primarily in the tubules.[31]

It is known that administration of DA into the renal vasculature elevated circulating renin and angiotensin levels. In the dog, intraarterial infusion of DA induces renin release attenuated by DA receptor antagonists.[32,33] The release is also stimulated by fenoldopam, a DA receptor agonist, and attenuated by SCH 23390, a DA-1 receptor antagonist.[34]

Many conflicting results have been reported in man. It has been suggested that the stimulation of renin release was secondary to renal vasodilation rather than a direct juxtaglomerular effect. Moreover, the combined use of radioreceptor binding and autoradiographic techniques with [^{3}H]-SCH 23390 as a ligand has revealed a significant accumulation of [^{3}H]-SCH 23390 binding sites in juxtaglomerular cells.[35]

We have observed, in disagreement with some authors,[36] an accumulation of DA-1 binding sites in juxtaglomerular cells, giving anatomical support for a DA-1 influence on renin release.[34]

3. Cholinergic Receptors

Very little information is available on the presence of cholinergic receptors in the mammalian kidney. Using [^{3}H]-QNB (quinuclidinyl benzylate) as a ligand for muscarinic receptors, Pirola et al.[18] found two distinct types of receptor sites. A homogeneous population of high-affinity muscarinic binding sites was observed in the outer cortex. On the other hand, muscarinic receptors seem to be absent in the inner cortex and medulla. The Kd and B_{max} values are not affected by denervation, thus suggesting the existence of a renal cholinergic system.

6-OHDA noradrenergic denervation and the absence of the modifications of binding parameters indicated that muscarinic receptors are mainly postsynaptic.[37] However, no data are available on possible age-induced changes in the cholinergic renal system.

Spontaneous hypertension is accompanied by a decrease in the number of renal muscarinic cholinergic receptors in the vascular tree and tubules.[37] Although the meaning of the hypertensive changes is not clear, these data are consistent with the hypothesis of a reduced cholinergic tone in hypertension.[39,40]

C. NEUROTRANSMITTER METABOLISM

Many studies have examined the metabolism of catecholamines in relation to the activation of afferent renal sympathetic nerves in the rat. These studies have shown that 85 to 95% of the noradrenaline, dopamine, and methoxytryptamine injected into the proximal tubules is recovered in the urine, and less than 10% is chemically altered.

Dopamine and noradrenaline are methylated by the enzyme catechol-*O*-methyltransferase and secreted into the tubular lumen, with complete delivery into the final urine. The catecholamine concentration was measured in the renal artery and vein, and urine, in innervated and denervated kidneys in the rat.[41] Noradrenaline derives from circulating noradrenaline and the renal nerves, since renal denervation reduced the renal component of urinary noradrenaline excretion to zero. Only 30% of the circulating noradrenaline was excreted with urine.

Approximately 70% of the urinary adrenaline excretion derives from circulating adrenaline. The remainder derives from intrarenal adrenaline production from adrenergic nerve endings.

DA urinary excretion derives from renal nerves and nonneuronal sources in the kidney. The low circulating arterial concentration of DA cannot explain the rate of DA excretion. Both innervated and denervated kidney contain L-DOPA-decarboxylase, the enzyme that converts L-DOPA to DA in renal tubular epithelial cells.[42]

DA can be synthesized within proximal tubules, and L-DOPA filtered at the glomerulus can be taken up by these cells with a sodium-coupled transport system. Experimental evidence suggests that endogenous DA plays an important role in the regulation of Na excretion via activation of DA-1 tubular receptors.[28,36] Renal-derived DA appears to be important for the control of Na balance during Na intake.[28,36] The increase of Na intake is accompanied by an increase in renal Na and DA excretion in experimental animals and humans.

Recent studies in rats provide evidence of the role of DA in renal responses to acute volume expansion. In rats, the natriuretic and diuretic response to acute volume expansion was attenuated by SCH 23390 and haloperidol.[43-45] These results suggest a possible role of the remaining natriuretic response, or, more likely, some other factors were also responsible for the natriuresis during acute volume expansion. This method causes the release of atrial natriuretic factor that should be responsible for the natriuresis during acute volume expansion.[46-48]

The existence of a DA-1 receptor blockade that is correlated with a reduction of urine output and urinary sodium excretion, on days with a normal sodium diet,[49] may be related to a dopaminergic tone regulating, in basal conditions, renal sodium excretion. The same results were obtained by inhibiting DA synthesis with carbidopa.[49]

DA produced during high sodium intake may cause natriuresis by inhibiting Na^+-K^+-ATPase.[42] The increase in dietary sodium intake was responsible for the renal DA increase, which acts to inhibit Na^+-K^+-ATPase. This action is most likely responsible for elimination of the excess Na and contributes to the maintenance of Na balance.

II. AGE-RELATED CHANGES IN KIDNEY INNERVATION

Age-related changes in the innervation of several organs have been investigated in different mammalian species, including man.[50] Age-related

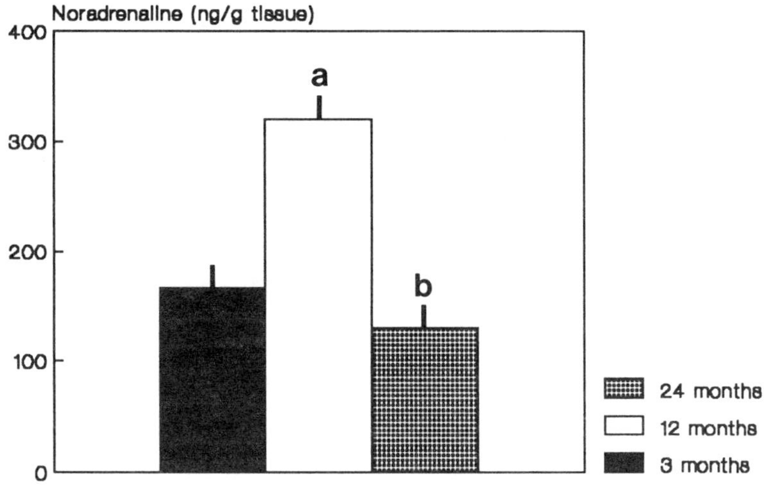

FIGURE 1. Noreadrenaline concentrations in the kidney of male Sprague-Dawley rats of different ages. Noradrenaline levels were measured by high-pressure liquid chromatography with electrochemical detection. Values are means ± SEM of data from eight animals per group. a, $p < 0.001$ vs. 3 months; b, $p < 0.01$ vs. 3 months and $p < 0.001$ vs. 12 months. (Derived from results in Vega, J. A., Ricci, A., and Amenta, F., *Mech. Ageing Dev.*, 54, 185, 1990.)

changes in the neuroeffector system of the renal artery and vein have been reported,[51,52] but many discrepancies exist among the various authors. Some of them report a nerve loss in the renal artery of adult rabbits, in accordance with the histochemical and analytical results obtained in the rat sympathetic ganglia,[51-54] and a decline in the β-adrenergic responsiveness of the aged rat kidney.[52]

On the other hand, some physiological and pharmacological studies revealed that [^{3}H]-noradrenaline accumulation does not change with age in rat renal blood vessels, and neither noradrenergic nerve density nor reactivity to noradrenaline in the same vessels declines with age.[55] Moreover, a decline in renal blood flow was described.[56]

Binding studies carried out using membrane preparations of young and old Sprague-Dawley rats demonstrated that high-affinity α_1- and β_2-adrenoceptors were significantly reduced in aged rats. Moreover, low-affinity α_1-, α_2-, and β_1-receptors were unchanged with age.[57]

The whole kidney noradrenaline concentration shows two kinds of developmental changes in male Sprague-Dawley rats: an increase at 12 months and a decrease at 24 months of age (Figure 1). The same is true for perivascular noradrenergic nerves (Figures 2 and 3), which show an increase in the number of bundles and/or varicosities around the circumference of four types of kidney arteries investigated in adult animals and a remarkable decrease in old animals[58] (Figure 4).

Similar observations have been done for acetylcholinesterase (AChE)-positive nerve fibers (Figure 5), although their density was rather low to allow

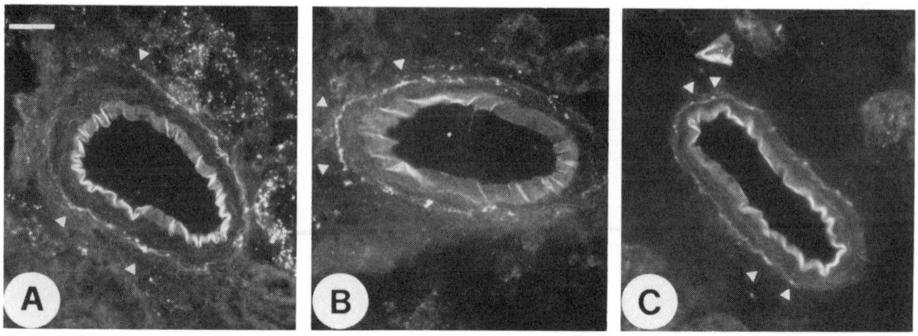

FIGURE 2. Glyoxylic acid histofluorescence of adrenergic nerve fibers (arrowheads) in sections of the arcuate artery of male Sprague-Dawley rats of 3 (A), 12 (B), and 24 (C) months of age. Note the highest expression of fluorescent axons in the 12-month-old animal and the lowest in the 24-month-old one. (Bar = 30 μm.)

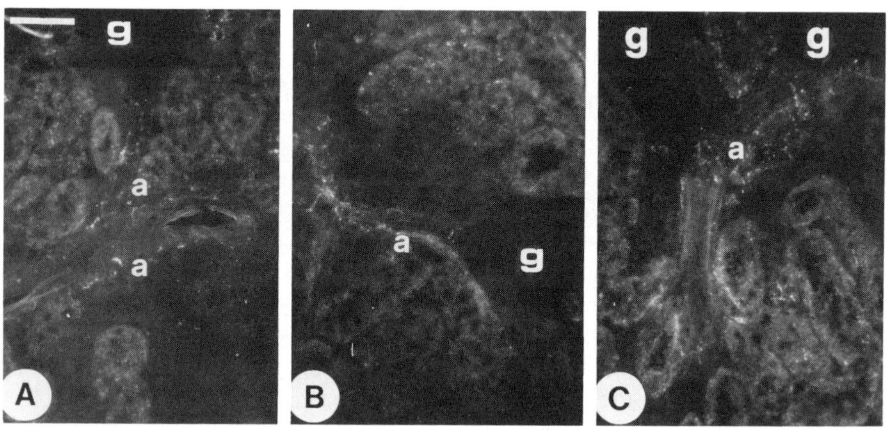

FIGURE 3. Glyoxylic acid histofluorescence of adrenergic axons supplying afferent arterioles (a) in the renal cortex of male Sprague-Dawley rats of 3 (A), 12 (B), and 24 (C) months of age. The heaviest density of axons is noticeable in the middle-aged animal; g, glomerulus (Bar = 25 μm.)

a detailed quantification of the age-dependent changes in the renal arterial tree. The different expression of noradrenergic nerves in the kidneys of adult and old rats is different to explain, and few reports have compared morphological, functional, and biochemical parameters not limited to young and old, or adult and old, animals.[52,57,59,60]

Since the main age-related changes in renal morphology and function were described between middle and old age, it is possible that increased expression of noradrenergic periarterial nerves may represent a compensatory mechanism for counteracting the impairment of renal function between young and adult rats.[52] The subsequent decrease from adult to old rats should be

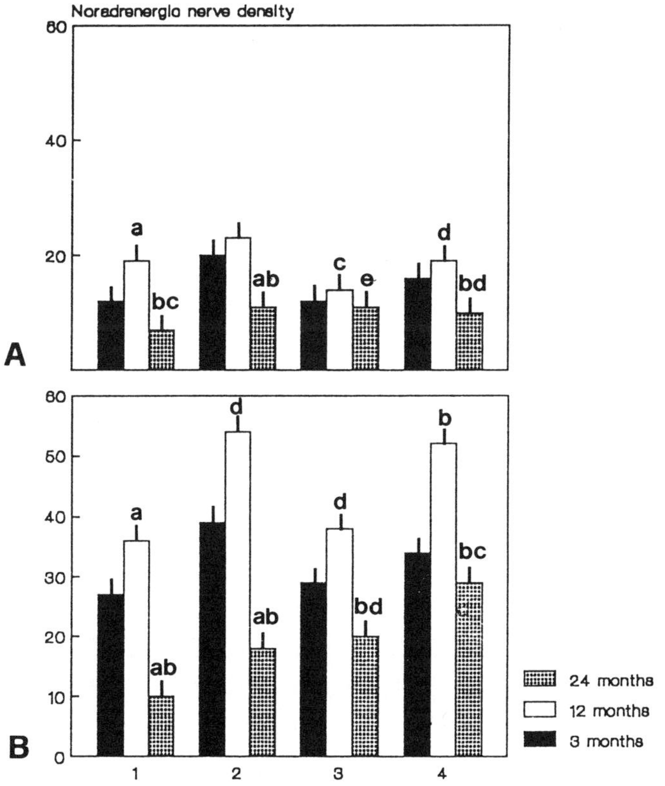

FIGURE 4. Density of fluorescent noradrenergic axons (A) and varicosities (B) in sections of lobar arteries (1), arcuate arteries (2), interlobular arteries (3), and afferent arterioles (4) of male Sprague-Dawley rats of 3, 12, and 24 months of age. The density of the axons is expressed as the number of bundles per 0.01 mm^{-1} of vessel circumference; the density of the varicosities is expressed as the number of fluorescent varicosities per 0.01 mm^{-1} of vessel circumference. Values are means ± SEM of data from eight animals per group. a, $p < 0.001$ vs. 3 months; b, $p < 0.001$ vs. 12 months; c, $p < 0.05$ vs. 3 months; d, $p < 0.01$ vs. 3 months; e, $p < 0.01$ vs. 12 months. (Derived from results in Vega, J. A., Ricci, A., and Amenta, F., *Mech. Ageing Dev.*, 54, 185, 1990.)

due to the progressive development of degenerative lesions that occur in old age.

III. NEUROTRANSMITTER RECEPTORS IN AGING

A. α-ADRENOCEPTORS

Using isolated perfused kidneys from male Fischer 344 rats of different ages, some authors have observed an age-related decrease in prejunctional α-adrenoceptors and an influence in the increasing noradrenaline release from renal sympathetic nerves, but no alterations in vascular response to sympathetic nerve activation.

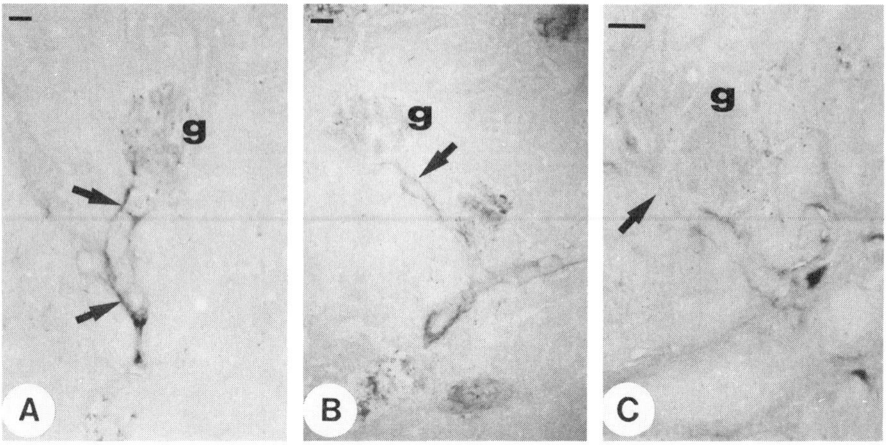

FIGURE 5. Acetylcholinesterase-positive nerve fibers of the afferent arteriole (arrow) in the renal cortex of male Sprague-Dawley rats of 3 (A), 12 (B), and 24 (C) months of age. g, glomerulus. (Bar = 20 μm.)

B. β-ADRENOCEPTORS

β-Adrenoceptors in the kidney were studied using ligand-affinity binding techniques, basal adenylate cyclase activity, and cAMP synthesis.[61] In male Wistar rats, a significant increase in β-receptor density was described. Moreover, a decrease in ligand affinity was observed. The age-related changes observed in the rat kidney were similar to those described in some other organs and seem to be related to a compensatory mechanism to ensure equal cAMP synthesis for a given agonist stimulation.

C. DOPAMINE RECEPTORS

Renal DA, DA-1, and DA-2 receptors evidence an age-dependent loss in number (B_{max}), but not affinity (Kd), suggesting a similarity to the age-dependent loss of D-1 and D-2 receptors in the central nervous system. In renal cortex membrane, the use of DA-sensitive adenylate cyclase activity as a marker of postjunctional DA-1 receptors revealed a decrease in DA-1 when mature animals were compared with young ones.[30] These data, compared with binding analysis, allow us to hypothesize that the number rather than the pharmacological profile of DA-1 receptor sites is impaired with age. A series of investigations revealed a DA-1 age-dependent reduction in rabbit renal artery.[30] The author described a 40% decrease in adenylate cyclase levels in old rabbits compared to young animals. In the rabbit renal artery, old age is also accompanied by changes in [^{3}H]-spiroperidol binding sites. Once again, the dissociation constant (Kd) was unchanged and the maximal binding capacity was reduced by more than 40%.[30] Renal prejunctional DA-2 receptors, negatively coupled with the cAMP-generating system,[26] are affected by age.

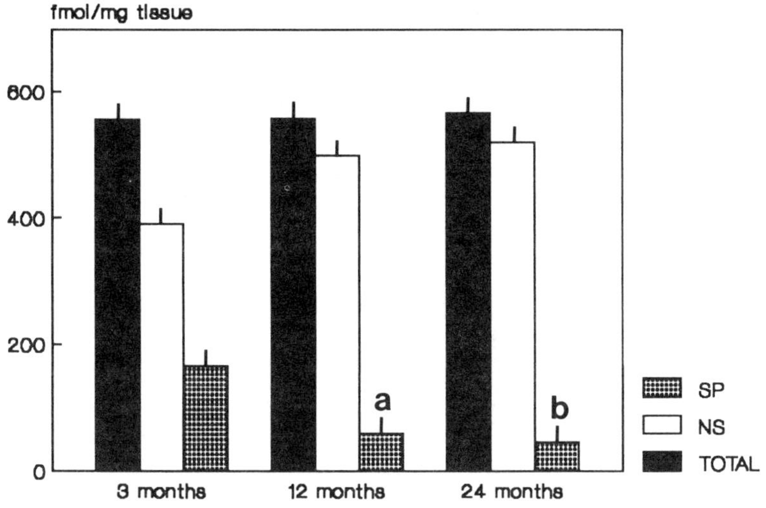

FIGURE 6. Values of total (TOTAL), nonspecific (NS), and specific (SP) binding values of [^{3}H]-spiroperidol to sections of kidney of male Wistar rats of different ages. Values are means ± SEM of binding data obtained by incubating sections with 3 n*M* [^{3}H]-spiroperidol alone (total binding) or plus 1 μ*M* (+)-butaclamol to generate nonspecific binding. Specific binding was obtained by subtracting nonspecific from total binding. a, $p < 0.001$ vs. 3 months; b, $p < 0.001$ vs. 3 months and $p < 0.01$ vs. 12 months.

Using Wistar rat kidney sections and [^{3}H]-spiroperidol as a ligand in 3-, 12-, and 24-month-old animals, we observed a decrease of [^{3}H]-spiroperidol binding sites in 12- vs. 3-month-old rats. A further decrease was observed in the oldest age group (Figure 6). Autoradiographic techniques revealed that [^{3}H]-SCH 23390 binding sites, which are located primarily within proximal tubules and to a lesser extent within smooth muscle cells in renal artery branches, decreased in a more evident way in the tubular than in the vascular bed (Figure 7). Hence, tubular rather than vascular DA-1 receptors are more sensitive to age.

IV. NEUROTRANSMITTER METABOLISM AND PHYSIOLOGICAL FUNCTIONS OF NEUROTRANSMITTERS IN AGING

Sympathetic neurons play a fundamental role in regulating the activity of various peripheral neuroeffector mechanisms. Hence, any failure of autonomic control and activity could have serious effects on the well-being of the organism. If these functions are altered, either through pathology or aging, the effects can be important. Two factors can trigger these alterations: intrinsic events and extrinsic changes. Aging seems to act by modifying neural intrinsic metabolism (protein, lipids, and cellular respiratory systems). The extrinsic

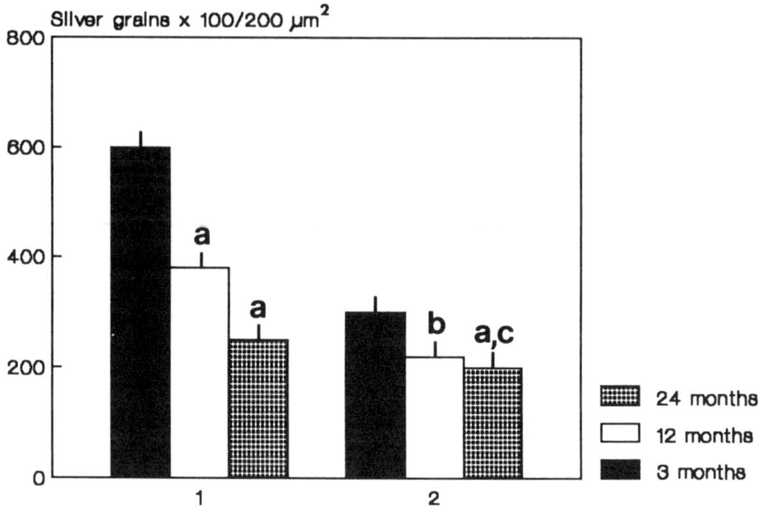

FIGURE 7. Specific silver grains (total, nonspecific binding) developed in proximal tubules of the renal cortex (1) and in the tunica media of the interlobular artery (2). Sections of the kidney of male Sprague-Dawley rats of different ages were exposed to 5 n*M* [^{3}H]-SCH 23390 and then processed for light-microscope autoradiography. Values are means ± SEM of data obtained from eight animals per age group. a, $p < 0.001$ vs. 3 or 12 months; b, $p < 0.01$ vs. 3 months; c, $p < 0.05$ vs. 3 months.

factors are the results of lifetime environmental challenges. Intrinsic and extrinsic factors are usually superimposed, and it is impossible to differentiate one from the other.

Intraneuronal lipopigment accumulation and the death of many cells suggest the existence of compensatory changes in the surviving neurons that seem to increase their activity.[62] The investigation of sympathetic noradrenaline levels in humans and rats shows that, with age, some organs become partially denervated at the periphery.[62]

Reports are not yet available concerning the influence of aging on DA levels and catabolism at the periphery. Some studies carried out using human material show a significant decrease of DA-containing nerve fibers in the kidney of old patients.[69] The loss of the dopaminergic system at the periphery seems to be similar to some events that occur in the central nervous system.

Although it is well known that rat kidney is particularly sensitive to aging, and several age-related changes in renal morphology and functions have been reported,[50] information is not yet available concerning the possible role of age in the innervation of vascular and nonvascular components of the kidney.

A significant loss of noradrenergic fibers in the renal artery of adult compared with young rabbits has been described[51] as well as a decline in the β-adrenergic responsiveness of aged kidney.[52]

V. PATHOPHYSIOLOGY AND CLINICAL CORRELATES

The renal nerves are involved in the efferent control system for the regulation of some renal functions. Many events are able to induce alterations in renal blood flow, renal tubule solute transport, water excretion, and release of vasoactive substances.

The functional absence of sympathetic cholinergic vasodilatory fibers in the kidney allows us to hypothesize that a decrease or increase of renal sympathetic nerve activity produces, directly or reflexively, changes in renal tubular solutes and water reabsorption (probably mediated by adrenergic or cholinergic nerves and receptors).

The cortical proximal convoluted tubule has been identified as the site of this neural control. It seems to control renal physiological responses to dietary sodium intake. Moreover, the renin responses to renal sympathetic stimulation seem to be due to the stimulation of β_1- and DA-1 receptors in relation to an intact macula densa receptor mechanism.

Renal nerve stimulation is also correlated with an increase in the secretion of renal prostaglandin and kallikrein, which are involved in a nonneuronal control of renal functions. All these systems are rich in adaptive and compensatory responses, particularly during renal failure. These responses seem to be mediated by afferent renal nerves responsible for the renal reflex. It is important to clarify whether the age-related changes observed in the kidney are a generalized phenomenon, and whether they are limited to laboratory animals or occur in humans as well.

These changes should contribute to an understanding of the role of senescence in renal function; both afferent and efferent nerves have homeostatic significance in the control of body fluids, electrolyte balance, arterial pressure, and the development and maintenance of hypertension.

Exact knowledge of these events can contribute to an understanding of the pathophysiological changes correlated with aging, and their possible treatment.

ACKNOWLEDGMENTS

The results on the aging of the renal dopaminergic system are part of an investigation supported by the Sandoz Foundation for Gerontological Research.

REFERENCES

1. **Pappenaim, S.,** Effect of controlateral nephrectomy and renal nerves on renal function, *Med. Ann. D.C.*, 35, 467, 1966.
2. **Mitchell, G. A. G.,** The nerve supply of the kidneys, *Acta Anat.*, 10, 1, 1950.
3. **Takeuchi, J., Iino, S., Hanada, S., Sakai, S., Oya, N., and Kitamura, K.,** Experimental studies on the nervous control of the renal circulation — effect of the electrical stimulation of the spinal cord on the renal circulation, *Jpn. Heart J.*, 5, 69, 1964.
4. **Norvell, J. E.,** The aorticorenal ganglion and its role in renal innervation, *J. Comp. Neurol.*, 133, 101, 1979.
5. **Barajas, L.,** Anatomy of the juxtaglomerular apparatus, *Am. J. Physiol.*, 237, F333, 1979.
6. **Zimmerman, H. D.,** Elektronenmikroskopische befunde zur innervation des nephron nach untersuchungen an der fetalen nachniere des menschen, *Z. Zellforsch.*, 129, 65, 1972.
7. **DiBona, G. F.,** Neurogenic regulation of renal tubular sodium reabsorption, *Am. J. Physiol.*, 233, F73, 1977.
8. **Pang, P. K. T., Uchiyama, M., and Sawyer, W. H.,** Endocrine and neural control of amphibian renal function, *Fed. Proc.*, 41, 2365, 1982.
9. **Weitsen, H. A. and Norvell, J. E.,** Cholinergic innervation of the autotransplanted canine kidney, *Circ. Res.*, 251, 535, 1969.
10. **Norvell, J. E.,** Renal nerves: are they essential?, *N. Engl. J. Med.*, 283, 261, 1970.
11. **Barajas, L. and Wang, P.,** Demonstration of acetylcholinesterase in the adrenergic nerves of the renal glomerular arterioles, *J. Ultrastruct. Res.*, 53, 244, 1975.
12. **Burn, J. H. and Rand, M. J.,** Acetylcholine in adrenergic transmission, *Annu. Rev. Pharmacol.*, 5, 163, 1965.
13. **Niijima, A.,** Observation on the localization of mechanoceptors in the kidney and afferent nerve fibres in the renal nerves in the rabbit, *J. Physiol.*, 254, 81, 1975.
14. **Ferguson, M. and Bell, C.,** Substance P-immunoreactive nerves in the rat kidney, *Neurosci. Lett.*, 60, 183, 1985.
15. **Su, H. C., Wharton, J., Polak, J. M., Mulderry, P. K., Ghatei, M. A., Gibson, S. J., Terenghi, G., Morrison, J. F. B., Bellesta, J., and Bloom, S. R.,** Calcitonin gene-related peptide immunoreactivity in afferent neurons supplying the urinary tract: combined retrograde tracing and immunochemistry, *Neuroscience*, 18, 727, 1986.
16. **Knight, D. S., Beal, J. A., Yuan, Z. P., and Fournet, T. S.,** Vasoactive intestinal peptide-immunoreactive nerves in the rat kidney, *Anat. Rec.*, 219, 193, 1987.
17. **Alvarez, A. L., Pirola, C. J., Bennarroch, E. E., Finkielman, S., and Nahmod, V. E.,** Muscarinic receptor involvement in the electrolytic changes of kidney tubular cells elicited by bradykinin and cholinergic agonists, *IRCS Med. Sci.*, 8, 197, 1980.
18. **Pirola, C. J., Balda, M. S., Alvarez, A. L., Finkielman, S., and Nahmod, V. E.,** Interaction between acetylcholine and bradykinin in the lateral septal area of the rat brain: involvement of muscarinic receptors in cardiovascular responses, *Neuropharmacology*, 25, 1387, 1986.
19. **Insel, P. A. and Snavely, M. D.,** Catecholamines and the kidney: receptors and renal function, *Annu. Rev. Physiol.*, 43, 625, 1981.
20. **Young, W. S. and Kuhar, M. J.,** Alpha-2 adrenergic receptors are associated with renal proximal tubules, *Eur. J. Pharmacol.*, 67, 493, 1981.
21. **Felder, R., Pelyao, J., Schoelkopf, L., Cooke, M., Jose, P., and Eisner, G.,** Glomerular adrenergic receptors, *Physiologist*, 23, 162, 1980.
22. **Woodcock, E. A., Funder, J. W., and Johnston, C. I.,** Decreased cardiac beta adrenoceptor in hypertensive rats, *Clin. Exp. J. Pharmacol.*, 67, 493, 1980.
23. **Gavendo, S., Kapuler, S., Serban, I., Iania, I., Ben-David, E., and Eliahou, H.,** Beta-1 adrenergic receptors in kidney tubular cell membrane in the rat, *Kidney Int.*, 17, 764, 1980.

24. **Felder, R., Schoekopf, L., Pelazo, J., Blecher, M., Calcagno, P., Eisner, G., and Jose, P.,** Adrenergic and dopaminergic receptors in glomeruli and cortical tubules, *Proc. Am. Soc. Nephrol.,* 13, 132, 1980.
25. **Bell, C.,** Dopamine and the kidney, in *Peripheral Dopamine Pathophysiology,* Amenta, F., Ed., CRC Press, Boca Raton, FL, 1990, 87.
26. **Lokhandwala, M. F. and Hegde, S. S.,** Cardiovascular pharmacology of dopamine receptor agonists, in *Peripheral Dopamine Pathophysiology,* Amenta, F., Ed., CRC Press, Boca Raton, FL, 1990, 63.
27. **Hegde, S. S., Ricci, A., Amenta, F., and Lokhandwala, M. F.,** Evidence from functional and autoradiographic studies for the presence of tubular effects of fenoldopam, *J. Pharmacol. Exp. Ther.,* 251, 1237, 1989.
28. **Felder, R. A., Blecher, M., Eisner, G. M., and Jose, P. A.,** Cortical tubular and glomerular dopamine receptors in the rat kidney, *Am. J. Physiol.,* 246, F557, 1984.
29. **Felder, R. A.,** Autoradiographic localization of DA-1 dopamine receptors in microdissected rat proximal convoluted tubule, *Kidney Int.,* 31, 432, 1987.
30. **Amenta, F.,** Biochemistry and autoradiography of peripheral dopamine receptors, in *Peripheral Dopamine Pathophysiology,* Amenta, F., Ed., CRC Press, Boca Raton, FL, 1990, 39.
31. **Ricci, A., Rossodivita, A., and Amenta, F.,** Dopamine DA-2 receptor sites in the rat renal cortex: a light microscope study, *Naunyn-Schmiedeberg's Arch. Pharmacol.,* 344, 259, 1991.
32. **Bell, C., Lang, W. J., and Laska, F.,** Dopamine containing vasomotor nerves in the dog kidney, *J. Neurochem.,* 31, 77, 1978.
33. **Imbs, J. L., Schmidt, M., and Schwartz, J.,** Effect of dopamine on renin secretion in the anaesthetized dog, *Eur. J. Pharmacol.,* 33, 151, 1975.
34. **Montier, F., Walrant, P., Pratz, J., and Cavero, I.,** Studies on the fenoldopam-evoked renin release in dogs, *Clin. Exp. Hypertens.,* A9, 1091, 1987.
35. **Amenta, F. and Ricci, A.,** Autoradiographic localization of dopamine DA-1 receptors in the rat renal vasculature using [3H]-SCH 23390 as a ligand, *J. Auton. Pharmacol.,* 10, 373, 1990.
36. **Huo, T. and Healy, P. H.,** Autoradiographic localization of dopamine DA-1 receptors in rat kidney with [3H]-SCH 23390, *Am. J. Physiol.,* 257, F414, 1989.
37. **Yamada, S., Hayashi, E., Yamamura, H. I., and Roeske, W. R.,** Presynaptic muscarinic cholinergic and postsynaptic beta-adrenergic receptors in splenic tissue, in *Molecular Pharmacology of Neurotransmitter Receptors,* Segawa, T., Ed., Raven Press, New York, 1983, 43.
38. **De Michele, M., Amenta, F., and Cavallotti, C.,** Autoradiographic localization of muscarinic receptors within the rat kidney, *Eur. J. Pharmacol.,* 169, 297, 1989.
39. **Julius, S., Pascual, A. V., and London, R.,** Role of the parasympathetic inhibition in the hyperkinetic type of borderline hypertension, *Circulation,* 44, 413, 1971.
40. **Korner, P. I., Shaw, J., Uther, J. B., West, M. J., McRitchie, R. J., and Richards, J. G.,** Autonomic and non-autonomic circulatory components in essential hypertension in man, *Circulation,* 48, 107, 1973.
41. **Baines, A. D., Craan, A., Chan, W., and Morgunov, N.,** Tubular secretion and metabolism of dopamine, norepinephrine, methoxytyramine and normetanephrine by the rat kidney, *J. Pharmacol. Exp. Ther.,* 208, 144, 1979.
42. **Aperia, A., Bertorello, A., and Seri, I.,** Dopamine causes inhibition of Na^+-K^+-ATPase activity in rat proximal convoluted tubule segments, *Am. J. Physiol.,* 252, F39, 1987.
43. **Martin-Grez, M., Angchanpen, M., Gambaro, P., Schnermann, J., Schbert, G., and Biggs, J. P.,** Evidence for an involvement of dopamine receptors in the natriuretic response to atrial natriuretic peptide, *Klin. Wochenschr.,* 65 (Suppl. 8), 97, 1987.
44. **Jose, P. A., Halloway, R. R., Campbell, T. W., and Elsner, G. M.,** Dopamine blockade attenuates the natriuresis of saline loading in the adrenalectomized rat, *Nephron,* 48, 54, 1988.

45. **Angchanpen, P., Marin-Grez, M., and Scnermann, J.,** Effect of dopamine antagonists on the urine flow of rats infused with hypotonic saline, *Br. J. Pharmacol.,* 93, 151, 1988.
46. **Hirth, C., Stac, J. P., Jon, A., Kazda, S., Morich, F., Neuser, D., and Wohlfeil, S.,** The renal response to acute hypervolemia is caused by atrial natriuretic peptides, *J. Cardiovasc. Pharmacol.,* 8, 268, 1986.
47. **Kraibi, A. A., Granger, J. P., Burnett, J. C., Jr., Walker, K. R., and Knox, F. G.,** Role of atrial natriuretic factor in the natriuresis of acute volume expansion, *Am. J. Physiol.,* 252, R921, 1987.
48. **Barbee, R. W. and Trippodo, N. C.,** The contribution of atrial natriuretic factor to acute volume natriuresis in rats, *Am. J. Physiol.,* 253, F1129, 1987.
49. **Frederickson, E. D., Bradley, T., and Goldberg, L. I.,** Blockade of renal effects of dopamine in the dog by the DA-1 antagonist SCH 23390, *Am. J. Physiol.,* 249, F236, 1985.
50. **Masoro, E. J.,** *Handbook of Physiology in Aging,* CRC Press, Boca Raton, FL, 1981, 175.
51. **Cowen, T., Haven, A. J., Wen-Quin, C., Gallen, D. D., Franc, F., and Burnstock, G.,** Development and aging of perivascular adrenergic nerves in the rabbit. A quantitative fluorescent histochemical study using image analysis, *J. Auton. Nerv. Syst.,* 5, 317, 1982.
52. **Simpkins, J. W., Fields, F. P., and Ress, R. J.,** Age-related decline in adrenergic responsiveness of the kidney, heart and aorta of male rats, *Neurobiol. Aging,* 4, 233, 1983.
53. **Santer, R. M.,** Fluorescence histochemical evidence for decreased noradrenaline synthesis in sympathetic neurons of aged rats, *Neurosci. Lett.,* 15, 177, 1979.
54. **Santer, R. M., Partanen, M., and Hervonen, A.,** Glyoxylic acid fluorescence and ultrastructural studies of neurons in the coeliac-superior mesenteric ganglion of the aged rat, *Cell Tissue Res.,* 211, 475, 1980.
55. **Dukles, S. P., Carter, B. J., and Williams, C. L.,** Vascular adrenergic neuroeffector function does not decline in aged rats, *Circ. Res.,* 56, 109, 1985.
56. **Tuma, R. F., Irion, G. L., Vasthare, U. S., and Heinel, L. A.,** Age-related changes in regional blood flow in the rat, *Am. J. Physiol.,* 249, H485, 1985.
57. **Galbusera, M., Garattini, S., Remuzzi, G., and Mennini, T.,** Catecholamine receptor binding in rat kidney: effect of aging, *Kidney Int.,* 33, 1073, 1988.
58. **Vega, J. A., Ricci, A., and Amenta, F.,** Age-dependent changes of the sympathetic innervation of the rat kidney, *Mech. Ageing Dev.,* 54, 185, 1990.
59. **Barrows, C. H., Jr., Falzone, J. A., and Shock, N. W.,** Age differences in the succinoxidase activity of homogenates and mitochondria from the livers and kidneys of rats, *J. Gerontol.,* 15, 130, 1960.
60. **Beuchene, R. E., Fanestil, D. D., and Barrow, C. H., Jr.,** The effect of age on active transport and sodium-potassium-activated ATPase activity in renal tissues of rats, *J. Gerontol.,* 20, 306, 1965.
61. **Vanscheeuwijck, P., Van de Velde, E., and Fraeyman, N.,** The beta-adrenergic transduction system in kidneys from young and senescent rats, *Eur. J. Pharmacol.,* 188, 129, 1990.
62. **Hervonen, A., Partanen, M., Helen, P., Koistinaho, J., Alho, H., Baker, D. M., Johnson, J. E., Jr., and Santer, R.,** The sympathetic neuron as a model of neuronal ageing, in *Neurochemistry: Modern Methods and Applications,* Hervonen, A. and Partanen, M., Eds., Alan R. Liss, New York, 1986, 569.
63. **Bell, C.,** personal communication.

Chapter 12

NORADRENERGIC SYMPATHETIC INNERVATION OF LYMPHOID ORGANS DURING DEVELOPMENT, AGING, AND IN AUTOIMMUNE DISEASE

Denise L. Bellinger, Suzanne Y. Felten, Kurt D. Ackerman, Dianne Lorton, Kelley S. Madden, and David L. Felten

TABLE OF CONTENTS

0-8493-6981-9/93/$0.00 + $.50

I. INTRODUCTION

In the early literature, noradrenergic sympathetic innervation of lymphoid organs was described in a variety of mammals,[1-4] but was considered mainly as vascular innervation functioning to regulate blood flow and to contract capsular/trabecular smooth muscle. The possibility that cells of the immune system also were targets of sympathetic nerves in lymphoid organs was not investigated until recently. Studies from our laboratories[5-15] and others[16-21] have demonstrated noradrenergic sympathetic innervation not only along smooth muscle of the vasculature and capsular/trabecular system, but also extensive innervation of the parenchyma of specific functional compartments of primary and secondary lymphoid organs. In young adult rodents, noradrenergic innervation of lymphoid compartments is regional and specific, generally distributing extensively into zones of T-lymphocytes and macrophages, with minimal direct distribution into nodular or follicular zones of B-lymphocytes in adults. The presence of adrenoceptors on a variety of cell types constituting the host defense and immune systems, including T- and B-lymphocytes, monocytes and macrophages, and granulocytes, also has been demonstrated[22-33] (reviewed in Reference 34). Functional studies in young adult experimental animals have demonstrated predictable alterations in immune responses following manipulation of noradrenergic innervation, noradrenaline (NA) itself, and its receptor[8,9,11,12,26,35-40] (reviewed in Reference 41). Collectively, these findings indicate a role for autonomic sympathetic nerves in lymphoid organs in immunomodulation, and provide an additional route besides the hypothalamo-pituitary neuroendocrine outflow available to the brain for influencing cells of the immune system.

It is well documented that during development, maturation, and aging, marked changes occur in both the organization of lymphoid organs and regulation of immunocompetence. Findings from our laboratories suggest that noradrenergic sympathetic nerves are responsive and possibly contributory to developmentally-, plasticity-, and age-related changes of the immune system. In this chapter, we describe the ontogeny, maturation, and aging of sympathetic neurotransmission in primary and secondary lymphoid organs, and discuss the possible implications of a dynamic noradrenergic sympathetic modulation of the immune system in immunosenescence and autoimmune disorders.

II. NORADRENERGIC INNERVATION OF PRIMARY LYMPHOID ORGANS

A. BONE MARROW

1. Maturation

In young adult rats, noradrenergic innervation to long bones courses with the appropriate branch of the spinal nerve supplying that region. These nerves further branch into bundles that enter the interior of the bone along with the vasculature via the nutrient foramena. Within the bone marrow (Plate 1*), fine, varicose noradrenergic nerve fibers form plexuses along the vasculature, and occasionally nerve fibers branch from these plexuses into the surrounding parenchyma among hemopoietic cells in the marrow[42] (reviewed in Reference 43). Some noradrenergic nerve fibers are present in the parenchyma of the bone marrow, as free linear profiles without association with the vasculature (Plate 2).

2. Ontogeny and Senescence

Very few developmental studies of innervation of bone marrow have been undertaken.[21,44,45] In the Wistar rat[44] and the New Zealand rabbit,[45] innervation of the bone marrow is first apparent during late fetal life. Miller and McCuskey[45] found that noradrenergic fibers are present prior to the onset of hemopoietic activity, leading these investigators to hypothesize a role for noradrenergic innervation in the subsequent development of hematopoiesis. We are aware of no studies that have examined noradrenergic innervation of bone marrow from aged animals.

B. THYMUS

1. Maturation

Noradrenergic nerve fibers that distribute to the thymus originate from postganglionic cell bodies in the upper paravertebral ganglia of the sympathetic chain, especially the superior cervical and stellate ganglia.[17,46,47] In young adult rodents, noradrenergic fibers enter the thymus with large blood vessels as dense nerve plexuses, and travel in the thick capsule and interlobular septa or continue with the vasculature into the cortex[5,6,9,17,46,48-52] (reviewed in Reference 43). From these plexuses, fibers branch into the cortical region, where they are found adjacent to thymocytes (Plate 3). Noradrenergic fibers also appear to associate with yellow autofluorescent cells in the cortex (Plate 3). In the septa, noradrenergic nerves, as well as neuropeptide-containing nerves, are found in close proximity to mast cells. Noradrenergic nerve fibers are predominantly located in the cortical region, with a slightly higher density near the corticomedullary junction. At the junction between the cortex and the medulla, noradrenergic fibers are found along the medullary venous sinuses, which are continuous with the system of innervated vessels in the septa. Noradrenergic fibers extend from these sinuses into the adjacent cortical region. Occasionally, noradrenergic fibers course through the medullary parenchyma. Neurochemical analysis has revealed that catecholamine innervation of the thymus is virtually all noradrenergic.

Immunohistochemical methods have revealed the presence of numerous neuropeptides in nerve-like profiles in the thymus, including neuropeptide Y (NPY), tachykinins (substance P, SP); neurokinin A and B, collectively), SP,

* Plates 1 through 32 follow page 274.

calcitonin gene-related peptide, and vasoactive intestinal polypeptide[9,43,53-57] (reviewed in Reference 43). While the origin of most neuropeptidergic nerve fibers has not been revealed, the distribution of NPY+ nerves in the rat thymus overlaps with the pattern of tyrosine hydroxylase (TH, the rate-limiting enzyme in the synthesis of NA) staining in the capsular/septal system and in cortical regions, and denervation of noradrenergic nerve fibers with 6-hydroxy-dopamine (6-OHDA) also depletes NPY, suggesting that NA and NPY may colocalize in sympathetic nerves. Like noradrenergic sympathetic innervation, NPY+ nerve fibers are present as varicose punctate and linear profiles, and distribute with the vasculature and into parenchymal fields of thymocytes and associated cells. Other peptidergic innervation of the thymus is not as robust as noradrenergic or NPY innervation of the thymus, and has a different pattern of distribution (see References 43 and 58 for reviews).

2. Ontogeny

The ontogeny of noradrenergic innervation of the thymus is largely a postnatal event[59] (reviewed in Reference 60). With fluorescence histochemistry for localization of catecholamines (CA), noradrenergic sympathetic nerves in the rodent thymus have been observed as early as days (d) 17 to 18 of gestation (full term is 20 to 21 d).[18,61] At this stage of development, only an occasional noradrenergic fiber is observed in the cortex and along arterioles at the corticomedullary junction. In thymuses from Fischer 344 (F344) rats at 1 d of age examined with fluorescence histochemistry, the density of fluorescent profiles is sparse, and the intensity of the fluorescence is dull compared with thymuses from adults.[59,60] Most of the noradrenergic innervation is present as nerve bundles traveling in the capsules and along interlobular septa that extend to the corticomedullary boundary. Few fibers are detected in the cortex beneath the capsule. At the corticomedullary junction, dense plexuses of fine fibers course adjacent to some of the small to medium-size arteries, and single fluorescent profiles extend a short distance from the septa into the surrounding parenchyma.

By postnatal day 7, the thymus is much larger and further subdivided into lobules by connective tissue septa.[59,60] The density of noradrenergic nerves in the interlobular septa is greater than that seen in the thymus of newborn rats, but has not attained the density observed in adulthood. Denser plexuses of noradrenergic nerve fibers also are present along arteries in the septa and their branches that extend deep into the cortex. Occasional fibers exit these arterial plexuses and arborize in the parenchyma of the cortex among thymocytes. These fibers are often associated with cortical autofluorescent (CAF) cells, which are scattered throughout the cortex. In the adult thymus, CAF cells, whose function is currently not known, form a zone near the corticomedullary boundary, which receives extensive noradrenergic innervation. Large venous sinuses also are innervated at 7 d of age.

Between 7 and 14 d of age, the density and intensity of noradrenergic innervation of the vasculature is greatly enhanced.[59,60] Septal and penetrating thymic arteries are surrounded by dense, tangled plexuses of varicose fibers, resembling those seen in adults. Individual fibers extend from arterial plexuses into the deep cortex, especially along the zone of CAF cells. This adult-like distribution of noradrenergic nerves corresponds to an adult-like concentration of thymic NA (expressed as NA per milligram of wet weight) that is maintained through 56 d of age (Figure 1A and C). Additional fibers also are present along venous sinuses and the interlobular septa.

From 14 to 56 d of age, noradrenergic innervation of septal and cortical arteries, sinuses, and capsule and interlobular septa keep pace with thymic growth.[59,60] The only major change in innervation is in the organization of the corticomedullary junction. CAF cells form a concentrated zone of cells in this region of the thymus which is accompanied by a dense plexus of noradrenergic nerve fibers coursing among these cells and along the blood vessels. In the medulla, noradrenergic fibers supply arterioles and venous sinuses; few parenchymal fibers are present.

These findings indicate that NA is available to interact with thymocytes throughout development and in adulthood. During early development, the presence of noradrenergic nerves in the thymus may be important for the entry of thymic stem cells into the thymus, either directly through an effect on lymphocyte chemotaxis and migration or indirectly through regulation of blood flow, and also may assist in the regulation of proliferation (diminished by CA) and differentiation (enhanced by CA). Later in the development of the thymus, the presence of noradrenergic nerve fibers may influence the maturation and emigration of mature thymocytes for the seeding of secondary lymphoid organs.

3. Senescence

In humans and experimental animals, the maximum size of the thymus is attained at sexual maturation, begins to atrophy shortly after puberty, and continues to degenerate with normal aging.[48,62] Prior to the onset of thymic involution, a decline in thymic hormone secretion also has been reported.[63] We have examined noradrenergic innervation of the thymus from 3-, 8-, 12-, 17-, 21-, and 27-month-old F344 rats with fluorescence histochemistry for localization of CA and neurochemical measurement of NA.[48] In thymuses from 3-month-old rats (Plate 3), the distribution of noradrenergic nerve fibers is consistent with earlier studies in young adult rodents,[5,6,9,17,46,52,64] as described above. By 8 months of age, thymic involution is evident by a decline in thymic wet weight (Figure 2). Fluorescence histochemical staining for CA reveals intact noradrenergic innervation of thymuses from 8-month-old rats, with the most remarkable difference in innervation at 8 months of age compared with 3 months of age being a slight increase in the density of noradrenergic nerve fibers in the thymic cortex and paracortex. Noradrenergic fibers in the cortex and paracortex are linear or punctate profiles associated with thymic blood vessels or coursing within the thymic parenchyma, most often in close proximity to yellow CAF cells (Plate 4). At the corticomedullary

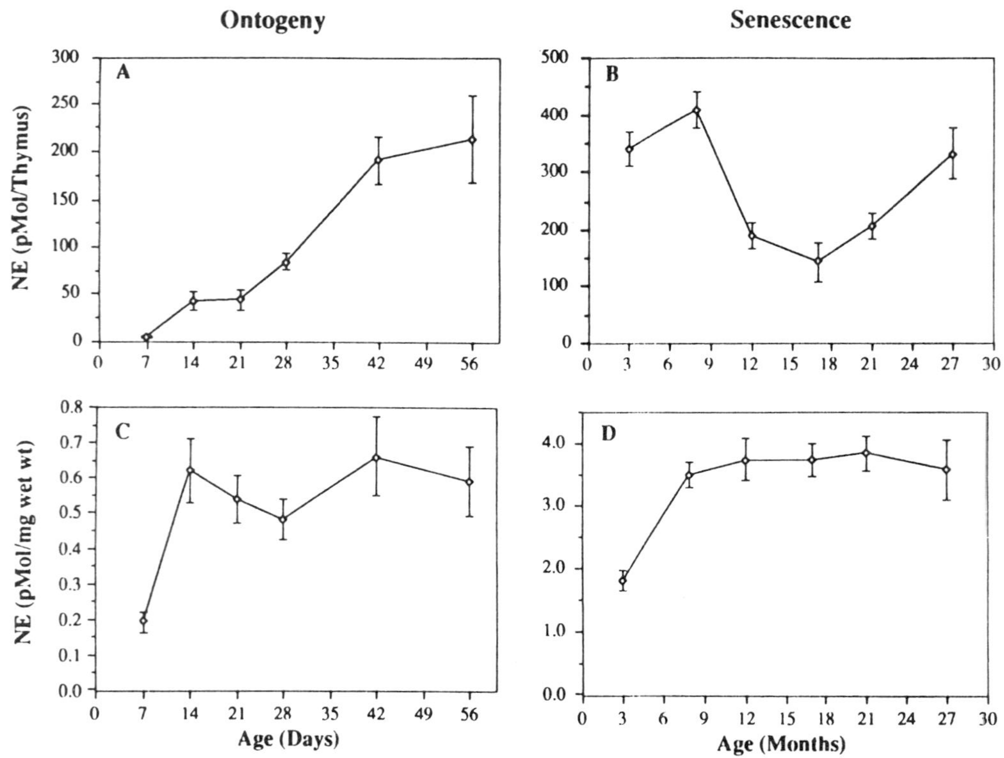

FIGURE 1. Noradrenaline content in the thymus expressed as picomoles of NA per whole thymus (A, B) and as picomoles of NA per milligram of wet weight (C, D) during ontogeny (A, C) and in senescence (B, D). Mean ± SEM. LCEC. (From Ackerman, K. D., Bellinger, D. L., Felten, S. Y., and Felten, D. L., in *Pseudoneuroimmunology II*, Ader, R., Felten, D. L., and Cohen, N., Eds., Academic Press, New York, 1990, 71. With permission.)

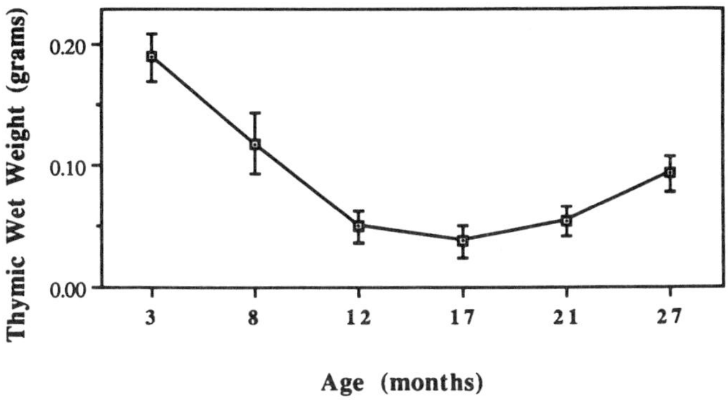

FIGURE 2. Thymic wet weight in grams from F344 rats at 3, 8, 12, 17, 21, and 27 months of age. Mean ± SD.

junction, noradrenergic vascular plexuses give rise to numerous noradrenergic fibers that extend into the paracortical region.

Between 8 and 27 months of age, the density of noradrenergic nerve fibers (Plates 4, 5, and 6) in all compartments of the thymus progressively increases.[48] Noradrenergic innervation is especially dense in the cortical and paracortical regions of the aging thymus (Plate 6), forming more extensive plexuses of sympathetic fibers than have been described anywhere else in the body, with the possible exception of the vas deferens. Dense tangles of noradrenergic fibers course along the vasculature and as free profiles through the parenchyma of the cortex and paracortex. These fibers generally form longitudinal arrays oriented parallel to the long axis of the thymic surface (Plates 5 and 6). The number of CAF cells in the thymus also increases with age, making the boundary between the cortex and medulla much more distinct than at younger ages. These cells, singly or in clusters, often are observed among dense tangles of noradrenergic fibers (Plate 6). The density of noradrenergic nerve fibers in the medulla is not as markedly altered with advancing age, making the boundaries between the medulla and cortex more distinct. We suggest that the noradrenergic innervation increases in density with age because the full complement of nerve fibers is retained in the ever-diminishing thymic cortex as it involutes. It also is noteworthy that in some cortical regions, where only minimal cortex remain and few CAF cells are present, noradrenergic innervation also is lost. It is tempting to speculate that at these cortical sites, the number of noradrenergic target cells falls below a threshold whereby noradrenergic innervation can no longer be maintained.

The wet weight of thymuses (Figure 2) from these same animals demonstrates a progressive degeneration of the thymus through 17 months of age, followed by a slight increase in weight at 21 and 27 months of age.[48] Hematoxylin and eosin staining reveals the hallmark signs of aging in the thymus, a loss in cellularity in the thymic cortex and an increase in the number of

Hassall's corpuscles. The slight increase in thymic weight observed at 21 and 27 months of age results from infiltration and deposition of fatty tissue between the thymic lobules. The venous sinuses display irregular contours, and the septa that separate the thymus into lobules also show marked degeneration. In spite of these changes, the morphology of the thymus, with relatively distinct cortical and medullary boundaries and the presence of interlobular septa dividing thymic lobules, is maintained through 27 months of age.

In the same animals studied with fluorescence histochemistry, NA content in the thymus (Figure 1B, D) has been measured using an aluminum extraction method.[48] NA content per whole thymus (Figure 1B) demonstrates relatively stable NA levels throughout the ages examined, while thymic NA concentration per milligram of wet weight (Figure 1D) shows a progressive increase in NA concentration from 3 to 17 months of age, followed by a decline to a second plateau from 21 to 27 months of age. These findings support the hypothesis that with advancing age, noradrenergic nerve fibers in the thymus accommodate to the collapsing thymic parenchyma. As the thymic parenchyma recedes, the remaining nerve fibers become packed into a smaller volume of tissue, yielding an increased NA concentration in the remaining parenchyma. The decreased concentration of NA per milligram of wet weight of thymus at 21 to 27 months of age likely reflects an infiltration of fat, which is not as well innervated as the remaining thymic tissue.

These findings demonstrate the remarkable plasticity of noradrenergic nerve fibers to accommodate to the loss of cellularity in the cortex and conform to the changing geometry of the aged thymus. The rise in NA concentration in the thymus with age suggests increased NA availability in the thymic cortex for interaction with the remaining thymocytes and other target cells of the immune system. Whether these target cells are capable of responding to enhanced NA availability has not been examined, but would depend on the continued presence of adrenoceptors on these cells and the ability of the ligand-receptor interaction to initiate an appropriate second-messenger response with functional consequences for the target cells.

III. NORADRENERGIC INNERVATION OF SECONDARY LYMPHOID ORGANS

A. SPLEEN

1. Maturation

Innervation of the spleen has been studied more widely than that of any other lymphoid organ. Fluorescence histochemical and immunocytochemical studies have demonstrated the presence of noradrenergic sympathetic nerve fibers in spleens from humans[65] and from dog,[66] cat,[2,67,68] beluga whale,[69] mouse,[5,49,70-73] rat[14,64,74-80] (reviewed in References 9-12, 43, 64, 81, and 82), lung fish,[83] dogfish,[84] cod,[85,86] and cane toad.[87] Immunohistochemical localization of noradrenergic nerve fibers in the spleen using antibodies to TH or NA also has been demonstrated.

Postganglionic cell bodies that innervate the rat spleen reside mainly in the superior mesenteric-celiac ganglionic complex (Plates 7 and 8),[15,77] and the vast majority of noradrenergic fibers reaching the spleen either originate there or pass through the plexus from an origin in the sympathetic trunk,[88] since ganglionectomy eliminates noradrenergic nerve fibers in the spleen (Plates 9 and 10). Noradrenergic fibers enter the spleen as a dense plexus associated with the splenic artery and its branches beneath the capsule of the spleen. Fibers of the subcapsular plexuses follow the trabeculae and continue deep into the spleen. Varicose fibers from vascular and trabecular plexuses follow the central arteries into the white pulp and branch within the white pulp (Plate 10). Noradrenergic fibers (Plate 10) branch from this vascular plexus into the surrounding periarteriolar lymphatic sheaths (PALS), regions composed primarily of T-lymphocytes. Noradrenergic fibers also are present along the marginal sinus, and in the marginal zone along blood vessels, in nerve bundles or as free fibers. Occasionally, noradrenergic fibers are found in the follicles or in the red pulp not associated with trabecular or venous sinuses. Noradrenergic plexuses also are associated with the large venous channels that drain splenic blood.

Single- and double-label immunocytochemical studies with TH, the rate-limiting enzyme in the synthesis of NA, and specific markers of immunocytes have demonstrated TH+ nerve fibers (Plate 11) in the same density and distribution in the spleen as observed with fluorescence histochemistry, and further reveal the anatomical relationship of TH+ nerve fibers with specific populations of cells of the immune system (Plates 12 to 14).[10,11,76] In the PALS, TH+ nerve fibers (Plate 12) reside in close proximity to T-lymphocytes of both the T-helper and T-cytotoxic/suppressor subsets. TH+ nerve fibers also course along the marginal sinus adjacent to ED3+ macrophages, and in the marginal zone in close association with ED3+ macrophages and IgM+ B-lymphocytes (Plates 13 and 14 respectively). Only an occasional TH+ nerve profile is observed among B-lymphocytes in follicles.

Electron microscopic (EM) examination of TH-immunoreactive profiles demonstrates TH+ nerve terminals adjacent to smooth muscle of the central arteriole (Figure 3), and indirect apposition to lymphocytes in the PALS, distant from the central arterioles (Figure 4).[13,64,81,89] These terminals possess long, smooth zones of contact with lymphocyte plasma membranes, separated by as little as 6 nm. In many cases, the TH+ terminals indent into lymphocytes. Similar close appositions are found in the marginal zone between TH+ terminals and lymphocytes and macrophages. No postsynaptic specializations are apparent; however, such central nervous system (CNS)-type synaptic specializations seldom are found at any peripheral sympathetic neuroeffector junction with target cells. These TH+ profiles are not present in spleens denervated by either 6-OHDA or ganglionectomy, or in sections immunocytochemically incubated with goat anti-rabbit IgG (the secondary antibody) alone.

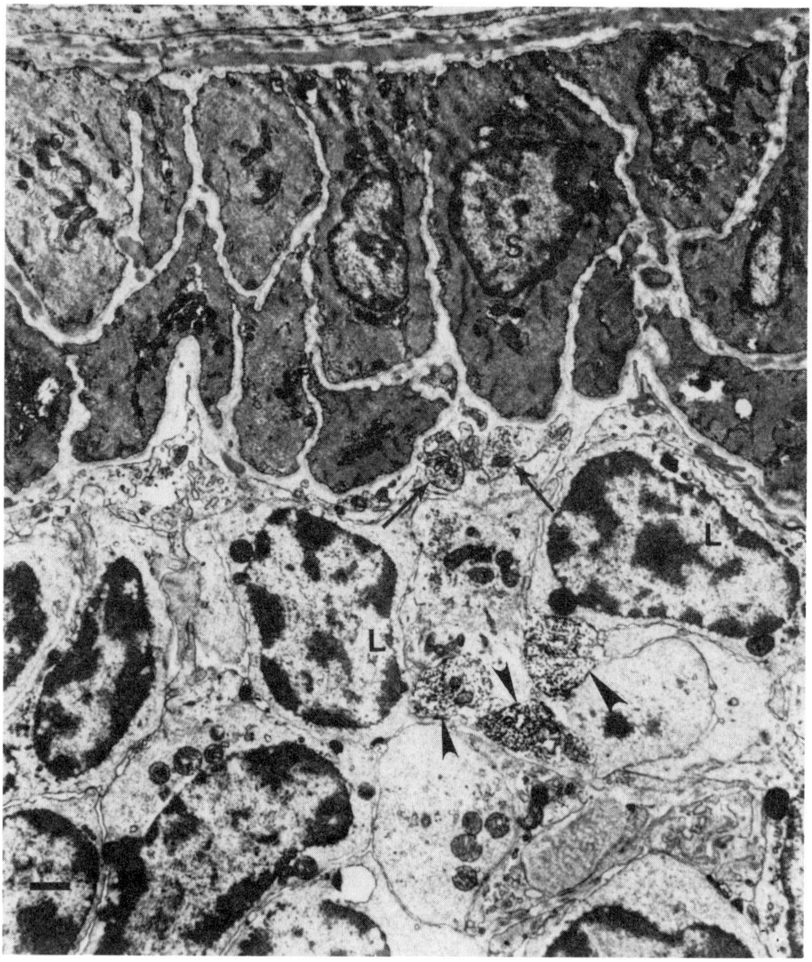

FIGURE 3. TH+ nerve terminals (arrows and arrowheads) present in the adventitia separating the smooth muscle of a central arteriole of the white pulp from the adjacent cells of the PALS. All of the TH+ terminals are separated from the smooth muscle cells (S) by at least a basement membrane, and sometimes by additional cell processes. Some of the TH+ terminals directly contact (arrowheads) lymphocytes (L) with flat or indented smooth appositions. Terminals that do not contact lymphocytes in this section are indicated with arrows. EM ICC for TH. (Bar = 1 μm.)

Like the rat thymus, the distribution of NPY+ nerve fibers closely overlaps the distribution of noradrenergic innervation of the rat spleen, and chemical sympathectomy with 6-OHDA destroys both NA- and NPY-containing nerves in the spleen.[54,90,91] Fried et al.[92] demonstrated NA and NPY colocalized in large, dense core vesicles with enkephalin in the bovine splenic nerve. Other investigators have shown NA and NPY to be colocalized in

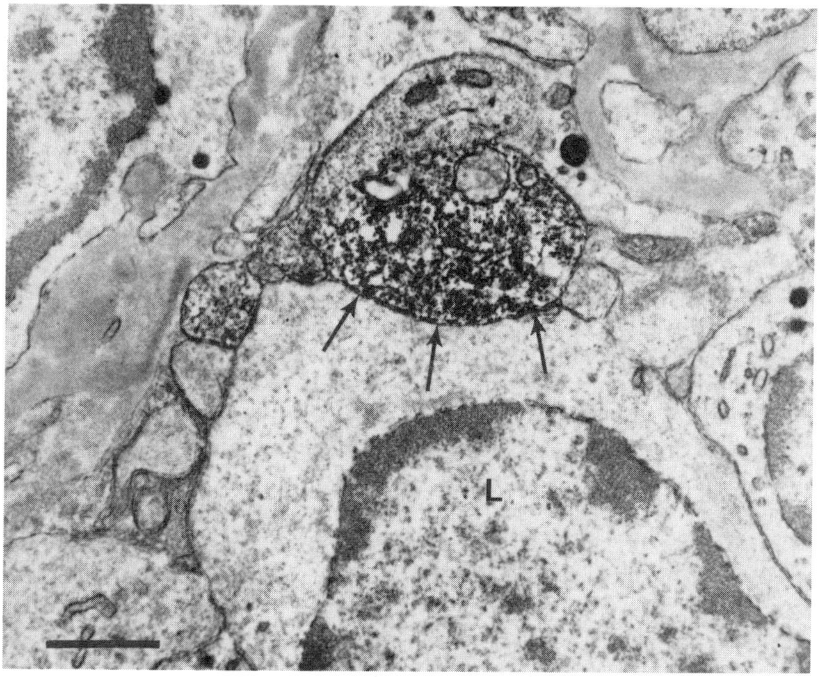

FIGURE 4. TH+ nerve terminal contacts a lymphocyte (L) within the PALS distant from the central arteriolar system. EM ICC for TH. (Bar = 1 μm.)

vascular trabecular nerves of spleens from cats[93] and pigs.[94] At the EM level, NPY-immunoreactive terminals similar to those positive for TH have been demonstrated. Nerve fibers that are immunoreactive for SP, calcitonin gene-related peptide, met-enkephalin, vasoactive intestinal polypeptide, cholecystokinin, neurotensin, and interleukin-1 also have been reported in the spleen.[9,92,93,95-97] The origin of these peptidergic nerve fibers has not been elucidated.

2. Ontogeny

Noradrenergic sympathetic innervation of the rat spleen is first detectable at 17 to 19 d of gestation and develops largely postnatally.[59,60,74-76,81,82,98,99] (reviewed in Reference 60). In newborn rats (postnatal d 1 to 3), noradrenergic nerves enter the spleen at the hilus in bundles traveling with the splenic artery and its branches, and are present in the white pulp of the spleen. Double-label immunocytochemical staining for TH and specific markers on lymphoid cells reveal TH+ nerve bundles in the white pulp among B-lymphocytes at the outer border of the PALS (Plate 15). These TH+ nerves encircle the zone containing T- and B-lymphocytes, and occasionally branch from this plexus into the developing PALS. The most striking feature of this pattern of innervation is the total lack of TH+ nerve fibers associated with the central artery (Plates 15 and 16). At the ultrastructural level, direct contacts between TH+ nerve terminals and lymphocytes in newborn rats[59,60,74-76,81,82,98] are observed that are morphologically similar to those observed in the PALS of adult rats,[13,64,89,99] as well as direct appositions between TH+ profiles and an unidentified reticular-like cell[59,60,74-76,81,82,98] not present within the inner PALS of the adult rat.

By 7 d of age, the four compartments of the PALS are distinguishable: (1) central PALS, (2) peripheral PALS, (3) marginal sinus, and (4) marginal zone. T-lymphocytes are present throughout the PALS, and are mixed with B-lymphocytes in the outer PALS (Plate 17). B-lymphocytes also are present in the marginal zone, separated from the PALS by the marginal sinus (Plate 17). The central arteriole is surrounded by a dense plexus of TH+ nerve fibers (Plates 17 and 18). Numerous TH+ profiles extend from these vascular plexuses into the PALS among T- and B-lymphocytes (Plates 17 and 18). Although noradrenergic innervation is not prominent in compartments of the spleen where B-lymphocytes reside in the adult spleen, TH+ nerve fibers arborize extensively among B-lymphocytes in the outer PALS and inner marginal zone of the neonatal spleen. ED3+ macrophages, scattered throughout the spleen at birth, are confined to the marginal zone by 7 to 10 d of age, often adjacent to bundles of nerve fibers in the marginal sinus (Plate 19). The density of nerve fibers and intensity of TH staining throughout the white pulp is greater at this age than at any other age examined.

Between 10 and 14 d of age, the PALS reorganizes to form adult-like boundaries between T- and B-lymphocytes, with T- and B-lymphocytes primarily in the inner and outer PALS, respectively (Plates 20 and 21).[59,60,74-76] B-lymphocytes also form the follicles at this time, and additional B-cells are present in the marginal zone adjacent to the marginal sinus. Concomitant with this cellular reorganization, noradrenergic nerve fibers redistribute to the inner PALS, forming dense, tangled plexuses of fibers (Plates 20 to 22). Numerous TH+ profiles course from this plexus into the inner portion of the PALS (Plate 21). TH+ nerves also are found in the parafollicular zone, and the marginal sinus at postnatal day 14. TH+ profiles (Plates 21 and 22) in the marginal sinus arborize among ED3+ macrophages and B-lymphocytes in the inner marginal zone. Noradrenergic innervation resembles the pattern of innervation seen in the adult spleen by 14 d of age. Between 14 and 28 d of age, the relative size of each splenic compartment shifts, with the inner PALS becoming thin and elongated, and growth and expansion of follicles and marginal zones. Noradrenergic nerve fibers accommodate to the changes in the organization of the white pulp. TH+ nerve fibers localize to the inner one third of the PALS, similar to the distribution in the adult spleen. Noradrenergic nerve fibers remain in the marginal zone among the thickening rim of ED3+ macrophages, and the outer portion of this zone receives additional fibers from noradrenergic plexuses that course along the venous sinuses and trabeculae. TH+ fibers also are found in the parafollicular zone, with some fibers occasionally extending from this zone into the follicle. Between 21 and 28 d of age, the most prominent change in innervation is the marked increase in the density of TH+ nerve fibers along the venous sinuses and trabeculae. Relatively few fibers extend from the venous sinuses and trabeculae into the surrounding red pulp. Between day 28 and adulthood, all compartments of the spleen increase in size, and are accompanied by parallel growth of noradrenergic innervation.

NA content in spleens (Figure 5A) has been measured from the same animals examined with fluorescence histochemistry.[59,60,74-76,81,82,98] At 1 d of age, in spite of the low density of noradrenergic nerve fibers associated with the PALS at this age, the concentration of NA in the spleen is approximately 25% of adult values. By day 7, the NA concentration per milligram of wet weight is markedly increased. By day 14, the splenic NA concentration approaches young adult levels. We have examined the turnover of NA in rat spleens during development by blocking the synthetic enzyme TH with α-methyl-*p*-tyrosine (200 mg/kg body weight, i.p.) and measuring the rate of decline of splenic NA as it is presumably used but not replaced by synthesis (Figure 5B).[59,60,74-76,81,82,98] These studies have been performed because methoxyhydroxyphenylglycol (MHPG)/NA ratios, often used to estimate turnover, are not easily obtainable in the rat. In the rat, MHPG is sulfated, making extraction more complicated. Estimates of NA turnover in the developing spleen demonstrate an increase in NA metabolism throughout development, parallel with increases in NA concentration and the density of noradrenergic nerve fibers. NA turnover is low at birth, increasing from 0 to 7 d of age, reaching a second plateau from 14 to 56 d of age, then gradually increasing to adult levels.

Equilibrium-binding studies utilizing the β-adrenergic antagonist ^{125}I-cyanopindolol (^{125}I-CYP) in spleens from developing F344 rats reveal two plateaus in the receptor density on splenocytes between postnatal day 1 and 90 (Figure 6A and B).[59,60,74-76,81,82,98] Splenocytes from 1- to 10-d-old rats display a B_{max} of 400 to 500 sites per cell, whereas splenocytes from 14- to 90-d-old rats contain adult levels of 800 to 1000 sites per cell. This shift in receptor density may reflect a change in spleen cell populations or the development of immunocompetent cells, with splenocytes developing toward a more activated state in which they possess an increased number of receptors.

The presence of noradrenergic fibers in developing compartments of the spleen from birth through adulthood indicate the availability of NA for interaction with target cells of the immune system at critical periods during the development of immunocompetence. The presence of β-adrenoceptors on cells of the immune system supports this hypothesis. Collectively, these findings suggest a possible role for NA in the formation and maturation of splenic compartments through selective migration of specific populations of cells into the spleen, modulation of proliferation and maturation of lymphocytes, and activation of specific cell functions at critical time points during splenic development. Conversely, lymphoid and nonlymphoid cells that reside near

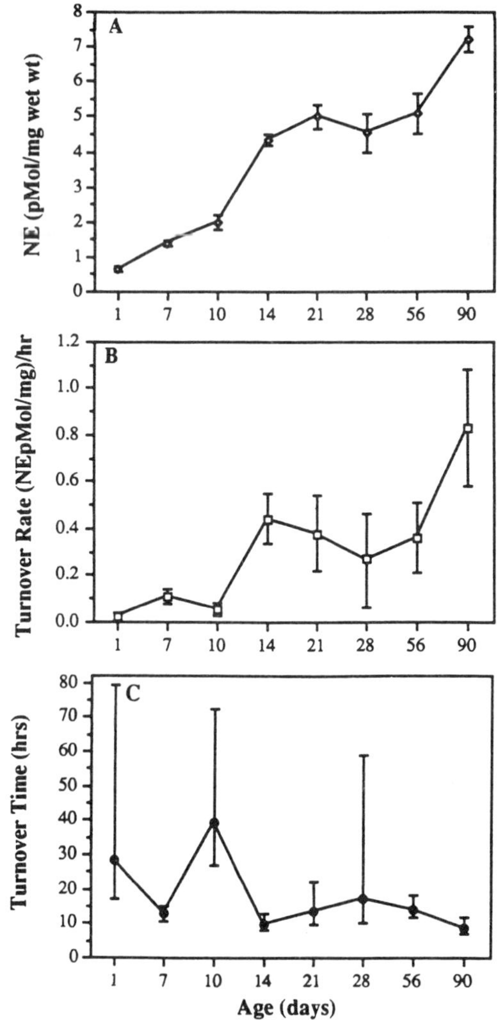

FIGURE 5. (A) Noradrenaline content in the rat spleen during ontogeny expressed as picomoles of NA per milligram of wet weight. (B) Ontogeny of noradrenaline turnover rate estimated from NA content in the rat spleen following administration of α-methyl-*p*-tyrosine, a NA synthesis blocker, at 1, 2, and 4 h prior to animal sacrifice, expressed as picomoles of NA per milligram of wet weight per hour. Splenic NA turnover was calculated from the rate of decrease of NA following blockade of TH, using regression analysis of the plot of log NA vs. time from injection. (C) NA turnover time in hours, calculated as the inverse of the rate constant for degradation of NA derived from the regression plot of log NA vs. time from injection. The error bars are unsymmetrical, since they represent +1 SEM for the rate constant calculated prior to taking its reciprocal. Mean ± SEM. LCEC. (From Ackerman, K. D., Bellinger, D. L., Felten, S. Y., and Felten, D. L., in *Psychoneuroimmunology II,* Ader, R., Felten, D. L., and Cohen, N., Eds., Academic Press, New York, 71. With permission.)

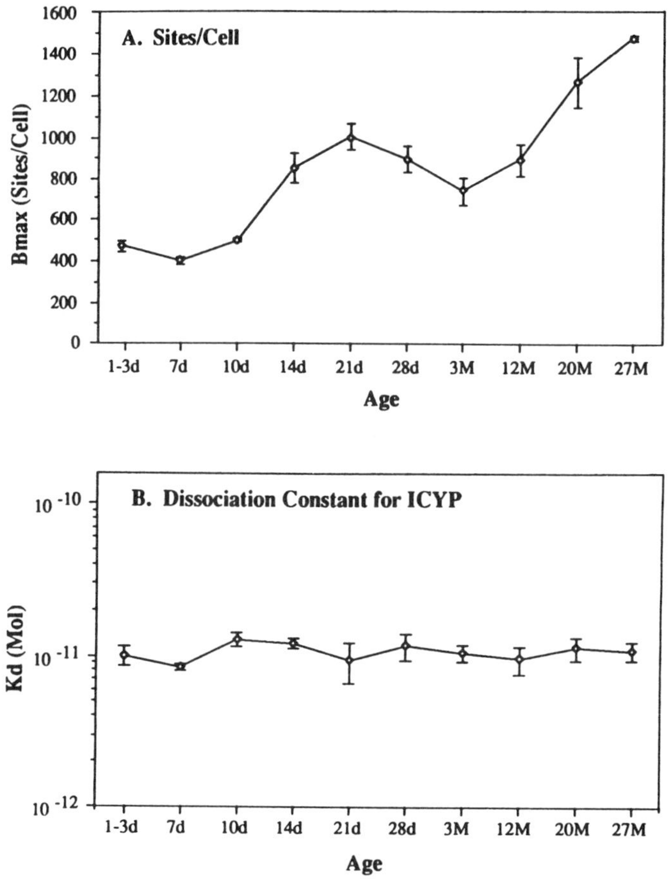

FIGURE 6. Equilibrium β-adrenoceptor analysis of intact spleen cells was modified from the method of Madden et al., 1989. Spleen cells were incubated for 60 min at 37°C with eight concentrations of ^{125}I-CYP ranging from 0.1 to 2 p*M*. Nonspecific binding was determined using parallel assays incubated in the presence of 10^{-6} *M* CGP-12177, a hydrophilic β-adrenoceptor antagonist. The data were analyzed using Scatchard analysis and represent the mean + SEM of at least three pools of animals. (A) β-Adrenoceptor density expressed as sites per cell; (B) dissociation constant (Kd) for ^{125}I-CYP binding to intact splenocytes. Mean ± SEM. (From Ackerman, K. D., Bellinger, D. L., Felten, S. Y., and Felten, D. L., in *Psychoneuroimmunology II*, Ader, R., Felten, D. L., and Cohen, N., Eds., Academic Press, New York, 71, 1990. With permission.)

nerve fibers in the marginal sinus may provide signals that influence the growth and maturation of noradrenergic nerves.

3. Senescence

The mature pattern of noradrenergic innervation of the rat spleen is observed by 28 d after birth, and adult proportions are attained by 56 d of age.[59,60,74-76,81,82,98] Early studies from our laboratories have examined age-related changes in innervation of the rat spleen using two age groups of male F344 rats, with 8-month-old rats representing mature young adults and 27-month-old rats representing aged animals[14,79] (reviewed in References 11, 60, 82, and 98). In this study, we reported noradrenergic innervation in 8-month-old rats that is similar to innervation described for the young adult rats above, and in previous reports from our laboratories and others[14,59,60,74-76,80,98] (reviewed in References 9, 11, 12, 64, 81, and 82). Further, spleens from 27-month-old rats show a loss of noradrenergic innervation as assessed by fluorescence histochemistry and neurochemical measures of NA.[14,79] NA content in the spleen is approximately 50% lower in 27-month-old rats compared with their 3-month-old counterparts. Quantitation of fluorescent varicosities reveals an age-related decrease in noradrenergic innervation in all splenic compartments of approximately 70 to 80%.[14] These studies demonstrate a greater loss in noradrenergic nerve terminals than is indicated from splenic NA content in the aged rat spleen. The reason for this smaller decline in splenic NA content compared with the extent of lost noradrenergic nerve fibers will require further investigation of the fate of lost noradrenergic nerve fibers, NA metabolism in age-resistant nerve fibers, and availability of NA from the microenvironment for uptake into these terminals. Perhaps the loss of some noradrenergic fibers leads to increased NA production and release in neighboring fibers, similar to what is seen in central dopaminergic nigrostriatal neurons.

Longitudinal studies using fluorescence histochemistry, double-label immunocytochemistry, and neurochemical measurement of NA have been performed to examine the time course and manner in which noradrenergic innervation is lost as a function of age, and to examine the relationship of specific populations of immunocytes that reside in compartments innervated by noradrenergic nerves.[78] Data from these studies indicate that noradrenergic innervation is maintained through 12 months of age (Plates 23 and 24) and then begins to gradually decline through 27 months of age. At 12 months of age, noradrenergic nerve fibers appear to be densely packed in the PALS (Plate 24). Double-label immunocytochemistry further reveals a gradual decline in the density of T-lymphocytes in the PALS (Plates 23 to 26) and ED3+ macrophages in the marginal zone (Plates 27 to 30) that parallels the decline in noradrenergic innervation. At 12 months of age, these immune cell compartments (PALS and marginal zone) are reduced as a result of cell loss. At this time point, noradrenergic innervation of these compartments is still robust (Plates 24 and 28). Noradrenergic nerve fibers appear to retract into smaller lymphoid compartments to maintain their anatomical distribution within these shrinking immune cell zones, giving the appearance of hyperinnervation. By 17 months of age, further loss of T-lymphocytes (Plate 25) and ED3+ macrophages (Plate 29) is apparent; at this time point there also is a decline in the density of noradrenergic nerve fibers associated with these compartments. This parallel decline in the density of specific populations of cells of the immune system and in noradrenergic innervation continues through 27 months of age (Plates 26 and 30).

The pattern of noradrenergic nerve fiber loss and decline in T-lymphocytes and ED3+ macrophages in these two splenic compartments occur in a precise manner, with nerve fibers first lost in lymphoid compartments that reside most distal from the hilar region, the site of entry of these nerve fibers. Noradrenergic innervation continues to decline in this distal-to-proximal fashion with respect to the hilus until the only remaining noradrenergic nerve fibers in the 27-month-old spleen are distributed near the hilus. The density of T-lymphocytes in the PALS and ED3+ macrophages in the marginal zone displays a similar pattern of loss that closely overlaps the loss of noradrenergic innervation; cell loss in these compartments occurs first at sites most distal from the hilar region, with progressive loss nearer to the hilus with continual aging.

Double-label immunocytochemistry for TH and IgM, a B-lymphocyte marker, indicates little change in the density and distribution of B-lymphocytes in the marginal zone or in the follicle of spleens from old rats (Plates 31 and 32). With age, there is a less distinct boundary between cells in the follicle and the marginal zone, and between the PALS and marginal zone. A decline in TH+ fibers is apparent in the marginal zone with aging; an occasional TH+ profile can be found in the follicle of spleens from 27-month-old rats.

Neurochemical measurement of NA in spleens from rats at 3, 8, 12, 17, 21, and 27 months of age reveals a progressive decline in splenic NA content with age (Figure 7). This decline is less severe than the density of noradrenergic innervation would indicate, similar to our previous findings performed in rats at 8 and 27 months of age. We also have preliminary evidence to suggest that the time course of splenic noradrenergic denervation with age is dependent upon the continued exposure of the adult rat to environmental antigens. Pilot data suggest that splenic noradrenergic denervation is accelerated in rats raised under standard vivarium conditions compared with rats raised in barrier facilities (pathogen free). A similar decline in NPYI nerve fibers that overlaps the distribution of noradrenergic nerves and parallels the loss of T-lymphocytes and ED3+ macrophages occurs with age, while SP innervation, based on immunocytochemical staining and direct measurement of splenic SP, does not appear to change with the normal aging process. These findings support the hypothesis that NPY is colocalized in noradrenergic sympathetic nerves that distribute to the spleen.

A loss in noradrenergic innervation of the spleen can result from (1) an inability of noradrenergic nerve terminals to synthesize enough NA to form observable fluorophore (or enough TH for immunostaining, since this decline in noradrenergic nerve fibers in the spleen also is observed with immunocytochemical staining for TH) or (2) an absence of the terminals. In preliminary studies to resolve this issue, we administered α-methylnoradrenaline (100 mg/kg body weight, i.p.), a compound that is taken up by the high-affinity carrier into noradrenergic terminals and persists because it cannot be

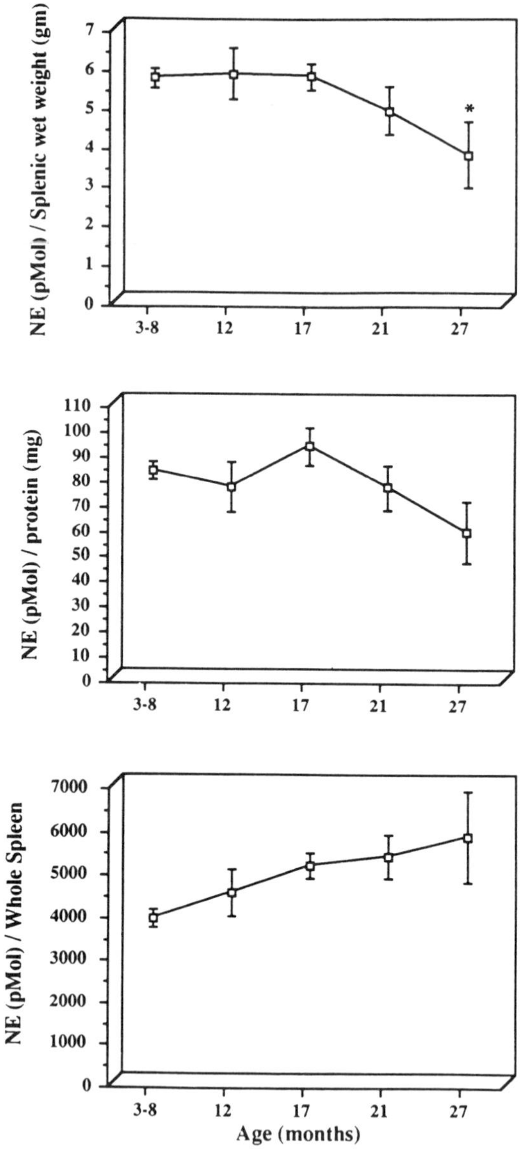

FIGURE 7. Noradrenaline content in the rat spleen expressed as (A) picomoles of NA per gram of wet weight, (B) picomoles of NA per milligram of protein, and (C) picomoles of NA per whole spleen at 3 to 8, 12, 17, 21, and 27 months of age. Mean ± SEM. LCEC. Significantly different from 3 to 8, 12, and 17 months of age ($p < 0.05$). (From Ackerman, K. D., Bellinger, D. L., Felten, S. Y., and Felten, D. L., in *Psychoneuroimmunology II,* Ader, R., Felten, D. L., and Cohen, N., Eds., Academic Press, New York, 1990, 71. With permission.)

catabolized by monoamine oxidase.[100] Preliminary data suggest that α-methylnoradrenaline is able to restore fluorescence in only a few nerve profiles in 27-month-old rat spleens compared with untreated and vehicle-treated controls, but not to the level of 3-month-old rats. These findings support an actual retraction and loss of noradrenergic fibers in the aging spleen; however, these findings also could be attributed to a deficit in the high-affinity uptake system in the aging spleen.

We have developed a perfused spleen model to examine the high-affinity uptake, release, and metabolism of [^{3}H]-NA in spleen slices, coupled with fluorescence histochemistry for CA to quantitate noradrenergic innervation in alternate slices.[100,101] In preliminary studies using this model with spleen slices from 3- and 21-month-old rats, we found that total uptake of [^{3}H]-NA in old rats is significantly less than the uptake in young rats, but that the efficiency of uptake of [^{3}H]-NA per nerve terminal is greatly enhanced in old rats. Kinetic analysis of the data suggests an increase in the density of high-affinity uptake binding sites for NA in age-resistant noradrenergic nerve terminals in the spleen, probably as a compensatory mechanism resulting from the decline in NA concentration and/or loss of noradrenergic nerve terminals that occur with aging. We have not yet examined the effect of age on the synthesis of NA in spleens from young and old rats. It is likely that synthesis in age-resistant terminals in the spleen also is enhanced, and may account for the higher splenic NA content compared with the density of nerve terminals. The increased circulating NA, reported to occur with aging,[102-105] which would be available for uptake into age-resistant terminals, also may be contributory.

We have measured β-adrenoceptor density on splenocytes of F344 rats as a function of age from the same spleens used for histofluorescence and measurements of splenic NA concentration. These animals have been carefully examined for pathological abnormalities. These studies revealed an increase in the density of β-adrenoceptors in splenocytes with age, consistent with upregulation of β-adrenoceptors in response to declining NA levels from an age-related loss of innervation (Figure 6A and B). No differences in receptor affinity were detected. Examination of β-adrenoceptor density in two rats with splenomegaly and one rat with a pituitary adenoma show a considerably lower number of receptors (300, 775, and 800 sites per cell, respectively) than in splenocytes taken from "healthy" aged rats (approximately 1000 sites per cell). These values have been excluded from analysis of the data from normal aging rats; inclusion of these values would have obscured the increase in receptor expression we observed in apparently healthy old rats.

De Blasi and co-workers[106] reported that aged rats respond to restraint stress (as a mechanism to increase NA availability) with a slower time course and diminished extent of downregulation of lymphocyte β-adrenoceptors compared with young adult controls. They suggested that animals may lose their ability to regulate β-adrenoceptor density in response to agonist availability;

however, the facts that restraint stress also increases circulating glucocorticoid levels, which also have been shown to modulate β-adrenoceptor expression, and that old rats may have exaggerated responses to stressors compared with young rats make this study difficult to interpret mechanistically. Measurements of β-adrenoceptors over time, following administration of exogenous adrenergic agonists, in old rats have not been performed to test this hypothesis.

It is clear that measurement of β-adrenoceptor density alone is not a sufficient indicator of the ability of cells to respond to noradrenergic stimulation, especially in lymphocytes from aged subjects. Based on several studies examining the β-adrenoceptor coupling with adenylate cyclase (AC),[107,108] in spite of the greater number of β-adrenoceptors, lymphocyte responsiveness to noradrenergic stimulation may be impaired. These studies indicate an age-related intrinsic defect in the activation of the AC system following β-adrenoceptor stimulation, the result of an inability of the receptor to activate AC and/or altered catalytic capacity of AC. Similar findings have been reported in nonlymphoid organs that receive sympathetic innervation, such as rat myocardium and lung.[109-112]

Extensive investigation of immune functions during aging in a variety of species, including man, have shown a decline in normal immune functions, primarily those involving the T-cell responses (Table 1), that parallels the decline in the peripheral noradrenergic system. Age-related immunologic dysfunction is associated with an increased incidence of autoimmune and immune complex diseases, certain types of cancer, and viral and fungal infections. Impairment of immune capacity with age is likely to result from a combination of both age-related changes in immunocytes and alterations in their microenvironment. The fact that the normal aging process is associated with a decline in both immunocompetence and noradrenergic innervation of secondary lymphoid organs makes it tempting to hypothesize a causal relationship. Conversely, noradrenergic denervation may be compensatory rather than causal to immune dysfunction. This hypothesis is currently being tested in our laboratory. The finding that the direction of change in immune responses following acute chemical sympathectomy with 6-OHDA is often similar to the direction of immune responses measured in aged rodents supports this hypothesis (Table 1). Preliminary examination of noradrenergic innervation of the murine spleen following lymphocyte depletion using hydrocortisone or cyclophosphamide in young adults did not result in a diminution of noradrenergic innervation acutely,[70] suggesting that the noradrenergic denervation may not be compensatory to the loss of T-lymphocytes in the white pulp of the spleen with age.

B. LYMPH NODES

1. Maturation

Fluorescence histochemical studies of lymph nodes in rats[9,11,16] and several strains of mice[8,12,98] demonstrate dense plexuses of noradrenergic nerves

TABLE 1
Effect of Aging and Acute Chemical Sympathectomy on Immune Responsiveness

Immune response	Age-related changes	Sympathectomy-induced changes
T-Cell-Mediated Responses		
Delayed-type hypersensitivity	Decreased	Decreased
Cytotoxic T-lymphocyte activity	Decreased	Decreased
Con A-induced proliferation	Decreased	Decreased
IL-2 receptors	Decreased	ND[a]
B-Cell-Mediated Responses		
LPS-induced proliferation	Mixed	Mixed
Ig secretion in response to polyclonal B-cell differentiation	Increased	Increased
Response to Antigen Stimulation		
Primary antibody response (T dependent)	Decreased	Decreased
Secondary antibody response (T dependent)	Mixed	Decreased[b]
Primary antibody response (T independent)	Mixed	Increased
IL-2 production	Decreased	Decreased
Lymph node cellular proliferation	Increased	Increased
Response to autologous antigen	Increased	Increased
Tumoricidal Activity		
Natural killer cell activity	Mixed	Increased
Resistance to tumor challenge	Decreased	Increased

Note: See Reference 60 for a review of the effects of aging and acute sympathectomy on immune responsiveness.

[a] Not determined.
[b] Sympathectomy prior to boosting.

From Ackerman, K. D., Bellinger, D. L., Felten, S. Y., and Felten, D. L., in *Psychoneuroimmunology II,* Ader, R., Felten, D. L., and Cohen, N., Eds., Academic Press, New York, 1990, 71. With permission.

along blood vessels at the hilus, the presumed point of entry of noradrenergic nerve fibers. In lymph nodes, noradrenergic nerve fibers distribute either into a subcapsular plexus or continue in vascular plexuses through the medullary cords. These fibers course adjacent to both vascular and lymphatic channels in the medulla and continue with small vessels into the paracortical regions, which are rich in T-lymphocytes. Individual varicose fibers branch from these vascular plexuses and travel in the paracortical parenchyma. Noradrenergic

TABLE 2
Noradrenergic Innervation of Functional Compartments of Secondary Lymphoid Organs

Compartments with noradrenergic innervation	Functional significance of compartment
Spleen	
Marginal zone (macrophages/B-lymphocytes)	Site of lymphocyte entry
Marginal zone (macrophages/B-lymphocytes)	Site of antigen capture
PALS (T-lymphocytes)	Site of antigen presentation/lymphocyte activation
Parafollicular/marginal zone (macrophages/B-lymphocytes)	Site of antigen presentation/lymphocyte activation
Outer marginal zone	Site of lymphocyte egress
Lymph Node	
Corticomedullary junction	Site of lymphocyte entry
Subcapsular sinus	Site of antigen capture
Paracortex	Site of antigen presentation/lymphocyte activation
Medullary cords	Site of antigen presentation/lymphocyte activation
Medullary sinuses	Site of lymphocyte egress

fibers from the subcapsular plexus continue along small vessels into the cortex, and single varicose profiles course in the cortical parenchyma. Parenchymal fibers are present in cortical and paracortical regions, among T-lymphocytes, but are absent from nodular regions and germinal centers where B-lymphocytes predominate.

The compartmentation of noradrenergic nerve fibers in lymph nodes shows numerous similarities with noradrenergic innervation of the spleen, suggesting a common functional role in both organs. Innervated compartments of both lymph node and spleen are compared in Table 2, with a summary of the major functional significance of each site (reviewed in Reference 11). A role for NA at sites of antigen capture and antigen presentation and lymphocyte activation is suggested from studies demonstrating a decline in primary and secondary spleen and lymph node plaque-forming cell responses following sympathetic denervation.[12] Egress of activated lymphocytes into the circulation from spleen and lymph nodes occurs following infusion of CA,[11,36,37] indicating an additional functional role for NA at this site.

2. Senescence

We have examined cervical, mesenteric, and popliteal lymph nodes from several strains of aged mice, including C3H, C57Bl/6, NZW, NZB, and NZB × NZW F1, and in aged F344 rats with fluorescence histochemistry for CA. Noradrenergic innervation in all compartments of lymph nodes was diminished in all lymph nodes examined.[98] The time course of diminished noradrenergic innervation varies, depending on the strain, and correlates with the expected lifespan of the animals.

IV. PLASTICITY OF NORADRENERGIC INNERVATION OF LYMPHOID ORGANS

Our approach to assessing the functional significance of noradrenergic innervation of lymphoid organs on immune functions has been to examine the immune system following sympathetic denervation with 6-OHDA, a neurotoxin specific for noradrenergic nerve fibers, in neonates and in mature adult and aging rodents. Prior to evaluating the impact of denervation on immune function, it is first necessary to determine the completeness of chemical sympathectomy and the duration of noradrenergic denervation, and to rule out direct nonneuronal effects of 6-OHDA or toxic products generated during its oxidation.

A. ONTOGENY

Neonatal administration of 6-OHDA produces a profound central and peripheral noradrenergic denervation. We have found that the degree and time course of denervation following administration of 6-OHDA differs in the spleen and thymus from F344 rats (Figure 8A and B).[59] Chemical sympathectomy by 6-OHDA treatment at 1, 2, and 3 d of age results in a virtual depletion of NA content in thymuses from sympathectomized rats at 7 d of age (greater than 95% depletion compared with vehicle controls), followed by a progressive increase in thymic NA concentration from postnatal day 7 to 56 (Figure 8A). At 14 d of age, the thymic NA concentration in such sympathectomized rats is greater than 60% of vehicle control values. From 21 to 56 d of age, no difference in thymic NA concentration is present between sympathectomized and vehicle control groups, suggesting full neurochemical recovery.

Measurement of splenic NA following neonatal 6-OHDA treatment (Figure 8B) reveals a more pronounced and longer-lasting denervation than in the thymus.[59,113] Acute injection of young adult rats with 6-OHDA results in a 70% reduction of thymic NA relative to vehicle controls, while splenic NA content in acutely sympathectomized rats is virtually depleted (greater than 95%) through day 28 postnatally, and by 56 d of age is still depleted by greater than 80% of vehicle control values. NA content in spleens from 6-OHDA-treated animals appears to increase slightly between 28 and 56 d of age. Although this increase may result from reinnervation by noradrenergic fibers, reinjection of neonatally sympathectomized rats with 6-OHDA on days 3 and 5 prior to sacrifice at 56 d of age does not alter the splenic NA concentration. Further, this slight rise in splenic NA concentration is not statistically significant compared with any other treatment group.

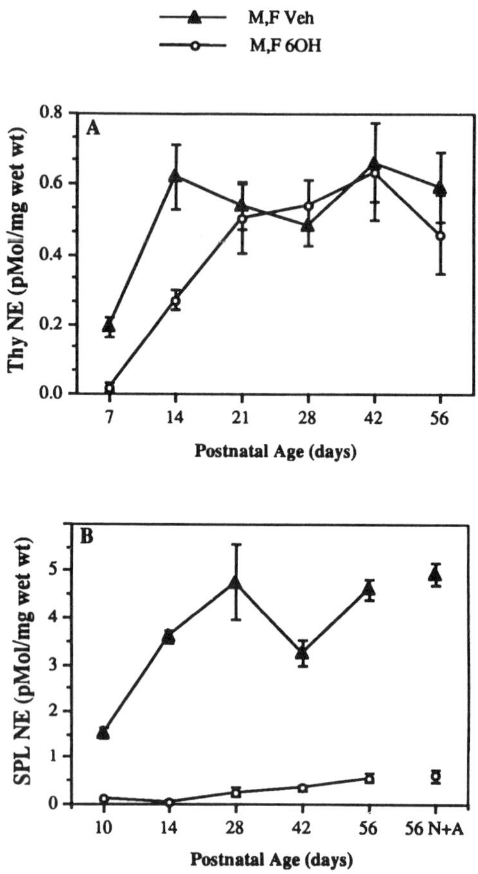

FIGURE 8. Ontogeny of thymic (A) and splenic (B) NA content in intact and chemically sympathectomized F344 rats, expressed as picomoles of NA per milligram of NA wet weight. Sympathectomized rats received 100 mg/kg 6-OHDA in 0.1% ascorbate vehicle, i.p., on postnatal days 1, 2, and 3. Additional 56-d-old rats injected with 6-OHDA or vehicle at days 1, 2, and 3 were reinjected at 51 and 53 d of age (56 N + A). M, male; F, female; Veh, vehicle; 60H, 6-OHDA. Mean ± SEM. LCEC. (Figure B from Ackerman, K. D., Madden, K. S., Livnat, S., Felten, S. Y., and Felten, D. L., *Brain Behav. Immunol.*, 5, 235, 1991. With permission.)

The ontogeny of *in vitro* spleen cell proliferation to T- and B-cell mitogens (concanavalin A [Con A] and a mixture of equal concentrations of *Salmonella typhimurium* and dextran sulfate [STM/DxS], respectively) and immunoglobulin secretion has been examined in F344 rats at birth following administration of 6-OHDA on days 1 through 3 (Table 3).[59,113] A shift in the proliferative response to Con A is observed in splenocytes from sympathectomized rats sacrificed between postnatal days 7 and 14 compared with age-matched controls, with reduced responsiveness to low levels of Con A and increased responsiveness with higher doses. Between postnatal days 7 and

TABLE 3
***In Vitro* Lymphocyte Activity Following Neonatal Sympathectomy in the Developing F344 Rat**

Time of 6-OHDA[a]	Age[b]	Lymphocyte proliferation			B-cell differentiation	
		Background	T-Cell (Con A)	B-Cell (S/D)	Background IgM	Net IgM
Neonatal (days 1–3)	7–14	No change[c]	Shift in dose response	No change	Reduced	ND[d]
Neonatal (days 1–3)	21	No change	Reduced (f)	Reduced	Reduced	Reduced
Neonatal (days 1–3)	28	No change	Reduced	Reduced	Reduced	Reduced
Neonatal (days 1–3)	42	Reduced (f) Increased (m)	Reduced (m)	Reduced	No change	Reduced
Neonatal (days 1–3)	56	No change	No change	No change	Reduced (trend f > m)	No change (f) Increased (trend, m)
Neonatal (days 1–3) + reinjected (days 51, 53)	56	No change (m)	No change (m)	Reduced (m)	No change (m)	Increased (m)
Days 51, 53 (acute)	56	No change	Reduced	Reduced	Reduced (f) No change (m)	Reduced

Note: f, female; m, male.

[a] 100 mg/kg 6-OHDA or 0.1% ascorbate vehicle was administered at the days indicated after birth.
[b] Age at which spleen cell responsiveness was assessed.
[c] If not stated, effects are independent of gender.
[d] Not detectable.

From Ackerman, K. D., Madden, K. S., Livnat, S., Felten, S. Y., and Felten, D. L., *Brain Behav. Immunol.*, 5, 235, 1991. With permission.

14, B-cell spontaneous proliferation was suppressed by STM/DxS in all groups of rats examined, and no effect of sympathectomy was observed. From 21 to 42 d of age, the mature patterns of mitogen responsiveness develop. Proliferative responses to both T- and B-cell mitogens are reduced at this time period, and return to vehicle age-matched control values by 56 d of age.

Polyclonal immunoglobulin (Ig) production by B-cells *in vitro* following chemical sympathectomy at birth shows a reduction in spontaneous IgM production throughout development compared with age-matched controls (Table 3).[59,113] The magnitude of mitogen-induced IgM secretory response is lower in neonatally sympathectomized rats from 28 to 42 d of age, an effect that is not observed at day 56. The time course of denervation-induced alteration in splenocyte proliferation and differentiation is affected by the sex of the animal, but the magnitude and direction of these changes are similar in both males and females. Overall, these findings suggest an important role for NA in the mediation of lymphoid cell development and maturation.

B. MATURATION

Acute administration of 6-OHDA also produces a profound noradrenergic denervation of peripheral organs. Denervation of central noradrenergic neural systems in neonatal rats occurs because the blood-brain barrier has not yet formed, allowing the neurotoxin to gain entry to the CNS; this phenomenon is not seen in adult animals. After a period of time following acute denervation with 6-OHDA, noradrenergic neuronal regrowth occurs, peripheral organs become reinnervated, and NA levels increase. In adults, noradrenergic nerve terminals are destroyed by 6-OHDA, while most ganglionic cell bodies are resistant to the destructive effects of the drug, allowing nerve terminals to arise from the original preterminal bundle.[114,115] The degree of NA depletion after 6-OHDA treatment varies, depending on age, strain of animal, drug regimen, and tissue. In some tissues such as vas deferens, sympathetic ganglia and thymus, the noradrenergic terminals appear to be more resistant to 6-OHDA treatment than those in the spleen.[6,115,116] Acute chemical sympathectomy in young adult rats, 1 d after treatment, produces approximately a 70 and 95% reduction in thymic and splenic NA concentration, respectively, compared with vehicle controls, indicating that noradrenergic innervation of the thymus is more resistant to 6-OHDA and that virtually all NA in the spleen is present in sympathetic nerves that are susceptible to the neurotoxic effects of 6-OHDA.[59,77,117] In spleens from young adult sympathectomized rats, profound loss of NA content persists through day 10 posttreatment, followed by a gradual increase in splenic NA concentration that returns to normal values by 56 d after treatment (Figure 9).[117]

The time course and pattern of reinnervation of noradrenergic sympathetic nerve fibers into the spleen following acute chemical sympathectomy in young adult F344 rats has been examined using glyoxylic acid histofluorescence for localization of CA (Figure 10).[117] The presence of a few 6-OHDA-resistant noradrenergic nerve fibers is observed in spleens from rats sacrificed 1 to 3

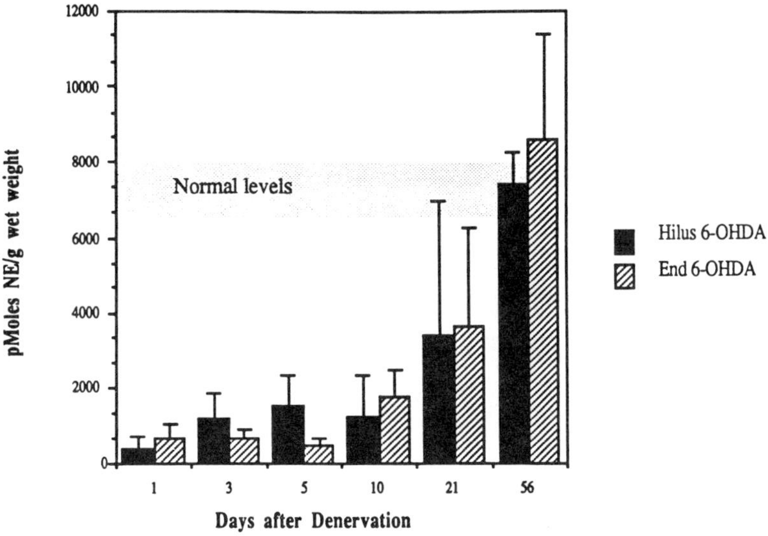

FIGURE 9. NA content in hilar and end pieces of spleen from 6-OHDA-treated adult F344 rats on posttreatment days 1, 3, 5, 10, 21, and 56 expressed as picomoles of NA per gram wet weight. Shaded region indicates NA content of hilar and end pieces of spleen from age-matched, vehicle-treated control rats. Mean ± SEM. LCEC. (From Lorton, D., Hewitt, D., Bellinger, D. L., Felten, S. Y., and Felten, D. L., *Brain Behav. Immunol.*, 4, 198, 1990. With permission.)

d after denervation. By day 5, reinnervation of the spleen by an ingrowth of noradrenergic nerve fibers proceeds initially along the splenic artery as it enters the hilar region of the spleen. By 10 d after the treatment, abundant ingrowing noradrenergic nerve fibers extend into the hilar region of the spleen, and between days 21 and 56 posttreatment, the ingrowth of noradrenergic nerve fibers is apparent in regions of the spleen distal from the hilar regions. Noradrenergic nerves reinnervating the spleen distribute to the same splenic compartments as seen in saline-injected and nontreated controls; however, the density of noradrenergic nerve fibers in these compartments display a high-to-low gradient from hilus to regions distal from the hilus throughout the course of reinnervation.

At 56 d after chemical sympathectomy, splenic compartments distal from the hilar region of the spleen contain fewer noradrenergic profiles than their age-matched nondenervated controls, suggesting that the anatomical process of reinnervation does not restore the density of nerve fibers in a fashion identical to that of nondenervated controls. In contrast, splenic NA concentrations at 56 d postdenervation does not differ from the concentration seen in nondenervated control spleens, either at the hilar region or distal to it. These findings suggest that functional restoration of splenic sympathetic innervation may involve metabolic and receptor upregulation in returning noradrenergic nerve fibers as a compensatory mechanism for the lack of complete

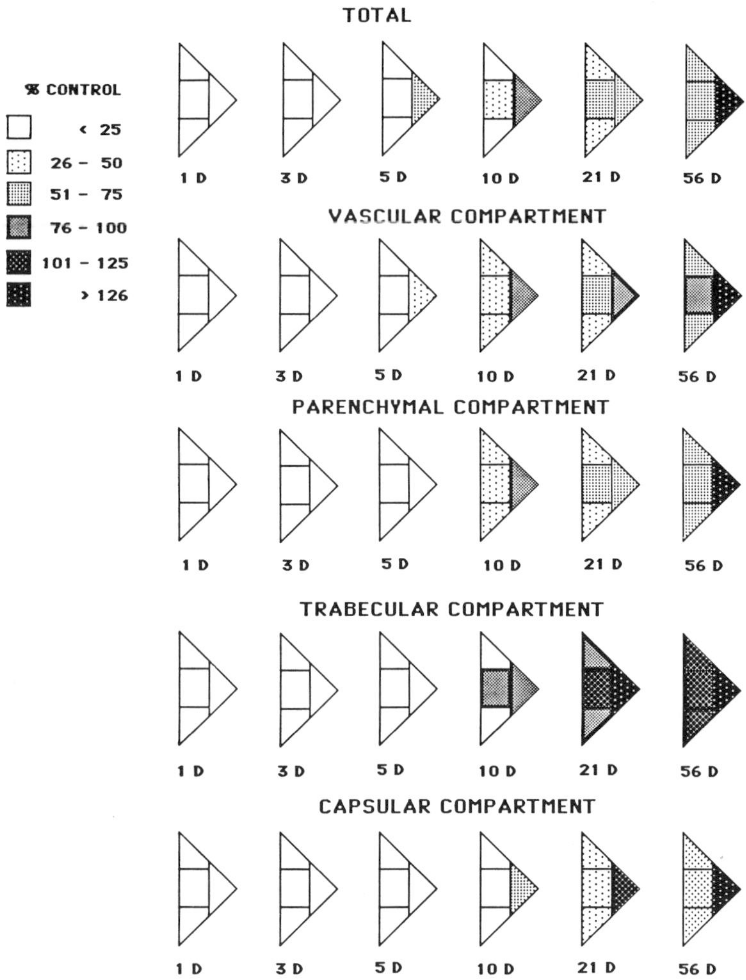

FIGURE 10. Schematic representation of a cross section of the spleen through the hilar region that summarizes the pattern of reinnervation of noradrenergic fibers into specific zones at days 1, 3, 5, 10, 21, and 56 after denervation with 6-OHDA. Sections were cut from the hilar region of the spleen and processed for glyoxylic acid histochemistry. To map the distribution of catecholamine fibers within the spleen at each posttreatment time point, a section from each group was divided into four regions on the basis of the distance from the entrance of the splenic artery into the spleen: (1) a region nearest the splenic artery and half the distance from the entrance of the splenic artery to the opposite side of the spleen, the proximal zone; (2) a region adjacent to the proximal zone and the medial one third of the base of the section, the medial zone; and (3) two regions distal to the entrance of the splenic artery and lateral to the medial zone, each consisting of one third the baseline of the section, the lateral distal zone. Quantitation of noradrenergic fiber density was carried out in each of these regions. (From Lorton, D., Hewitt, D., Bellinger, D. L., Felten, S. Y., and Felten, D. L., *Brain Behav. Immunol.*, 4, 198, 1990. With permission.)

fiber regrowth into regions of white pulp distal from the hilus. In addition, adrenal compensation has been reported with long-term chemical sympathectomy, resulting in increased TH activity, greater CA turnover from chromaffin cells, and higher plasma CA levels in 6-OHDA-treated animals.[118-120] Adrenal compensation, through the increased availability of NA for uptake by noradrenergic nerve terminals in the spleen, might explain the difference between neuroanatomical and neurochemical analysis at posttreatment day 56; however, regional stereologic analysis of noradrenergic reinnervation in the spleen reveal an increase in noradrenergic fiber density above control values in regions proximal to the hilar region. These data suggest a hyperinnervation of regions of the spleen near the hilus, while more distal regions of the spleen remain relatively denervated. Further work is necessary to determine at what time point, if ever, the pattern of noradrenergic innervation in both the hilar region and the distal region returns to that of controls.

Temporal discrepancies between recovery of noradrenergic fibers, NA content, and effector function have been reported for other peripheral organs after treatment with 6-OHDA. Following chemical sympathectomy of the cardiovascular system, investigators have reported that recovery of peripheral sympathetic effector function precedes both complete noradrenergic fiber regeneration[119,121] and recovery of NA to control levels in the effector organ.[122] These findings support our hypothesis that the recovery of effector function prior to the return of complete innervation or normal levels of NA may be the result of enhanced secretion of NA coupled with adrenoceptor upregulation from the relative state of denervation, perhaps augmented by enhanced adrenal medullary secretion of CA. Studies need to be undertaken to determine if restoration of function in the immune system precedes recovery of control noradrenergic levels.

A large body of evidence clearly indicates that sympathetic denervation in young adult rodents results in altered immune responses (reviewed in Reference 41) (Tables 4 and 5). Chemical sympathectomy with 6-OHDA in adults generally results in diminished primary immune responses,[11,12,40,123,124] in contrast to neonatal sympathectomy that results in enhanced primary immune responses.[6,31,32,125] Acute denervation of adult rodents with 6-OHDA also results in diminished secondary immune responses, diminished delayed-type hypersensitivity responses, diminished cytotoxic T-cell responses and interleukin-2 secretion, enhanced natural killer cell activity, and complex alterations in proliferative responses.[8,9,11,40] NA appears to affect the inductive phase, proliferative phase, and effector phase of the immune response, possibly through exerting enabling or enhancing effects in the inductive phase, mixed responses that depend on the concentration and cell subset at the proliferative phase, and inhibitory effects at the effector phase (Figure 11). In many of the denervation studies using 6-OHDA, the extent of denervation and possible reinnervation are not assessed, making cross comparisons of these experimental observations difficult at best. Another factor making assessment of the effect of sympathectomy on immune function difficult is the

TABLE 4
Changes in Immune Responses and Natural Killer Cell Activity after Chemical Sympathectomy in Adult Mice

Response	Lymphoid organ	Antigen	Change in activity
	In Vivo		
Delayed-type hypersensitivy	NA	TNCB	Reduced
Natural killer cell activity	Lungs[a]	NA	Enhanced
Primary antibody production	Spleen	SRBC	Reduced
Secondary antibody production	Spleen	SRBC	Reduced/enhanced[b]
Cytotoxic T-lymphocyte activity	Spleen/lymph nodes	Alloantigen	Reduced
	In Vitro		
Natural killer cell activity	Spleen	NA	Enhanced
Primary responses[c]			
Mixed lymphocyte reaction	Spleen	Alloantigen	Reduced
Cytotoxic T-lymphocyte activity	Spleen	Alloantigen	Reduced
Secondary responses[d]			
IL-2 production	Lymph nodes	TNP-self	Reduced
Cytotoxic T-lymphocyte activity	Lymph nodes	TNP-self	Reduced
	Spleen	Alloantigen	No change

Note: For sympathectomy, 100 or 150 mg/kg 6-OHDA (dissolved in 0.01% ascorbate in saline) or vehicle alone was injected 1 and 3 d prior to immunization (*in vivo* responses and secondary *in vitro* responses) or sacrificed (natural killer cell activity and primary *in vitro* responses). See text for references. NA, not applicable.

[a] Natural killer cell activity was assessed *in vivo* by clearance of radiolabeled YAC-1 cells from the lungs after intravenous injection. *In vitro,* natural killer cell activity was measured by lysis of ^{51}Cr-labeled YAC-1 cells.

[b] Secondary IgG responses were reduced when 6-OHDA was injected intraperitoneally at the time of secondary immunization with sheep red blood cells (SRBC). Sympathectomy prior to primary immunization with SRBC led to enhanced secondary IgG responses. (See Reference 12).

[c] C3H/HeJ spleen cells were cultured with irradiated BALB/c spleen cells for 5 d prior to assessment of proliferation or CTL activity.

[d] For lymph node IL-2 and CTL activity, mice were immunized epicutaneously with trinitrochlorobenzene. Five days later, draining inguinal and axillary lymph node cells were stimulated *in vitro* with haptenated, irradiated, syngeneic spleen cells. For generation of spleen cell CTL activity, C3H/HeJ mice were immunized intraperitoneally and stimulated *in vitro* with BALB/c spleen cells.

From Madden, K. S. and Livnat, S., in *Psychoneuroimmunology II,* Ader, R., Felten, D. L., and Cohen, N., Eds., Academic Press, New York, 1990, 283. With permission.

colocalization of regulatory neuropeptides, such as NPY, in noradrenergic sympathetic nerve terminals, which also are depleted along with NA following chemical sympathectomy.[54,91] The role or influence of these neuropeptides on altered immune function observed following sympathectomy has not been investigated.

TABLE 5
Sympathectomy in Adult Mice Alters *In Vivo* Cellular Proliferation and *In Vitro* Mitogen-Induced Responses

	Percentage of control responses		
	Lymph nodes	**Spleen**	**Bone marrow**
		***In Vivo* Proliferation**[a]	
Day 3	160–438[b]	89–132	138–158[b]
Day 7	600–3000[b]	135–141[b]	104–123
		***In Vitro* Proliferation**[c]	
Con A	35–60[b]	92–110	ND
LPS	150–160[b]	93–107	ND
		Immunoglobulin Production[c]	
IgM	50–75[b]	88–102	ND
IgG	230–500[b]	140–155[b]	ND

Note: Adult BALB/c mice were injected with 100 mg/kg 6-OHDA or 0.01% ascorbate in saline. Results represent the range of responses from two to five experiments. ND, not determined.

[a] *In vivo* proliferation was measured as ^{125}I-UdR uptake 3 or 7 d after sympathectomy.
[b] Significant difference (calculated separately for each experiment) based on Student's *t* test (two-tailed, $p \leq 0.05$).
[c] Spleen and lymph node cells were harvested 7 d after sympathectomy. *In vitro* proliferation induced by concanavalin A or by lipopolysaccharide was measured as [^{3}H]thymidine uptake. Immunoglobulin present in supernatants from LPS-stimulated cultures was determined by ELISA.

From Madden, K. S. and Livnat, S., in *Psychoneuroimmunology II,* Ader, R., Felten, D. L., and Cohen, N. Eds., Academic Press, New York, 1990, 283. With permission.

C. SENESCENCE

Using the same regimen of 6-OHDA treatment as described above for young adult F344 rats, we have begun a similar study in aged F344 rats. Preliminary qualitative analysis of our data indicates that noradrenergic nerve fibers are still capable of ingrowth back into the spleen following 6-OHDA treatment; however, initial ingrowth of the noradrenergic profile is delayed until day 15 after the last dose, ingrowth occurs over a slower time course, and the density of the noradrenergic nerve fibers that return to the spleen is reduced at day 56 compared with their young adult counterpart and aged-vehicle-treated controls.

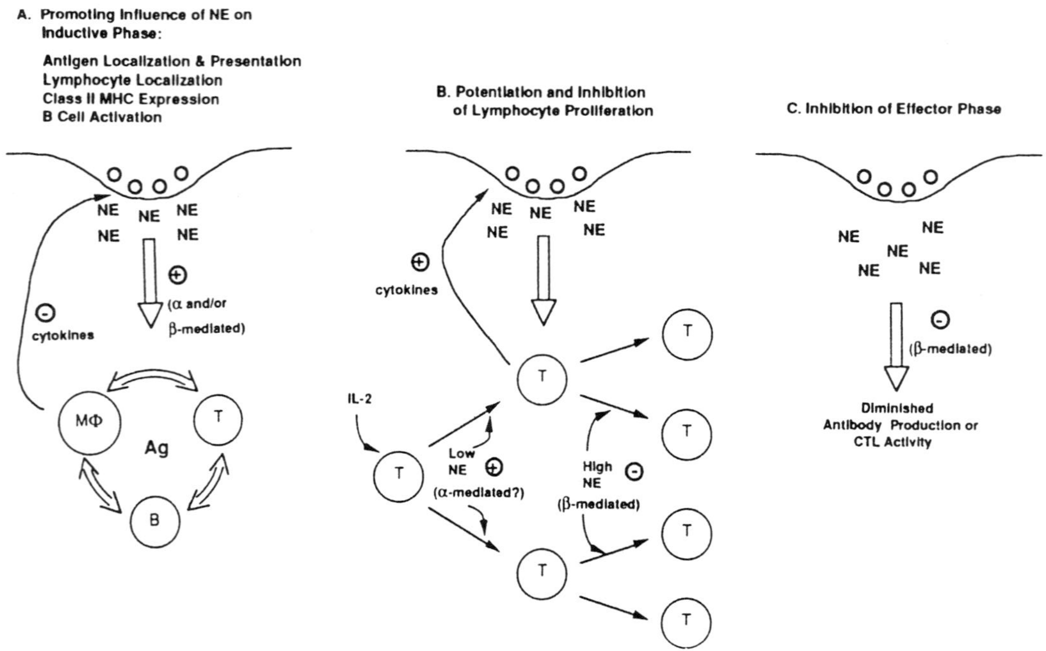

FIGURE 11. Working model for bidirectional communication between NA and cells of the immune system. This schematic diagram depicts the potential modulatory effects of NA during different phases of an immune response. The capacity for cellular responsiveness to NA will depend upon the presence of adrenoceptors, cell type, and activation state. (A) In the presence of antigen, NA may have an enabling influence on lymphocyte and accessory cell function. To achieve the proper balance between α- and β-mediated effects, regulation of NA levels may be critical. (B) Low levels of NA may stimulate α-adrenoceptors and thus potentiate lymphocyte proliferation. (B, C) With increasing NA, β-adrenoceptor-mediated inhibition of lymphocyte proliferation and effectory function may occur. The response to NA will be influenced by, and may alter the production of, cytokines (such as IL-s) and neuroendocrine peptides (not shown for clarity). Finally, products of the immune system may modulate NA availability by altering release or turnover. Direct and indirect evidence for these interactions is reviewed by Madden and colleagues.[39,41] (From Madden, K. S. and Livnat, S., in *Psychoneuroimmunology II,* Ader, R., Felten, D. L., and Cohen, N., Eds., Academic Press, New York, 1990, 283. With permission.)

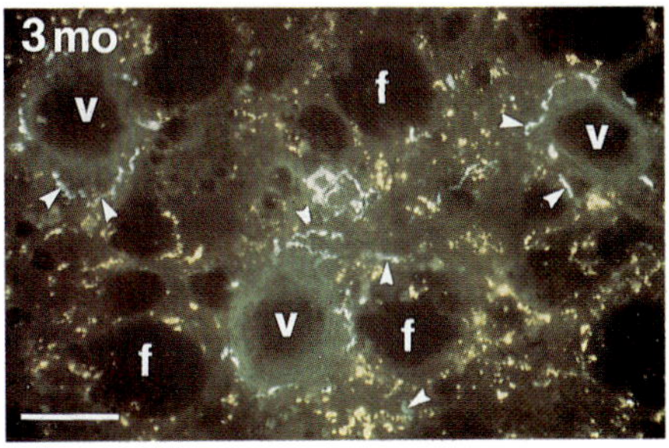

PLATE 1. Fluorescent noradrenergic nerves course along several blood vessels (v) in the bone marrow. Some fluorescent fibers (arrowheads) radiate away from these plexuses into the surrounding parenchyma. Glyoxylic acid histofluorescence; 3-month-old (3 mo) F344 rat. f, fat cells. (Bar = 100 μm.)

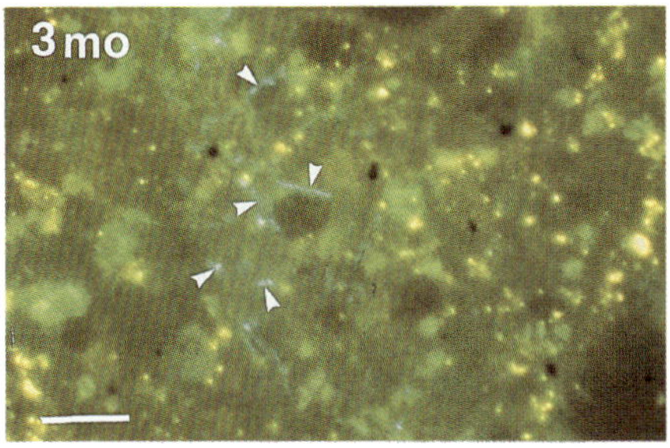

PLATE 2. Very fine fluorescent fibers (arrowheads) reside in the parenchyma of the bone marrow. Glyoxylic acid histofluorescence; 3-month-old (3 mo) F344 rat. (Bar = 50 μm.)

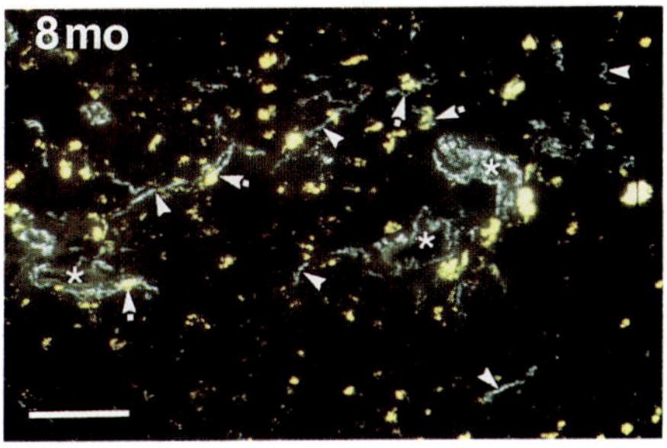

PLATE 3. Noradrenergic varicosities form dense arterial plexuses (*) in the thymic cortex, or course as free linear profiles among thymocytes (arrowheads) and in close proximity to cortical yellow autofluorescent cells (solid box, arrowheads). Glyoxylic acid histofluorescence; 8-month-old (8 mo) F344 rat. (Bar = 100 μm.)

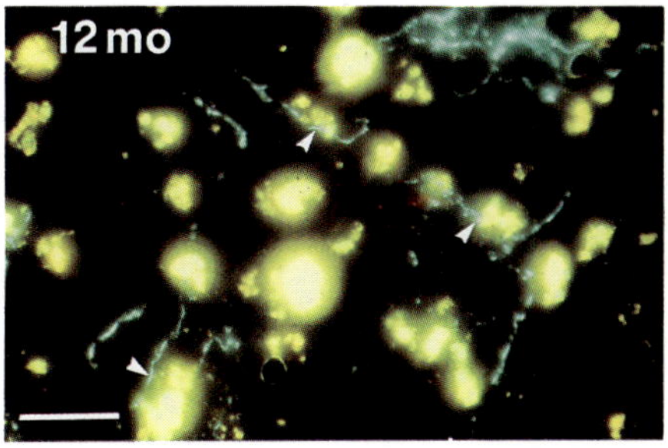

PLATE 4. High magnification demonstrating the close association of noradrenergic nerve fibers (arrowheads) with cortical yellow autofluorescent cells in the thymic cortex. These fibers also reside among thymocytes in the thymic cortex. Glyoxylic acid histofluorescence; 12-month-old (12 mo) F344 rat. (Bar = 50 μm.)

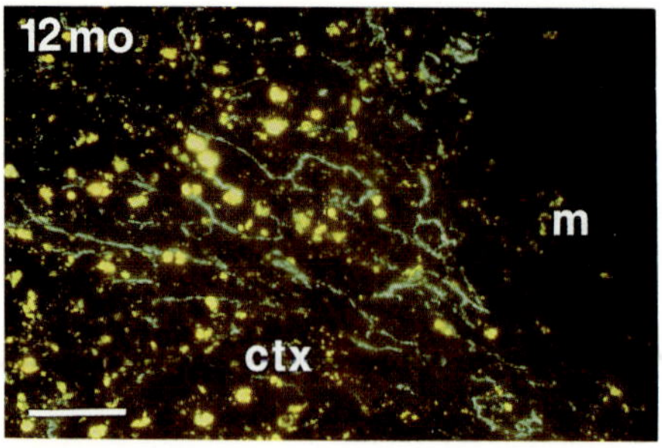

PLATE 5. Fine varicose noradrenergic fibers course in linear arrays among thymocytes and cortical yellow autofluorescent cells in the thymic cortex (ctx). Glyoxylic acid histofluorescence; 12-month-old (12 mo) F344 rat. m, medulla. (Bar = 100 μm.)

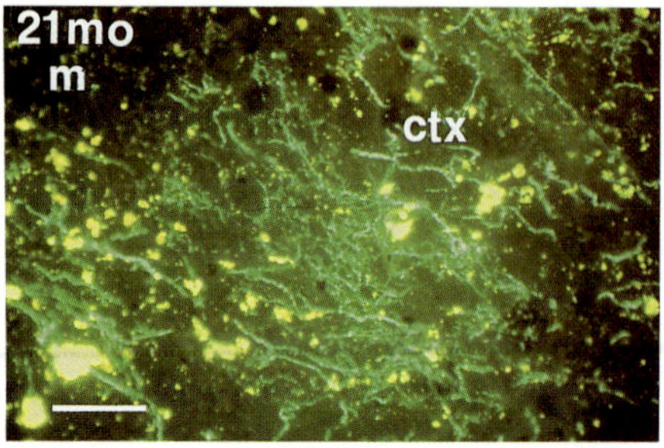

PLATE 6. A dense tangle of noradrenergic fibers in the parenchyma of the thymic cortex (ctx) among thymocytes and clusters of cortical yellow autofluorescent cells. Glyoxylic acid histofluorescence; 21-month-old (21 mo) F344 rat. m, medulla. (Bar = 100 μm.)

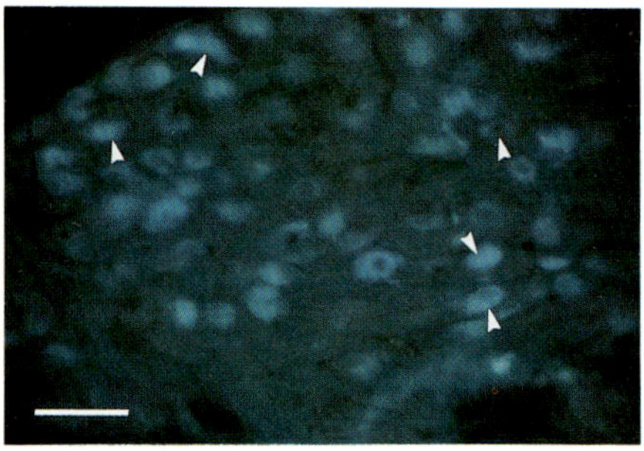

PLATE 7. Retrograde tracing following injection of true blue into the spleen demonstrates true blue-positive ganglion cells in the superior mesenteric-celiac ganglion of the F344 rat. Arrowheads indicate neurons that are positive for both true blue and TH (see Plate 8). Retrograde tracing of true blue. (Bar = 100 μm.)

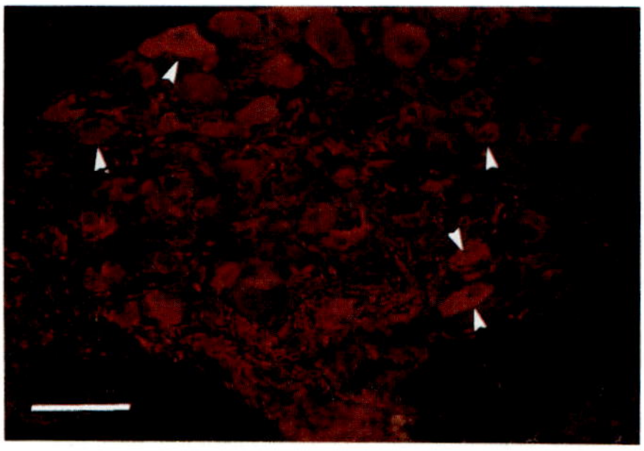

PLATE 8. Many of the TH-positive ganglion cells (arrowheads) in the superior mesenteric-celiac ganglion of the F344 rat also are positive for true blue (see Plate 7). Immunocytochemistry for TH; rhodamine-labeled secondary antibody. (Bar = 100 μm.)

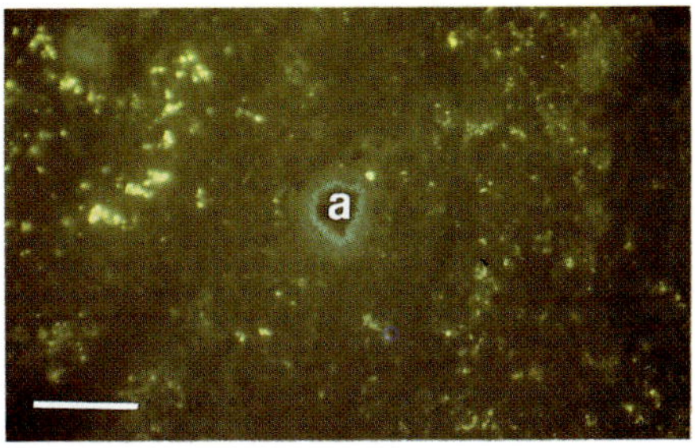

PLATE 9. Fluorescence histochemistry in spleens from F344 rats following surgical removal of the superior mesenteric-celiac ganglion demonstrates a lack of noradrenergic fibers along the central artery (a) of the white pulp of the spleen. Glyoxylic acid histofluorescence. (Bar = 100 μm.)

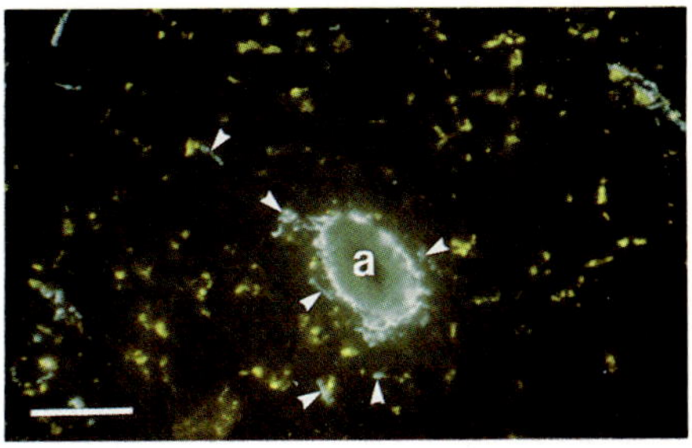

PLATE 10. Abundant noradrenergic varicosities surround the central artery (a) in the white pulp of the spleen from a 3-month-old F344 rat. Many fluorescent profiles (arrowheads) radiate away from this vascular plexus into the surrounding PALS. Glyoxylic acid histofluorescence. (Bar = 100 μm.)

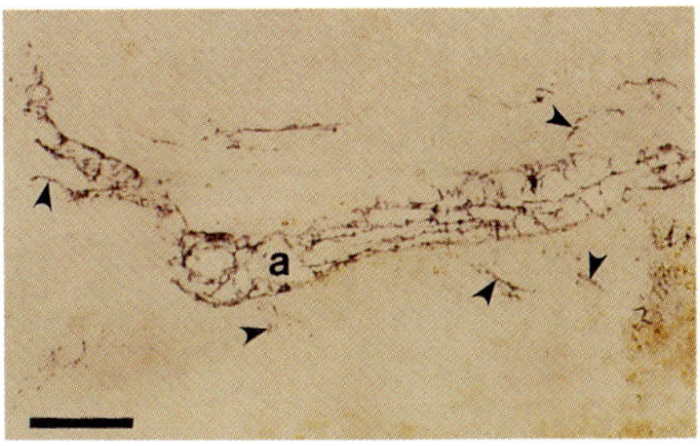

PLATE 11. Noradrenergic nerve fibers (black) coursing with the central arteriole (a) of the white pulp of the spleen and in the surrounding white pulp (arrowheads) are revealed with single-label immunocytochemistry for TH. Single-label ICC for TH. F344 rat. (Bar = 100 μm.)

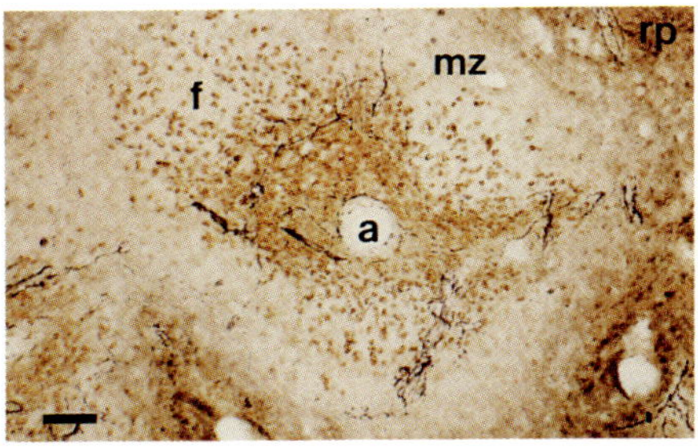

PLATE 12. TH+ nerve fibers (black) present along the central arteriole (a) of the white pulp of the spleen and in the surrounding PALS among OX19+ T-lymphocytes (brown). Double-label ICC for TH and OX19, a pan-T-lymphocyte marker. F344 rat. f, follicle; mz, marginal zone; rp, red pulp. (Bar = 100 μm.)

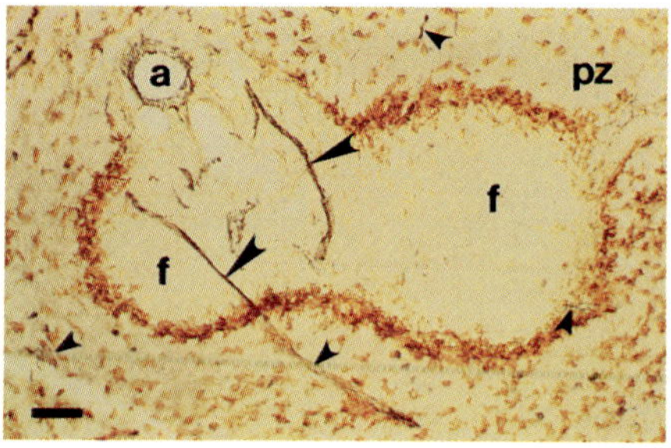

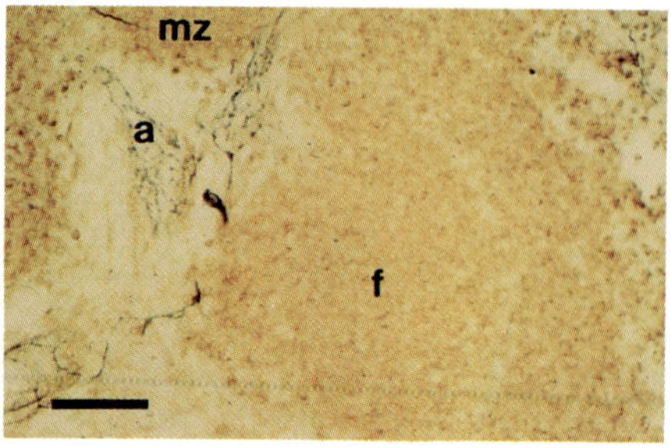

PLATE 13. TH+ nerves (black) course along the central arteriole (a) and in the surrounding white pulp of spleen. Two large nerve bundles (large arrowheads) traverse through the white pulp. TH+ profiles (small arrowheads) are found among ED3+ macrophages (brown) of the inner and outer marginal zone. ED3+ macrophages form a dense band of cells in the inner marginal zone/parafollicular (pz) along the marginal sinus and have a more diffuse distribution in the outer marginal zone/parafollicular. Double-label ICC for TH and ED3, a marker for macrophages that reside largely in the marginal zone. F344 rat. f, follicle. (Bar = 100 μm.)

PLATE 14. IgM+ B lymphocytes (brown) are present in the follicle (f) and in the marginal zone (mz). TH+ nerves (black) reside adjacent to the central arteriole (a), the surrounding PALS, and in the marginal zone among IgM+ B-lymphocytes. TH+ are occasionally observed in the follicle. Double-label ICC for TH and IgM, a marker for B-lymphocytes. F344 rat. (Bar = 100 μm.)

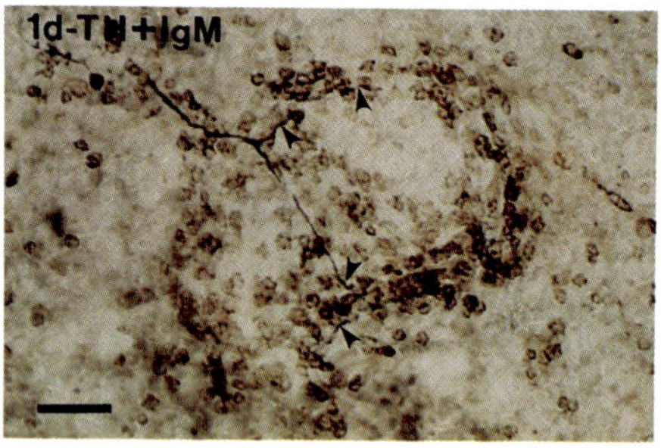

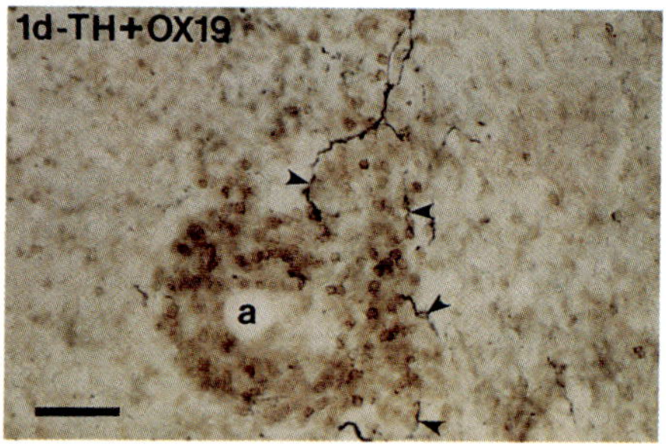

PLATE 15. One-day-old (1d) rat spleen stained for TH+ noradrenergic nerves (black fibers) and IgM+ B lymphocytes (brown cells). TH+ nerve fibers (arrowheads) course among IgM+ B-lymphocytes in the external portion of the PALS. Double-label ICC. (Bar = 50 μm.) (From Ackerman, K. D., Felten, S. Y., Dijkstra, C. D., Livnat, S., and Felten, D. L., *Exp. Neurol.*, 103, 239, 1989. With permission.)

PLATE 16. One-day-old (1d) rat spleen stained for TH+ noradrenergic nerves (black fibers) and OX19+ T-lymphocytes (brown cells). TH+ fibers in linear profile contact the outer limit of the region delineated by OX19+ T-lymphocytes (arrowheads). TH+ fibers enter this compartment (arrowheads), but do not contact the central artery (a). Double-label ICC. (Bar = 50 μm.) (From Ackerman, K. D., Felten, S. Y., Dijkstra, C. D., Livnat, S., and Felten, D. L., *Exp. Neurol.*, 103, 239, 1989. With permission.)

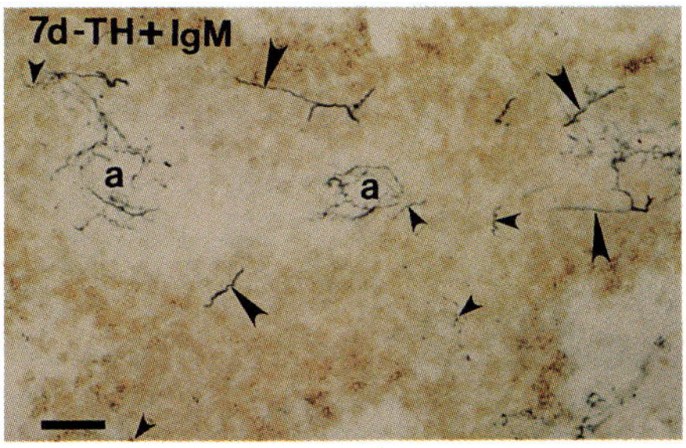

PLATE 17. Seven-day-old (7d) rat spleen stained for TH+ noradrenergic nerves (black fibers) and IgM+ B-lymphocytes (brown cells). Long linear profiles of TH+ nerves (large arrowheads) separate lightly stained IgM+ cells of the PALS from the darkly stained B-cells of the marginal zone. Additional TH+ fibers form a dense plexus within the PALS, arborizing among B-lymphocytes within the inner portion of the PALS (small arrowheads). Double-label ICC. a, central arteriole. (Bar = 100 μm.) (From Ackerman, K. D., Felten, S. Y., Dijkstra, C. D., Livnat, S., and Felten, D. L., *Exp. Neurol.*, 103, 239, 1989. With permission.)

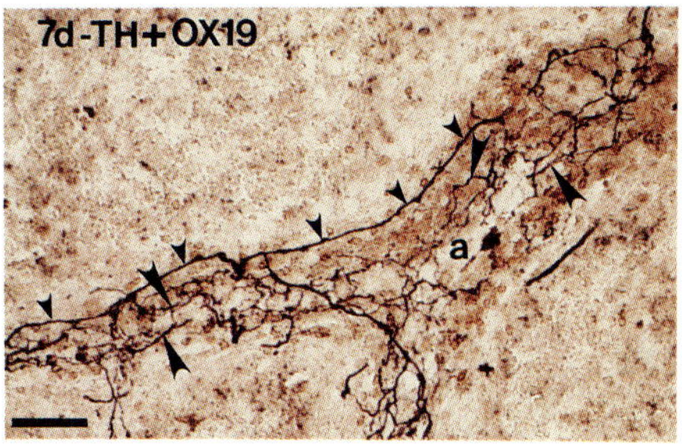

PLATE 18. Seven-day-old (7d) rat spleen stained for TH+ noradrenergic nerves (black fibers) and OX19+ T-lymphocytes (brown cells). TH+ fibers form a loose, tangled plexus along the central artery and its branches (large arrowheads), arborizing among OX19+ T-lymphocytes of the PALS. Additional TH+ fibers form long linear bundles at the external limit of the PALS defined by OX19+ lymphocytes (small arrowheads). Double-label ICC. (Bar = 50 μm.) (From Ackerman, K. D., Felten, S. Y., Dijkstra, C. D., Livnat, S., and Felten, D. L., *Exp. Neurol.*, 103, 239, 1989. With permission.)

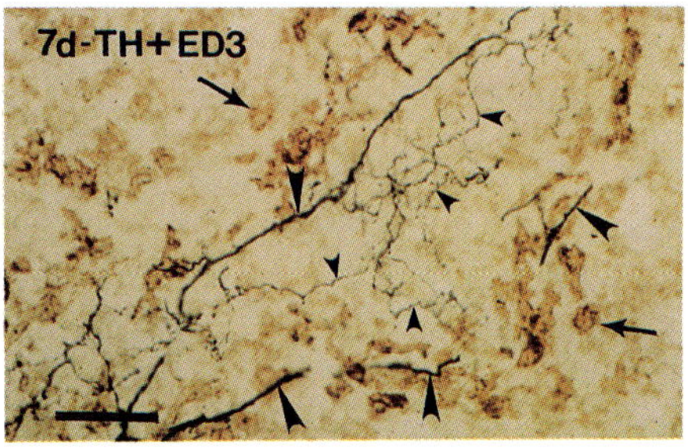

PLATE 19. Seven-day-old (7d) rat spleen stained for TH+ noradrenergic nerves (black fibers) and ED3+ macrophages (brown cells). TH+ nerve fibers (small arrowheads) form a dense plexus throughout the inner portion of the PALS devoid of ED3+ cells. Additional TH+ fibers in linear profile (large arrowheads) course among ED3+ cells with long, blunt processes at the site of the marginal sinus. Large, round ED3+ macrophages (arrows) are found in low density within the marginal zone. Double-label ICC. (Bar = 50 μm.) (From Ackerman, K. D., Felten, S. Y., Dijkstra, C. D., Livnat, S., and Felten, D. L., *Exp. Neurol.*, 103, 239, 1989. With permission.)

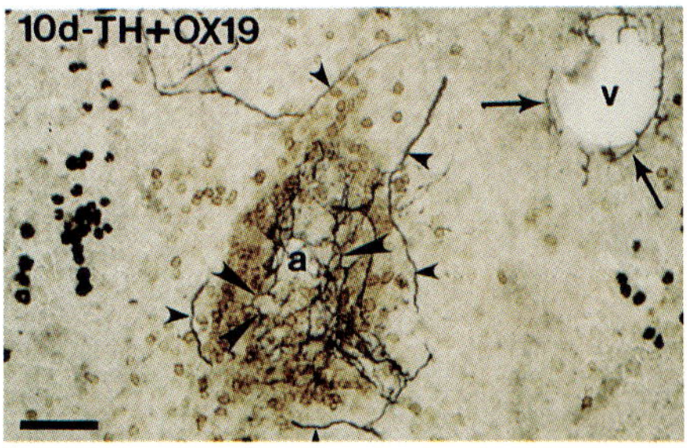

PLATE 20. Ten-day-old (10d) rat spleen stained for TH+ noradrenergic nerves (black fibers) and OX19+ T-lymphocytes (brown cells). TH+ fibers (large arrowheads) course throughout the inner PALS formed by OX19+ T-lymphocytes. Additional bundles, in long, linear arrays, encircle this region (small arrowheads). The red pulp contains a small vein (v, arrow) surrounded by TH+ fibers and numerous dark black cells (probably eosinophils) which stain nonspecifically with avidin-biotin complexes. Double-label ICC. a, central arteriole. (Bar = 50 μm.) (From Ackerman, K. D., Felten, S. Y., Dijkstra, C. D., Livnat, S., and Felten, D. L., *Exp. Neurol.*, 103, 239, 1989. With permission.)

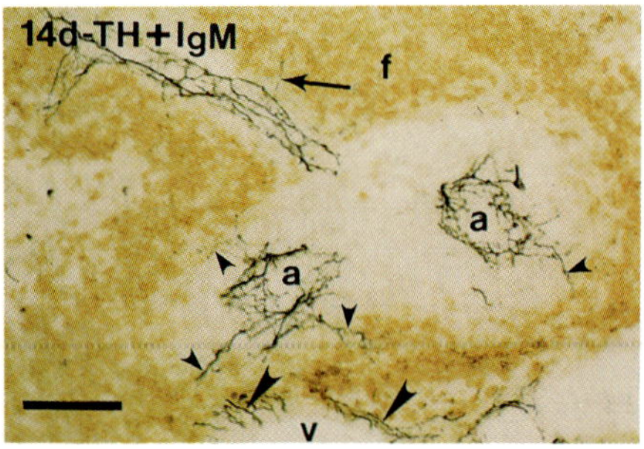

PLATE 21. Fourteen-day-old (14d) rat spleen stained for TH+ noradrenergic nerves (black fibers) and IgM+ B-lymphocytes (brown cells). TH+ fibers form a dense, tangled plexus within the inner PALS, an area devoid of B-lymphocytes. IgM+ cells are located within the outer PALS, marginal zone, and follicles. B-lymphocytes of the outer PALS and the inner marginal zone lie adjacent to TH+ fibers at the marginal sinus (small arrowheads), while additional TH+ fibers (large arrowheads), supplied by an adjacent vein (v), arborize among B-lymphocytes of the outer marginal zone. Finally, a delicate TH+ profile (arrow) is found within the follicle (f). Double-label ICC. (Bar = 100 μm.) (From Ackerman, K. D., Felten, S. Y., Dijkstra, C. D., Livnat, S., and Felten, D. L., *Exp. Neurol.*, 103, 239, 1989. With permission.)

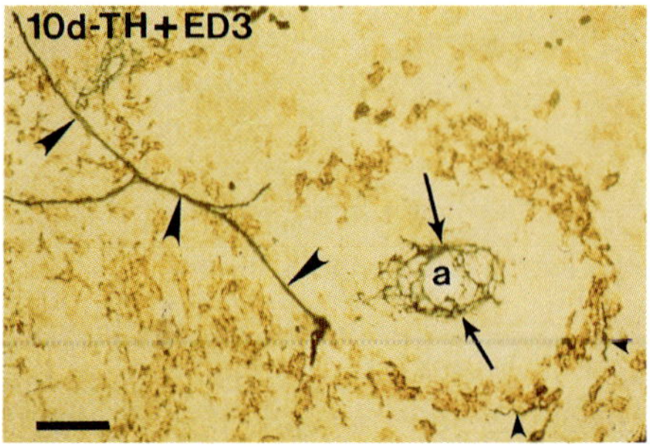

PLATE 22. Ten-day-old (10d) rat spleen stained for TH+ noradrenergic nerves (black fibers) and ED3+ macrophages (brown cells). A dense plexus of TH+ fibers (arrows) surrounds the central artery, associated with fibers which branch within the PALS. Additional TH+ fibers at the marginal sinus (large arrowheads) course among ED3+ cells with long processes. ED3+ macrophages with few processes (small arrowheads) are located external to the marginal sinus within the marginal zone. Double-label ICC. (Bar = 50 μm.) (From Ackerman, K. D., Felten, S. Y., Dijkstra, C. D., Livnat, S., and Felten, D. L., *Exp. Neurol.*, 103, 239, 1989. With permission.)

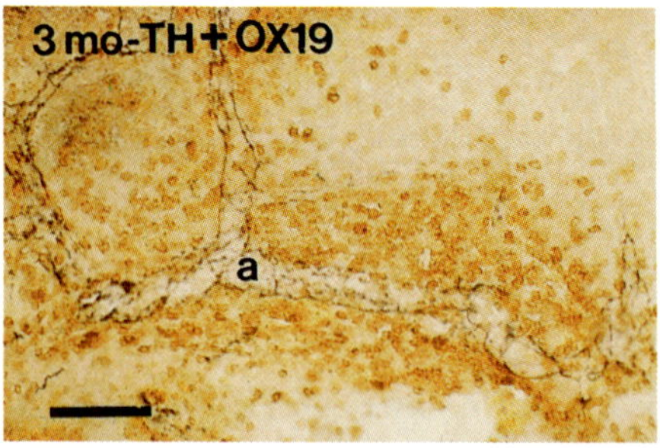

PLATE 23. Three-month-old (3 mo) rat spleen stained for TH+ noradrenergic nerves (black fibers) and OX19+ T-lymphocytes (brown cells). TH+ nerve fibers course along the central arteriole (a) cut in longitudinal section and in the surrounding PALS among OX19+ T-lymphocytes. Double-label ICC. (Bar = 100 μm.)

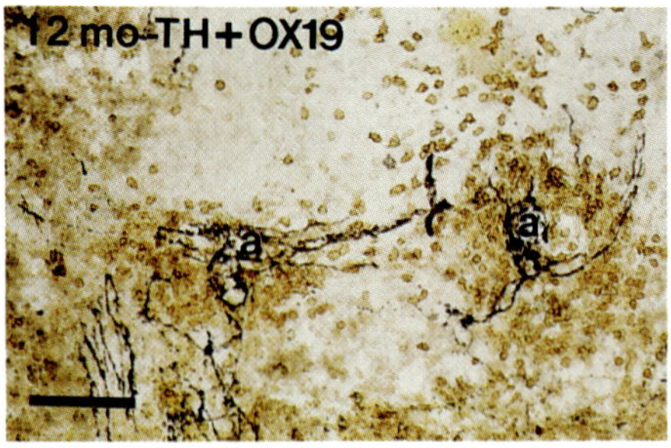

PLATE 24. Twelve-month-old (12 mo) rat spleen stained for TH+ noradrenergic nerves (black fibers) and OX19+ T-lymphocytes (brown cells). TH+ nerve fibers form a denser plexus along the central arteriole (a) and its branches compared with 3-month-old rats. From this dense vascular plexus, TH+ nerves extend into the surrounding PALS among OX19+ T-lymphocytes. The density of OX19+ T-lymphocytes that form the PALS is diminished compared with 3- to 8-month-old rats. Double-label ICC. (Bar = 100 μm.)

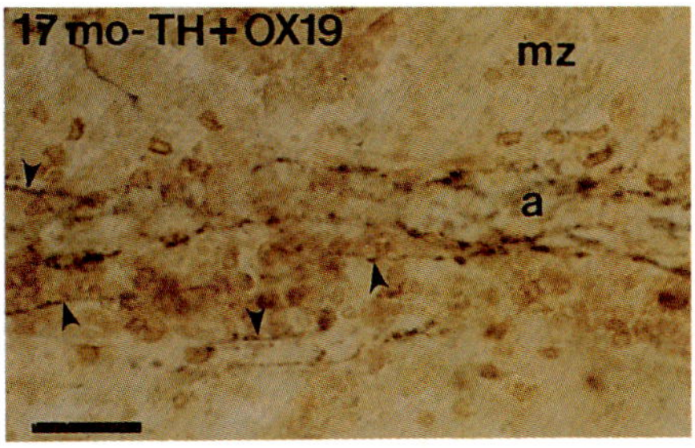

PLATE 25. Seventeen-month-old (17 mo) rat spleen stained for TH+ noradrenergic nerves (black fibers) and OX19+ T-lymphocytes (brown cells). Fewer TH+ nerve fibers traverse along the central arteriole (a). An occasional TH+ nerve fiber (small arrowheads) is found in the adjacent PALS among OX19+ T-lymphocytes, and in the marginal zone (mz) that does not contain T-lymphocytes. Double-label ICC. (Bar = 50 μm.)

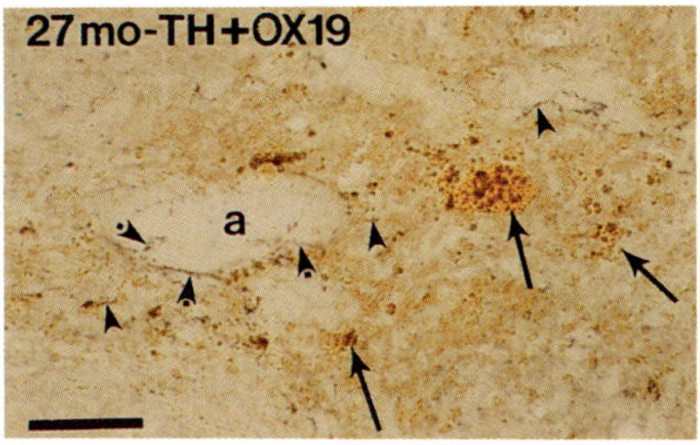

PLATE 26. Twenty-seven-month-old (27 mo) rat spleen stained for TH+ noradrenergic nerves (black fibers) and OX19+ T-lymphocytes (brown cells). The density of TH+ nerve fibers (solid box, small arrowheads) associated with the central arteriole (a) continue to decline. An occasional TH+ nerve profile (small arrowheads) is present in the PALS. Very few OX19+ cells stain in the PALS. In addition to OX19+ cells that appear as small rings of brown reaction product, large unstained gold granular cells (small arrows) are present in the PALS. Double-label ICC. (Bar = 100 μm.)

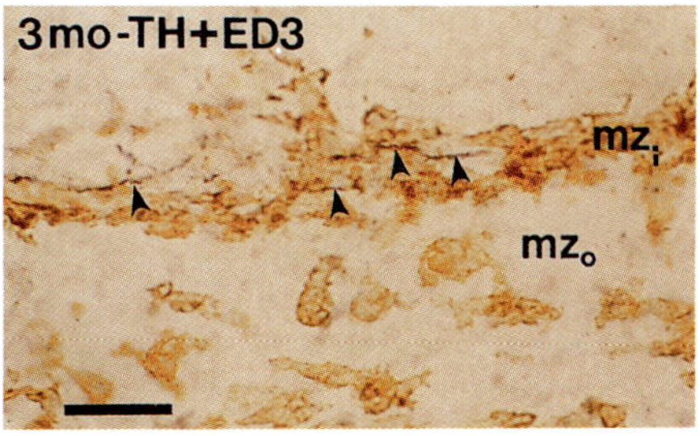

PLATE 27. Three-month-old (3 mo) rat spleen stained for TH+ noradrenergic nerves (black fibers) and ED3+ macrophages (brown cells). TH+ nerve fibers (small arrowheads) are present among ED3+ macrophages in the inner marginal zone (mz_i). Double-label ICC. mz_o, outer marginal zone. (Bar = 50 μm.)

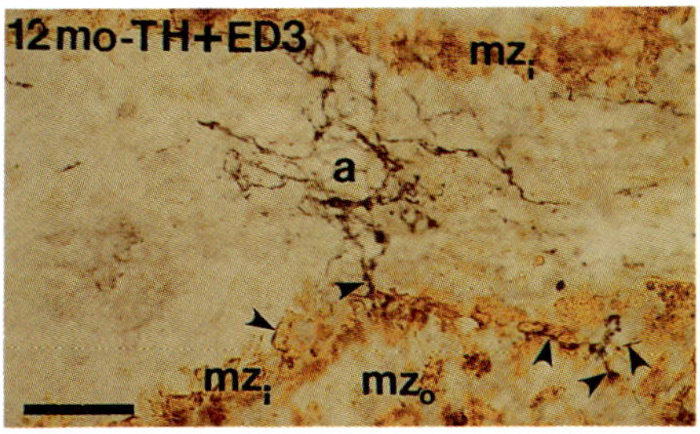

PLATE 28. Twelve-month-old (12 mo) rat spleen stained for TH+ noradrenergic nerves (black fibers) and ED3+ macrophages (brown cells). The density of ED3+ macrophages in the inner marginal zone (mz_i) of the PALS appears to be slightly reduced, and the packing of ED3+ cells in the outer marginal zone (mz_o) may be more dense. A dense plexus of TH+ nerves surround the central arteriole (a) in the unstained PALS. TH+ nerves radiate away from this plexus into the inner (mz_i) and outer marginal zone (small arrowheads). Double-label ICC. (Bar = 100 μm.)

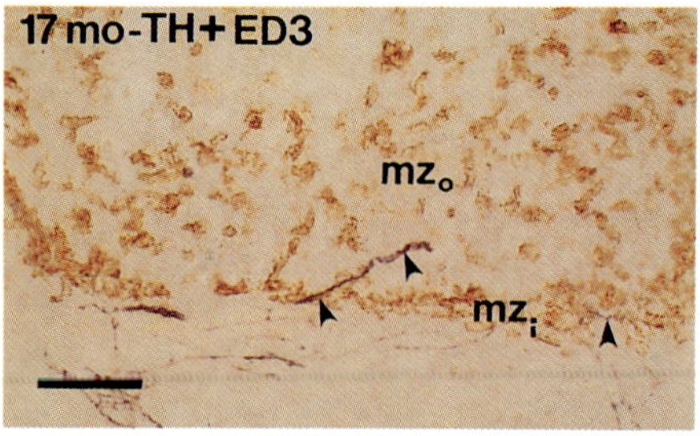

PLATE 29. Seventeen-month-old (17 mo) rat spleen stained for TH+ noradrenergic nerves (black fibers) and ED3+ macrophages (brown cells). The density of ED3+ macrophages in the inner (mz_i) and outer (mz_o) marginal zone is diminished and the delineation between these two regions becomes less distinct. TH+ nerve fibers (small arrowheads) are present in the inner and outer marginal zone among ED3+ macrophages. Double-label ICC. (Bar = 100 μm.)

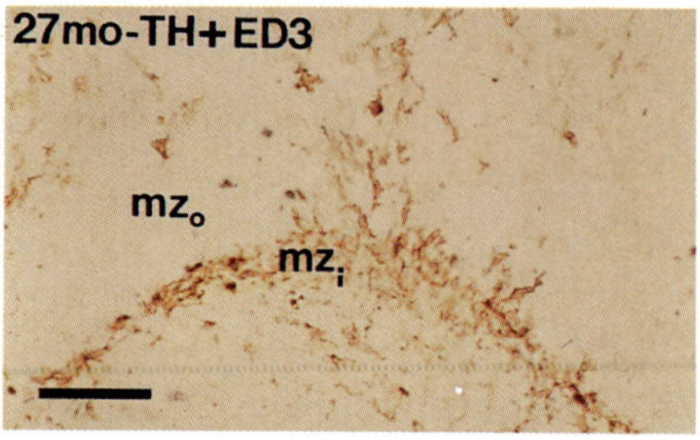

PLATE 30. Twenty-seven-month-old (27 mo) rat spleen stained for TH+ noradrenergic nerves (black fibers) and ED3+ macrophages (brown cells). Sparse accumulation of ED3+ macrophages for the inner (mz_i) and outer (mz_o) marginal zone at regions distal from the hilus. No TH+ nerves are present among ED3+ macrophages in this region of the spleen. Double-label ICC. (Bar = 100 μm.)

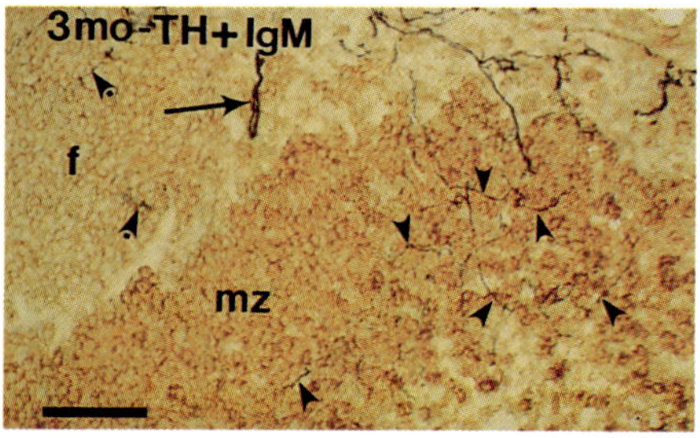

PLATE 31. Three-month-old (3 mo) rat spleen stained for TH+ noradrenergic nerves (black fibers) and IgM+ B-lymphocytes (brown cells). IgM+ B-lymphocytes reside in the follicle (f) and in the marginal zone (mz). A small bundle of TH+ nerve fibers (arrow) is present at the edge of a follicle. Long linear (TH+ profiles are present in the follicle (solid box, small arrowheads) and in the marginal zone (small arrowheads) among IgM+ B-lymphocytes. Double-label ICC. (Bar = 100 μm.)

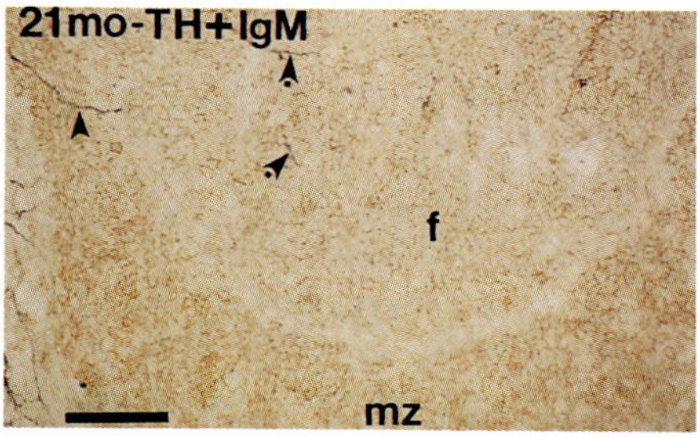

PLATE 32. Twenty-one-month-old (21 mo) rat spleen stained for TH+ noradrenergic nerves (black fibers) and IgM+ B-lymphocytes (brown cells). IgM+ B-lymphocyte staining remains relatively robust in the marginal zone (mz) and follicle (f). The density of TH+ nerve fiber (solid box, small arrowheads) in the follicle appears to be comparable with staining in spleens from 3- and 8-month-old rats. The density of TH+ nerve fibers (small arrowheads) in the marginal zone declines with age. Double-label ICC. (Bar = 100 μm.)

V. NORADRENERGIC INNERVATION AND AUTOIMMUNITY

Evidence from both animal research[78,98,101,126-138] (reviewed in Reference 139) and clinical studies[140,141] point toward involvement of the nervous system in the development and progression of autoimmune disorders in a variety of animal models for autoimmune disorders, including adjuvant-induced experimental arthritis (EA), New Zealand mice strains as a model for hemolytic anemia and lupus-like syndrome, and the experimental allergic encephalomyelitis (EAE) model for multiple sclerosis. The expression of EAE can be attenuated following lesions in the anterior hypothalamus prior to immunization.[142] Direct involvement of the sympathetic nervous system in the expression of autoimmune disease is indicated by the work of Chelmicka-Schorr and co-workers[128,129] and by studies from our laboratories.[126,127] Chemical sympathectomy in Lewis rats exacerbates EAE,[128] while administration of isoproterenol, an adrenergic agonist, to susceptible rats protects against the development of EAE.[129] Evidence from our laboratories suggests a relationship between the decline in noradrenergic innervation of secondary lymphoid organs (spleen and lymph nodes) and immunosenescence in F344 rats,[14,78,79,98,101] and with the expression of hemolytic anemia and lupus-like syndrome in New Zealand mice.[98,143] Clinical and animal studies also implicate neural involvement in the inflammation and joint destruction in rheumatoid arthritic patients, and in rats with EA. Neonatal treatment with capsaicin, a neurotoxin that destroys small nonmyelinated fibers, including SP-containing fibers, reduces the severity of joint injury in arthritic rats.[132,135,136] NA- and SP-containing nerves in the joint can mediate the extent of inflammation and subsequent destruction of affected joints in laboratory animals with EA, and in arthritic patients.[140,141]

Recent studies from our laboratory have examined the role of noradrenergic and peptidergic innervation of secondary lymphoid organs, without manipulation of the innervation of the joint tissue, in the development and progression of EA in young adult Lewis rats.[101,126,127,144] Local denervation of noradrenergic nerves in draining popliteal and inguinal lymph nodes by application of 6-OHDA prior to challenge with Freund's complete adjuvant into the hind footpad exacerbates the subsequent inflammatory responses and joint destruction following adjuvant-induced arthritis. Local denervation of SP-containing nerves in draining popliteal and inguinal lymph nodes by local application of capsaicin 1 d after immunization delays the onset and attenuates the severity of arthritis.

VI. SUMMARY

We and others have described the presence of noradrenergic, as well as peptidergic, innervation of both primary and secondary lymphoid organs.

Further, we have demonstrated the dynamic changes that occur in sympathetic innervation of lymphoid organs throughout the lifespan of the animal, and the plasticity of noradrenergic nerves in the spleen following insult. An abundance of evidence has accumulated to indicate a role for sympathetic innervation, and neuroendocrine outflow, in mediation of the immune system across the lifespan of the individual. There is evidence supporting involvement of noradrenergic sympathetic innervation during development in age-related changes in immune function and diseases processes such as autoimmune disorders. Collectively, these findings indicate that the development of disease states, including autoimmune disorders, depends upon highly complex interactions between the nervous and immune systems at multiple sites (i.e., at all effector tissues, at the site of antigen presentation and processing, and at the central nervous system). The mechanisms of action, routes of communication between these sites, and the neural and immune mediators that determine the overall health status of an individual await further investigation. Studies described in this chapter demonstrate that innervation of lymphoid organs can play an important role in health and illness, and under conditions of marginal functioning of the immune system, the outflow of neural signals may be a critical factor. Future research to elucidate pathways for neural-immune interactions, and the identity of neural and immune mediators and their mechanism of action in models of aging, disease, and autoimmunity, may ultimately result in unique pharmacological approaches to the maintenance of immune system homeostasis and host defense mechanisms throughout the lifespan of an individual, and to intervention in autoimmune disorders such as rheumatoid arthritis.

REFERENCES

1. **Ayers, A. B., Davies, B. N., and Withrington, P. G.,** Responses of the isolated, perfused human spleen to sympathetic nerve stimulation, catecholamines and polypeptides, *Br. J. Pharmacol.*, 44, 17, 1972.
2. **Fillenz, M.,** The innervation of the cat spleen, *Proc. R. Soc. London,* 174, 459, 1970.
3. **Gillespie, J. S. and Kirpekar, S. M.,** The localization of endogenous and infused noradrenaline in the spleen, *J. Physiol.*, 179, 46P, 1965.
4. **Reilly, F. D.,** Innervation and vascular pharmacodynamics of the mammalian spleen, *Experientia,* 41, 187, 1985.
5. **Williams, J. M. and Felten, D. L.,** Sympathetic innervation of murine thymus and spleen: a comparative histofluorescence study, *Anat. Rec.*, 199, 531, 1981.
6. **Williams, J. M., Peterson, R. G., Shea, P. A., Schmedtje, J. F., Bauer, D. C., and Felten, D. L.,** Sympathetic innervation of murine thymus and spleen: evidence for a functional link between the nervous and immune systems, *Brain Res. Bull.*, 6, 83, 1981.
7. **Felten, D. L., Overhage, J. M., Felten, S. Y., and Schmedtje, J. F.,** Noradrenergic sympathetic innervation of lymphoid tissue in the rabbit appendix: further evidence for a link between the nervous and immune systems, *Brain Res. Bull.*, 7, 595, 1981.

8. **Felten, D. L., Livnat, S., Felten, S. Y., Carlson, S. L., Bellinger, D. L., and Yeh, P.,** Sympathetic innervation of lymph nodes in mice, *Brain Res. Bull.,* 13, 693, 1984.
9. **Felten, D. L., Felten, S. Y., Carlson, S. L., Olschowka, J. A., and Livnat, S.,** Noradrenergic and peptidergic innervation of lymphoid tissue, *J. Immunol.,* 135, 755s, 1985.
10. **Felten, D. L., Ackerman, K. D., Wiegand, S. J., and Felten, S. Y.,** Noradrenergic sympathetic innervation of the spleen. I. Nerve fibers associate with lymphocytes and macrophages in specific compartments of the splenic white pulp, *J. Neurosci. Res.,* 18, 28, 1987.
11. **Felten, D. L., Felten, S. Y., Bellinger, D. L., Carlson, S. L., Ackerman, K. D., Madden, K. S., Olschowka, J. A., and Livnat, S.,** Noradrenergic sympathetic neural interactions with the immune system: structure and function, *Immunol. Rev.,* 100, 225, 1987.
12. **Livnat, S., Felten, S. Y., Carlson, S. L., Bellinger, D. L., and Felten, D. L.,** Involvement of peripheral and central catecholamine systems in neural-immune interactions, *J. Neuroimmunol.,* 10, 5, 1985.
13. **Felten, S. Y. and Olschowka, J. A.,** Noradrenergic sympathetic innervation of the spleen. II. Tyrosine hydroxylase (TH)-positive nerve terminals form synaptic-like contacts on lymphocytes in the splenic white pulp, *J. Neurosci. Res.,* 18, 37, 1987.
14. **Bellinger, D. L., Felten, S. Y., Collier, T. J., and Felten, D. L.,** Noradrenergic sympathetic innervation of the spleen. IV. Morphometric analysis in adult and aged F344 rats, *J. Neurosci. Res.,* 18, 55, 1987.
15. **Felten, S. Y., Carlson, S. L., Bellinger, D. L., and Felten, D. L.,** Overview of the efferent autonomic nervous system, in *Neuroregulation of Autonomic, Endocrine and Immune Systems,* Frederickson, R., Hendrie, H. C., Hingtgen, J. N., and Aprison, M. H., Eds., Martinus-Nijhof, New York, 1986, 109.
16. **Giron, L. T., Crutcher, K. A., and Davis, J. N.,** Lymph nodes — a possible site for sympathetic neuronal regulation of immune responses, *Ann. Neurol.,* 8, 520, 1980.
17. **Bullock, K. and Pomerantz, W.,** Autonomic nervous system innervation of thymic related lymphoid tissue in wild-type and nude mice, *J. Comp. Neurol.,* 228, 57, 1984.
18. **Singh, U.,** Sympathetic innervation of fetal mouse thymus, *Eur. J. Immunol.,* 14, 757, 1984.
19. **Walcott, B. and McLean, J. R.,** Catecholamine-containing neurons and lymphoid cells in a lacrimal gland of the pigeon, *Brain Res.,* 328, 129, 1985.
20. **Calvo, W.,** The innervation of the bone marrow in laboratory animals, *Am. J. Anat.,* 123, 315, 1968.
21. **Calvo, W. and Forteza-Vila, J.,** On the development of bone marrow innervation in newborn rats as studied with silver impregnation and electron microscopy, *Am. J. Anat.,* 126, 355, 1969.
22. **Bishopric, N. H., Cohen, H. J., and Lefkowitz, R. J.,** Beta adrenergic receptors in lymphocyte subpopulations, *J. Allergy Clin. Immunol.,* 65, 29, 1980.
23. **Hadden, J. W., Hadden, E. M., and Middleton, E., Jr.,** Lymphocyte blast transformation. I. Demonstration of adrenergic receptors in human peripheral lymphocytes, *Cell. Immunol.,* 1, 583, 1970.
24. **Williams, L. T., Snyderman, R., and Lefkowitz, R. J.,** Identification of β-adrenergic receptors in human lymphocytes by (− −)[^{3}H]-alprenolol binding, *J. Clin. Invest.,* 57, 149, 1976.
25. **Pochet, R., Delespesse, G., Gausset, P. W., and Collet, H.,** Distribution of beta-adrenergic receptors on human lymphocyte subpopulations, *Clin. Exp. Immunol.,* 38, 578, 1979.
26. **Singh, U.,** Effect of catecholamines on lymphopoiesis in fetal mouse thymic explants, *Eur. J. Immunol.,* 14, 757, 1979.
27. **Johnson, D. L. and Gordon, M. A.,** Characteristics of adrenergic binding sites associated with murine lymphocytes isolated from spleen, *J. Immunopharmacol.,* 2, 435, 1980.

28. **Loveland, B. E., Jarrott, B., and McKenzie, I. F. C.,** The detection of beta adrenoceptors on murine lymphocytes, *Int. J. Immunopharmacol.,* 3, 45, 1981.
29. **Bidart, J. M., Motte, Ph., Assicot, M., Bohuon, C., and Bellett, D.,** Catechol-O-methyltransferase activity and aminergic binding site distribution in human peripheral blood lymphocyte subpopulations, *Clin. Immunol. Immunopathol.,* 26, 1, 1983.
30. **Brodde, O. E., Engel, G., Hoyer, D., Block, K. D., and Weber, F.,** The beta-adrenergic receptor in human lymphocytes — subclassification by the use of a new radioligand (±)[^{125}iodo]cyanopindolol, *Life Sci.,* 29, 2189, 1981.
31. **Miles, K.,** The Sympathetic Nervous System and the Immune Response in Mice. Ph.D. thesis, University of Chicago, 1984.
32. **Miles, K., Quintáns, J., Chelmicka-Schorr, E., and Arnason, B. G. W.,** The sympathetic nervous system modulates antibody response to thymus-independent antigens, *J. Neuroimmunol.,* 1, 101, 1981.
33. **Miles, K., Atweh, S., Otten, G., Arnason, B. G. W., and Chelmicka-Schorr, E.,** Beta-adrenergic receptors on splenic lymphocytes from axotomized mice, *Int. J. Immunopharmacol.,* 6, 171, 1984.
34. **Hall, N. R., McGillis, J. P., Spangelo, B. L., Henly, D. L., Chrousos, G. P., Schulte, H. M., and Goldstein, A. L.,** Thymic hormone effects on the brain and neuroendocrine circuits, in *Neural Modulation of Immunity,* Guillemin, R., Cohn, M., and Melnechuk, T., Eds., Raven Press, New York, 1985, 179.
35. **Singh, U. and Owen, J. J. T.,** Studies on the maturation of thymus stem cells. The effects of catecholamines, histamine, and peptide hormones on the expression of T alloantigens, *Eur. J. Immunol.,* 6, 59, 1976.
36. **Ernström, U. and Sandberg, G.,** Stimulation of lymphocyte release from the spleen by theophylline and isoproterenol, *Acta Physiol. Scand.,* 90, 202, 1974.
37. **Ernström, U. and Söder, O.,** Influence of adrenaline on the dissemination of antibody-producing cells from the spleen, *Clin. Exp. Immunol.,* 21, 131, 1975.
38. **Webber, R. H., DeFelice, R., Ferguson, R. J., and Powell, J. P.,** Bone marrow response to stimulation of the sympathetic trunks in rats, *Acta Anat.,* 77, 92, 1970.
39. **Madden, K. S., Ackerman, K. D., Livnat, S., Felten, S. Y., and Felten, D. L.,** Patterns of noradrenergic innervation of lymphoid organs and immunological consequences of denervation, in *Neuroimmune Networks: Physiology and Diseases,* Goetzl, E. and Spector, N. H., Eds., Alan R. Liss, New York, 1989, 1.
40. **Madden, K. S., Felten, S. Y., Felten, D. L., and Livnat, S.,** Sympathetic neural modulation of the immune system. I. Depression of T cell immunity in vivo and in vitro following chemical sympathectomy, *Brain Behav. Immunol.,* 3, 72, 1989.
41. **Madden, K. S. and Livnat, S.,** Catecholaminergic influences on immune reactivity, in *Psychoneuroimmunology II,* Ader, R., Felten, D. L., and Cohen, N., Eds., Academic Press, New York, 1990, 283.
42. **DePace, D. M. and Webber, R. H.,** Electrostimulation and morphologic study of the nerves to the bone marrow of the albino rat, *Acta Anat.,* 93, 1, 1975.
43. **Felten, S. Y. and Felten, D. L.,** The innervation of lymphoid tissue, in *Psychoneuroimmunology II,* Ader, R., Felten, D. L., and Cohen, N., Eds., Academic Press, New York, 1990, 27.
44. **Calvo, W. and Haas, R. J.,** Die Histogenese des Knochemarks der Ratte, *Z. Zellforsch.,* 95, 377, 1969.
45. **Miller, M. L. and McCuskey, R. S.,** Innervation of bone marrow in the rabbit, *Scand. J. Haematol.,* 10, 17, 1973.
46. **Nance, D. M., Hopkins, D. A., and Bieger, D.,** Re-investigation of the innervation of the thymus gland in mice and rats, *Brain Behav. Immunol.,* 1, 134, 1987.
47. **Tollefson, L. and Bulloch, K.,** Dual-label retrograde transport: CNS innervation of the mouse thymus distinct from other mediastinum viscera, *J. Neurosci. Res.,* 25, 20, 1990.
48. **Bellinger, D. L., Felten, S. Y., and Felten, D. L.,** Maintenance of noradrenergic sympathetic innervation in the involuted thymus of the aged Fischer 344 rat, *Brain Behav. Immunol.,* 2, 133, 1988.

49. **Besedovsky, H. O., del Rey, A. E., Sorkin, E., Burri, R., Honegger, C. G., Schlumpf, M., and Lichtensteiger, W.,** T lymphocytes affect the development of sympathetic innervation of mouse spleen, *Brain Behav. Immunol.*, 1, 185, 1987.
50. **Motulsky, H. J. and Insel, P. A.,** Adrenergic receptors in man. Direction identification, physiologic regulation, and clinical alterations, *N. Engl. J. Med.*, 307, 18, 1982.
51. **Felten, D. L. and Felten, S. Y.,** Innervation of the thymus, in *Thymus Update,* Kendall, M. D. and Ritter, M. A., Eds., Harwood, London, 1989, 73.
52. **Sergeeva, V. E.,** Histotopography of catecholamines in the mammalian thymus, *Bull. Exp. Biol. Med.*, 77, 456, 1974.
53. **Weihe, E., Müller, S., Fink, T., and Zentel, H. J.,** Tachykinins, calcitonin gene-related peptide and neuropeptide Y in nerves of the mammalian thymus: interactions with mast cells in autonomic and sensory neuroimmunomodulation, *Neurosci. Lett.*, 100, 77, 1989.
54. **Romano, T. A., Felten, S. Y., Felten, D. L., and Olschowka, J. A.,** Neuropeptide-Y innervation of the rat spleen: another potential immunomodulatory neuropeptide, *Brain Behav. Immunol.*, 5, 116, 1991.
55. **Lorton, D., Bellinger, D. L., Felten, S. Y., and Felten, D. L.,** Substance P innervation of the rat thymus, *Peptides,* 11, 1269, 1990.
56. **Geppetti, P., Maggi, C. A., Zecchi-Orlandini, S., Santicioli, P., Meli, A., Frilli, S., Spillantini, M. G., and Amenta, F.,** Substance P-like immunoreactivity in capsaicin-sensitive structures of the rat thymus, *Regul. Pept.*, 18, 321, 1987.
57. **Geppetti, P., Frilli, S., Renzi, D., Santicioli, P., Maggi, C. A., Theodorsson, E.,and Fanciullacci, M.,** Distribution of calcitonin gene-related peptide-like immunoreactivity in various rat tissues: correlation with substance P and other tachykinins and sensitivity to capsaicin, *Regul. Pept.*, 23, 289, 1988.
58. **Bellinger, D. L., Lorton, D., Romano, T., Olschowka, J. A., Felten, S. Y., and Felten, D. L.,** Neuropeptide innervation of lymphoid organs, *Ann. N.Y. Acad. Sci.*, 594, 17, 1990.
59. **Ackerman, K. D.,** Noradrenergic Sympathetic Neurotransmission in the Adult and Neonatal Rat Spleen, Ph.D. thesis, University of Rochester, Rochester, NY, 1989.
60. **Ackerman, K. D., Bellinger, D. L., Felten, S. Y., and Felten, D. L.,** Ontogeny and senescence of noradrenergic innervation of the rodent thymus and spleen, in *Psychoneuroimmunology II,* Ader, R., Felten, D. L., and Cohen, N., Eds., Academic Press, New York, 1990, 71.
61. **Bulloch, K., Cullen, M. R., Schwartz, R. H., and Longo, D. L.,** Development of innervation within syngeneic thymus tissue transplanted under the kidney capsule of the nude mouse: a light and ultrastructural microscope study, *J. Neurosci. Res.*, 18, 16, 1987.
62. **Boyd, E.,** The weight of the thymus gland in health and in disease, *Am. J. Dis. Child.*, 43, 1162, 1932.
63. **Hall, N. R. S., O'Grady, M. P., and Farah, J. R., Jr.,** Thymic hormones and immune function: mediation via neuroendocrine circuits, in *Psychoneuroimmunology II,* Ader R., Cohen, N., and Felten, D. L., Eds., Academic Press, New York, 1990, 515.
64. **Felten, S. Y., Felten, D. L., Bellinger, D. L., Carlson, S. L., Ackerman, K. D., Madden, K. S., Olschowka, J. A., and Livnat, S.,** Noradrenergic sympathetic innervation of lymphoid organs, *Prog. Allergy,* 43, 14, 1988.
65. **Kudoh, G., Hoshi, K., and Murakami, T.,** Fluorescence microscopic and enzyme histochemical studies of the innervation of the human spleen, *Arch. Histol Jpn.*, 42, 169, 1979.
66. **Dahlström, A. B. and Zetterström, B. E. M.,** Noradrenaline stores in nerve terminals of the spleen: changes during hemorrhagic shock, *Science,* 147, 1583, 1965.
67. **Fillenz, M.,** Innervation of the cat spleen, *Proc. Physiol. Soc.*, 25–26, 2P, 1966.
68. **Fillenz, M.,** Innervation of blood vessels of lung and spleen, in *Sympathetic Electric Activation and Innervation of Blood Vessels,* Karger, B., Ed., Cambridge University Press, London, 1966, 56.

69. **Romano, T., Olschowka, J. A., Felten, S. Y., and Felten, D. L.,** Tyrosine hydroxylase and neuropeptide Y positive nerve fibers in the spleen of the beluga whale, *Assoc. Aquat. Anim. Res.*, 20, 82, 1989.
70. **Carlson, S. L., Felten, D. L., Livnat, S., and Felten, S. Y.,** Noradrenergic sympathetic innervation of the spleen. V. Acute drug-induced depletion of lymphocytes in the target fields of innervation results in redistribution of noradrenergic fibers but maintenance of compartmentation, *J. Neurosci. Res.*, 18, 64, 1987.
71. **Reilly, F. D. and McCuskey, R. S.,** Studies of the hemopoietic microenvironment. VII. Neural mechanisms in splenic microvascular regulation in mice, *Microvasc. Res.*, 14, 293, 1977.
72. **Reilly, F. D., McCuskey, R. S., and Meineke, H. A.,** Studies of the hemopoietic microenvironment. VIII. Adrenergic and cholinergic innervation of the murine spleen, *Anat. Rec.*, 185, 109, 1975.
73. **Reilly, F. D., McCuskey, P. A., Miller, M. L., McCuskey, R. S., and Meineke, H. A.,** Innervation of the periarteriolar lymphatic sheath of the spleen, *Tissue Cell,* 11(1), 121, 1979.
74. **Ackerman, K. D., Felten, S. Y., Bellinger, D. L., and Felten, D. L.,** Noradrenergic sympathetic innervation of the spleen. III. Development of innervation in the rat spleen, *J. Neurosci. Res.*, 18, 49, 1987.
75. **Ackerman, K. D., Felten, S. Y., Bellinger, D. L., Livnat, S., and Felten, D. L.,** Noradrenergic sympathetic innervation of spleen and lymph nodes in relation to specific cellular compartments, *Prog. Immunol.*, 6, 588, 1987.
76. **Ackerman, K. D., Felten, S. Y., Dijkstra, C. D., Livnat, S., and Felten, D. L.,** Parallel development of noradrenergic innervation and cellular compartmentation in the rat spleen, *Exp. Neurol.*, 103, 239, 1989.
77. **Bellinger, D. L., Felten, S. Y., Lorton, D., and Felten, D. L.,** Origin of noradrenergic innervation of the spleen in rats, *Brain Behav. Immunol.*, 3, 291, 1989.
78. **Bellinger, D. L., Ackerman, K. D., Felten, S. Y., and Felten, D. L.,** Loss of noradrenergic nerves and lymphoid cells in the aged rat spleen, *Exp. Neurol.*, 116, 295, 1992.
79. **Felten, S. Y., Bellinger, D. L., Collier, T. J., Coleman, P. D., and Felten, D. L.,** Decreased sympathetic innervation of spleen in aged Fischer 344 rats, *Neurobiol. Aging,* 8, 159, 1987.
80. **Fillenz, M. and Pollard, R. M.,** Quantitative differences between sympathetic nerve terminals, *Brain Res.*, 109, 443, 1976.
81. **Felten, D. L., Felten, S. Y., Carlson, S. L., Bellinger, D. L., Ackerman, K. D., Romano, T. A., and Livnat, S.,** Development, aging, and plasticity of noradrenergic sympathetic innervation of secondary lymphoid organs: implications for neural-immune interactions, in *Progress in Catecholamine Research Part A: Basic Aspects and Peripheral Mechanisms,* Dahlstrom, A., Belmaker, R. M., and Sandler, M., Eds., Alan R. Liss, New York, 1988, 517.
82. **Felten, D. L., Felten, S. Y., Madden, K. S., Ackerman, K. D., and Bellinger, D. L.,** Development, maturation, and senescence of sympathetic innervation of secondary immune organs, in *Development, Maturation, and Senescence of Neuroendocrine Systems,* Schreibman, M. P. and Scanes, C. G., Eds., Academic Press, San Diego, 1989, 381.
83. **Abrahamsson, T., Holmgren, S., Nilsson, S., and Pettersson, K.,** Adrenergic and cholinergic effects on the heart, the lung and the spleen of the African lungfish, *Protopterus aethiopicus, Acta Physiol. Scand.*, 107, 141, 1979.
84. **Nilsson, S., Holmgren, S., and Grove, D. J.,** Effects of drugs and nerve stimulation on the spleen and arteries of two species of dogfish, *Scyliorhinus canicula* and *Squalus acanthias, Acta Physiol. Scand.*, 95, 219, 1975.
85. **Nilsson, S. and Grove, D. J.,** Adrenergic and cholinergic innervation of the spleen of the cod: *Gadus morhua, Eur. J. Pharmacol.*, 28, 135, 1974.

86. **Winberg, M., Holmgren, S., and Nilsson, S.,** Effects of denervation of 6-hydroxydopamine on the activity of choline acetyltransferase in the spleen of the cod, *Gadus morhua, Comp. Biochem. Physiol.,* 69C, 141, 1981.
87. **Nilsson, S.,** Sympathetic innervation of the spleen of the cane toad, Bufo marinus, *Comp. Biochem. Physiol.,* 61C, 133, 1978.
88. **Nance, D. M. and Burns, J.,** Innervation of the spleen in the rat: evidence for absence of afferent innervation, *Brain Res. Bull.,* 3, 281, 1989.
89. **Felten, S. Y. and Felten, D. L.,** Are lymphocytes targets of noradrenergic innervation?, in *Frontiers of Stress Research,* Weiner, H., Helhammer, D., Murison, R., and Florin, I., Eds., Hans Huber, Toronto, 1988, 56.
90. **Olschowka, J. A., Felten, S. Y., Bellinger, D. L., Lorton, D., and Felten, D. L.,** NPY-positive nerve terminals contact lymphocytes in the periarteriolar lymphatic sheath of the rat splenic white pulp, *Soc. Neurosci. Abstr.,* 14, 1280, 1988.
91. **Romano, T. A., Olschowka, J. A., Felten, S. Y., and Felten, D. L.,** Neuropeptide-Y involvement in neural-immune interactions in the rat spleen, *Soc. Neurosci. Abstr.,* 15, 714, 1989.
92. **Fried, G., Terenius, L., Brodin, E., Efendic, S., Dockray, G., Fahrenkrug, J., Goldstein, M., and Hökfelt, T.,** Neuropeptide Y, enkephalin and noradrenaline coexist in sympathetic neurons innervating the bovine spleen. Biochemical and immunohistochemical evidence, *Cell Tissue Res.,* 243, 495, 1986.
93. **Lundberg, J. M., Änggård, A., Pernow, J., and Hökfelt, T.,** Neuropeptide Y-, substance P- and VIP-immunoreactive nerves in cat spleen in relation to autonomic vascular and volume control, *Cell Tissue Res.,* 239, 9, 1985.
94. **Lundberg, J. M., Rudehill, A., Sollevi, A., and Hamberger, B.,** Evidence for co-transmitter role of neuropeptide Y in the pig spleen, *Br. J. Pharmacol.,* 96, 675, 1989.
95. **Schultzberg, M., Svenson, S. B., Unden, A., and Bartfai, T.,** Interleukin-1-like immunoreactivity in peripheral tissues, *J. Neurosci. Res.,* 18, 184, 1987.
96. **Lorton, D., Bellinger, D. L., Felten, S. Y., and Felten, D. L.,** Substance P innervation of spleen in rats: nerve fibers associate with lymphocytes and macrophages in specific compartments of the spleen, *Brain Behav. Immunol.,* 5, 29, 1991.
97. **Lorton, D., Bellinger, D. L., Felten, S. Y., and Felten, D. L.,** Substance P (SP) and calcitonin gene-related peptide (CGRP) innervation of the rat spleen, *Soc. Neurosci. Abstr.,* 15, 714, 1989.
98. **Bellinger, D. L., Ackerman, K. D., Felten, S. Y., Lorton, D., and Felten, D. L.,** Noradrenergic sympathetic innervation of thymus, spleen, and lymph nodes: aspects of development, aging and plasticity in neural immune interaction, in *Proceedings of a Symposium on Interactions Between the Neuroendocrine and Immune Systems,* Nistico, G., Ed., Pythagora Press, Rome, 1988, 35.
99. **Felten, S. Y., Olschowka, J. A., Ackerman, K. D., and Felten, D. L.,** Catecholaminergic innervation of the spleen: are lymphocytes targets of noradrenergic nerves?, in *Progress in Catecholamine Research, Part A: Basic Aspects and Peripheral Mechanisms,* Dahlström, A., Belmaker, R. H., and Sandler, M., Eds., Alan R. Liss, New York, 1988, 525.
100. **Bellinger, D. L., Lorton, D., Felten, S. Y., and Felten, D. L.,** Age-related alteration in norepinephrine uptake in the rat spleen, *Soc. Neurosci. Abstr.,* 16, 1210, 1990.
101. **Bellinger, D. L., Felten, S. Y., and Felten, D. L.,** Neural-immune interactions: neurotransmitter signaling of cells of the immune system, *Annu. Rev. Psychiatry,* 11, 127, 1992.
102. **Esler, M., Skews, H., Leonard, P., Jackman, G., Bobik, A., and Korner, P.,** Age dependence of noradrenaline kinetics in normal subjects, *Clin. Sci.,* 60, 217, 1981.
103. **Krall, J. F., Connelly, M., Weisbart, R., and Tuck, M. L.,** Age-related elevation of plasma catecholamine concentration and reduced responsiveness of lymphocyte adenylate cyclase, *J. Clin. Endocrinol. Metab.,* 52, 863, 1981.

104. **Lake, C. R., Zeigler, M. G., Coleman, M. D., and Kopin, I. J.,** Age related norepinephrine levels are similar in normotensive and hypertensive subjects, *N. Engl. J. Med.*, 296, 208, 1977.
105. **Ziegler, M. G., Lake, C. R., and Kopin, I. J.,** Plasma noradrenaline increases with age, *Nature (London)*, 261, 333, 1976.
106. **De Blasi, A., Lipartiti, M., Algeri, S., Sacchetti, G., Costantini, C., Fratelli, M., and Cotecchia, S.,** Stress induced desensitization of lymphocyte β-adrenoceptors in young and aged rats, *Pharmacol. Biochem. Behav.*, 24, 991, 1986.
107. **Feldman, R. D., Limbird, L. E., Nadeau, J., Robertson, D., and Wood, A. J. J.,** Alterations in leukocyte β-receptor affinity with aging. A potential explanation for altered β-adrenergic sensitivity in the elderly, *N. Engl. J. Med.*, 310, 815, 1984.
108. **O'Hara, N., Daul, A. E., Fesel, R., Siekmann, U., and Brodde, O.-E.,** Different mechanisms underlying reduced β_2-adrenoceptor responsiveness in lymphocytes from neonates and old subjects, *Mech. Ageing Dev.*, 31, 115, 1985.
109. **Kusiak, J. W. and Pitha, J.,** Decreased response with age of the cardiac catecholamine sensitive adenylate cyclase system, *Life Sci.*, 33, 1679, 1983.
110. **O'Connor, S. W., Scarpace, P. J., and Abrass, I. B.,** Age-associated decrease in the catalytic unit activity of rat myocardial adenylate cyclase, *Mech. Ageing Dev.*, 21, 357, 1983.
111. **Kostrouch, Z., Raska, I., Felt, V., Nedvídková, J., and Holecková, E.,** Internalization of triidothyronine-bovine serum albumin-colloidal gold complexes in human peripheral leukocytes, *Experientia*, 43, 1119, 1987.
112. **Scarpace, P. J. and Abrass, I. B.,** Decreased beta-adrenergic agonist affinity and adenylate cyclase activity in senescent rat lung, *J. Gerontol.*, 38, 143, 1983.
113. **Ackerman, K. D., Madden, K. S., Livnat, S., Felten, S. Y., and Felten, D. L.,** Neonatal sympathetic denervation alters the development of in vitro spleen cell proliferation and differentiation, *Brain Behav. Immunol.*, 5, 235, 1991.
114. **Laverty, R., Sharman, D. F., and Vogt, M.,** Action of 2,4,5-trihydroxyphenylethylamine on the storage and release of noradrenaline, *Br. J. Pharmacol.*, 24, 549, 1965.
115. **Malmfors, T. and Sachs, D.,** Degeneration of adrenergic nerves produced by 6-hydroxydopamine, *Eur. J. Pharmacol.*, 3, 89, 1968.
116. **Kostrzewa, R. M. and Jacobwitz, D. M.,** Pharmacological actions of 6-hydroxydopamine, *Pharmacol. Rev.*, 26, 199, 1974.
117. **Lorton, D., Hewitt, D., Bellinger, D. L., Felten, S. Y., and Felten, D. L.,** Noradrenergic reinnervation of the rat spleen following chemical sympathectomy with 6-hydroxydopamine: pattern and time course of reinnervation, *Brain Behav. Immunol.*, 4, 198, 1990.
118. **De Champlain, J. and van Ameringen, M. R.,** Regulation of blood pressure by sympathetic nerve fibers and adrenal medulla in normotensive and hypertensive rats, *Circ. Res.*, 31, 617, 1972.
119. **Gauthier, P., Nadeau, R., and De Champlain, J.,** Acute and chronic cardiovascular effects of 6-hydroxydopamine in dogs, *Circ. Res.*, 31, 207, 1972.
120. **Mueller, R. A., Thoenen, H., and Axelrod, J.,** Adrenal tyrosine hydroxylase: compensatory increase in activity after chemical sympathectomy, *Science*, 163, 468, 1969.
121. **Finch, L., Haeusler, G., Kuhn, H., and Thoenen, H.,** Rapid recovery of vascular adrenergic nerves in the rat after chemical sympathectomy with 6-hydroxydopamine, *Br. J. Pharmacol.*, 48, 59, 1973.
122. **Nadeau, R. A., De Champlain, J., and Tremblay, G. M.,** Supersensitivity of the isolated rat heart after chemical sympathectomy with 6-hydroxydopamine, *Can. J. Physiol. Pharmacol.*, 49, 36, 1971.
123. **Hall, N. R., McClure, J. E., Hu, S.-K., Tare, N. S., Seals, C. M., and Goldstein, A. L.,** Effects of 6-hydroxydopamine upon primary and secondary thymus dependent immune responses, *Immunopharmacology*, 5, 39, 1982.

124. **Kasahara, K., Tanaka, S., Ito, R., and Hamashima, Y.,** Suppression of the primary immune response by chemical sympathectomy, *Res. Commun. Chem. Pathol. Pharmacol.,* 16, 687, 1977.
125. **Besedovsky, H. O., del Rey, A., Sorkin, E., Da Prada, M., and Keller, H. H.,** Immunoregulation mediated by the sympathetic nervous system, *Cell. Immunol.,* 48, 346, 1979.
126. **Felten, D. L., Lorton, D., Bellinger, D. L., Duclos, M., and Felten, S. Y.,** Substance P (SP) denervation of lymph nodes attenuates the expression of adjuvant induced arthritis, *Soc. Neurosci. Abstr.,* 16, 1210, 1990.
127. **Lorton, D., Bellinger, D. L., Duclos, M., Felten, S. Y., and Felten, D. L.,** Sympathectomy of lymph nodes exacerbates the expression of experimental arthritis, *Soc. Neurosci. Abstr.,* 16, 1210, 1990.
128. **Chelmicka-Schorr, E., Checinski, M., and Arnason, B. G. W.,** Chemical sympathectomy augments the severity of experimental allergic encephalomyelitis, *J. Neuroimmunol.,* 17, 347, 1988.
129. **Chelmicka-Schorr, E., Kwasniewski, M. N., Thomas, B. E., and Arnason, B. G. W.,** The β-adrenergic agonist isoproterenol suppresses experimental allergic encephalomyelitis in Lewis rats, *J. Neuroimmunol.,* 25, 203, 1989.
130. **Colpaert, F. C., Donnerer, J., and Lembeck, F.,** Effects of capsaicin on inflammation and on the substance P content of nervous tissue in rats with adjuvant arthritis, *Life Sci.,* 32, 1827, 1983.
131. **Costa, M. C., Sutter, P., and Gybel, J.,** Adjuvant-induced arthritis in rats: a possible animal model of chronic pain, *Pain,* 10, 173, 1985.
132. **Levine, J. D., Clark, R., Devor, H., Helms, C., Moskowitz, M. A., and Basbaum, A. I.,** Intraneuronal substance P contributes to the severity of experimental arthritis, *Science,* 226, 547, 1984.
133. **Levine, J. D., Coderre, T. J., Helms, C., and Basbaum, A. I.,** β_2-adrenergic mechanisms in experimental arthritis, *Proc. Natl. Acad. Sci. U.S.A.,* 85, 4553, 1988.
134. **Levine, J. D., Collier, D. H., Basbaum, A. I., Moskowitz, M. A., and Helms, C. A.,** Hypothesis: the nervous system may contribute to the pathophysiology of rheumatoid arthritis, *J. Rheumatol.,* 12, 406, 1985.
135. **Levine, J. D., Dardick, S. J., Basbaum, A. I., and Scipio, E.,** Reflex neurogenic inflammation. I. Contribution of the peripheral nervous system to spatially remote inflammatory responses that follow injury, *J. Neurosci.,* 5, 1380, 1985.
136. **Levine, J. D., Dardick, S. J., Roizen, M. F., Helms, C., and Basbaum, A. L.,** Contribution of sensory afferents and sympathetic efferents to joint injury in experimental arthritis, *J. Neurosci.,* 6, 3423, 1986.
137. **Levine, J. D., Fye, K., Heller, P., Basbaum, A. I., and Whiting-O'Keefe, Q.,** Clinical response to regional intravenous guanethidine in patients with rheumatoid arthritis, *J. Rheumatol.,* 13, 1040, 1986.
138. **Levine, J. D., Moskowitz, M. A., and Basbaum, A. I.,** The contribution of neurogenic inflammation in experimental arthritis, *J. Immunol.,* 135, 843s, 1985.
139. **Levine, J. D., Goetzl, E. J., and Basbaum, A. I.,** Contribution of the nervous system to the pathophysiology of rheumatoid arthritis and other polyarthritides, *Rheum. Dis. Clin. North Am.,* 13, 369, 1991.
140. **Kaplan, R., Robinson, C. A., Scavulli, J. F., and Vaughan, J. H.,** Propranolol and the treatment of rheumatoid arthritis, *Arthritis Rheum.,* 23, 253, 1980.
141. **Vyden, J. K., Groseth-Dittrich, M. F., Callis, G., Lars, M. M., and Weinberger, H.,** The effect of propranolol on peripheral hemodynamics in rheumatoid arthritis, *Arthritis Rheum.,* 14, 420, 1971.
142. **Wertian, E., Ovadia, H., Ferran, S., and Abramsky, O.,** Prevention of allergic experimental encephalomyelitis by anterior hypothalamic lesions in rats, *Neurology,* 35, 1468, 1985.

143. **Bellinger, D. L., Felten, S. Y., Coleman, P. D., Yeh, P., and Felten, D. L.,** Noradrenergic innervation and acetylcholinesterase activity of lymph nodes in young adult and aging mice, *Soc. Neurosci. Abstr.*, 11, 663, 1985.
144. **Lorton, D., Bellinger, D. L., Duclos, M., Felten, S. Y., and Felten, D. L.,** Sympathectomy of lymph nodes exacerbates the expression of adjuvant-induced arthritis, *Exp. Neurol.*, in press.

Chapter 13

AUTONOMIC NEUROEFFECTOR MECHANISMS IN THE REPRODUCTIVE AND ENDOCRINE SYSTEMS: AGE-RELATED CHANGES

Wade L. Collier

TABLE OF CONTENTS

0-8493-6981-9/93/$0.00 + $.50

I. INTRODUCTION

Aging has an enormous effect on the reproductive and endocrine systems. This chapter describes the effects of age on the reproductive and endocrine systems. The influence of the decline in autonomic nervous system function on the activity of the reproductive and endocrine systems is discussed.

II. THE FEMALE REPRODUCTIVE SYSTEM

Menopause is probably the best marker of aging of the female reproductive system. Similar changes are reported in other vertebrate species, but as their life span is shorter than that of *Homo sapiens*, the details can be markedly different.

A. MENSTRUAL CYCLES AND FERTILITY

It is well known that there are wide differences in the length of menstrual cycles in women. However, the normally accepted figure for the length of the female menstrual cycle is 28 d. As a woman approaches menopause, the length of the menstrual cycle becomes greatly irregular. A decreased menstrual cycle is usually the consequence of a decrease in the length of the follicular phase of the cycles, while an increased menstrual cycle occurs because of a failure of ovulation.[1-3] About 7 years before the onset of menopause, the woman may experience cycles which are normal, short, and long. Many women continue to ovulate until just prior to menopause.[2] As might be expected, important alterations occur in the endocrine system during this period of transition.

There appear to be various types of exogenous stimuli which can alter the length of a woman's menstrual cycle. Some of the stimuli reported in the literature are the frequency of sexual intercourse[4-6] and women being in close contact with each other.[4,7,8]

Geographic location does not appear to play an important role in the length of female reproductive cycles when cessation of fertility is defined by the age of the last birth. Supporting this is the observation that the average maternal age at last birth is 40.[9] Very few women become pregnant after the age of 50; in fact, the oldest documented age for a woman giving birth to a living child is 57 years.[10] This decrease in fertility is maternal rather than paternal: when 2000 women who had undergone artificial insemination were studied, a decrease in fecundity began around the age of 35 years.[11]

B. FEMALE SEXUALITY

Unlike fertility, females continue to experience sexual arousal and orgasm in postmenopausal years.[12] Even with the decrease of the sex steroids estrogen and progesterone, female sexuality was only slightly lower than that of young women.[13] However, this decrease can cause secondary consequences such as vaginal drying, leading to dyspareunia.

C. OOCYTE DECREASE

The number of oocytes and primordial ovarian follicles decreases exponentially after birth, leading to an almost complete loss of oocytes by the time menopause is reached. Research shows that by the age of 55 years, less than 1% of the original number of oocytes and primordial ovarian follicles remain.[14] A newer study by Richardson and co-workers[15] suggests that the ovaries become exhausted even faster than previously believed. In the women they studied (who ranged from 45 to 53 years of age), follicular reserves varied from 1000 to 2500 in women still menstruating to none in women who were postmenopausal. It now appears that ovarian exhaustion of follicles is the cause of reproductive aging in women. It is the brain, however, which is the cause of menstrual cycle irregularity immediately prior to the cessation of menstruation.[16] There are reports in the literature of an age-dependent decrease in the number of hypothalamic neurons and postmenopausal alteration of neuronal structure. It is now thought that a minimum number of ovarian follicles is needed to produce sufficient estrogen to cause the required preovulary spike of gonadotropins.[17] At the end of reproductive life, the number of functional follicles remaining causes random fluctuations in the amount of estrogen secreted, leading to cycles which are longer or shorter than what had been the average.

Following menopause, the ovarian production of estrogen and progesterone virtually stops so that the blood levels of these steroids fall to levels seen in women who have had their ovaries surgically removed. Before the end of menstruation, follicle-stimulating hormone (FSH) levels increase, with a decrease in this hypophyseal hormone occurring only during the final menstrual cycles.[18] It is now thought that the final FSH spike is caused by the decreased elaboration and secretion of the trophic hormone inhibin.[19] Following menopause, there is a similar spike in the secretion of luteinizing hormone (LH) which is probably due to a change in the structure of the molecule, causing its release from the pituitary.[20]

D. FERTILITY LOSS

The age-dependent loss of fertility has several causes. Age-dependent changes in the male partner cause decreased sperm production. Decreased sexual activity decreases the chance of fertilization.[21] It is recognized that older oocytes have a greater number of chromosomal disorders. For example, the risk of spontaneous abortion increases 100% just prior to menopause,[22] and the incidence of abnormal chromosome numbers in newborn babies increases 25 times as the mother ages from 30 to 45 years.

E. POSTMENOPAUSAL CHANGES

Postmenopausal changes are of two types. One type, which is not considered, does not show a straightforward linkage to the amount of steroids being produced by the ovaries. The second change, called ovariprival, is directly related to the levels of progesterone and estradiol being secreted by the ovary. Ovariprival changes at menopause cause many of the alterations which are troubling to women: hot flashes, degeneration of the reproductive and urinary tracts, and loss of mammary epithelial glands and the ducts which drain them. In fact, the tissues which atrophy are those which bloom at puberty when the ovarian steroid levels first begin to rise. Many of the aforementioned changes can be elicited in young women by the surgical removal of their ovaries, and as many may be ameliorated by hormonal replacement therapy. This suggests that an age-dependent loss or alteration in estrogen receptors may be the underlying cause.[23]

F. OVARIAN INNERVATION

In 1990, Ferrante and co-workers investigated the noradrenergic innervation of the ovary using 24-month-old rats, which are considered to be old, and high-pressure liquid chromatography with electrochemical detection and catecholamine histofluorescence.[24] They noted that old age was accompanied by a significant decrease in ovarian noradrenaline levels. In animals three months of age, which they defined as young, noradrenergic nerve fibers were distributed primarily to blood vessels and, to a smaller degree, the interstitial glands. In the 24-month-old animals, the perivascular noradrenergic fibers were reduced by more than 40%, while the interstitial gland nerve fibers were reduced by approximately 20%. It is still not known whether ovulation and endocrine ovarian functions in the rat are under direct autonomic nervous system control, but it is quite possible that there is a sympathetic noradrenergic regulation of ovarian blood flow and ovulation, subjects which have been investigated by Mohsin and Pennefather[25] and Unsicker.[26] Ferrante et al.,[24] using fluorescence histochemistry, showed that among the noradrenergic fibers supplying the ovary, the perivascular nerves were more sensitive to aging than those supplying the interstitial gland tissue (Figures 1 and 2).

G. UTERINE INNERVATION

Apparently, there is a relatively small number of nerve fibers innervating the uterine wall,[27] and there are large differences between species.[28] There is also an innervation of the wall of the blood vessels and the endometrium.[29]

In a 1990 histofluorescence study involving rats, Bronzetti et al.[30] reported an age-related change of noradrenergic innervation of the uterine vasculature of the uterine wall. Using animals 4, 12, and 24 months of age, they observed no changes in the density of noradrenergic innervation or in uterine noradrenaline levels in the 12-month-old animals compared to the 4-month-old rats. On the other hand, a significant decrease in the density of the noradrenergic perivascular plexus of the uterine artery and its intrauterine branches, and a decrease in the noradrenaline concentrations in the uterine horns and the cervix, were reported in the 24-month-old animals (Figures 1 and 2).

An age-dependent alteration in hypothalamic neurotransmission has been implicated as causing a decrease in female rat reproductive capability.[31] Both noradrenergic and dopaminergic functions decrease in the hypothalamus of

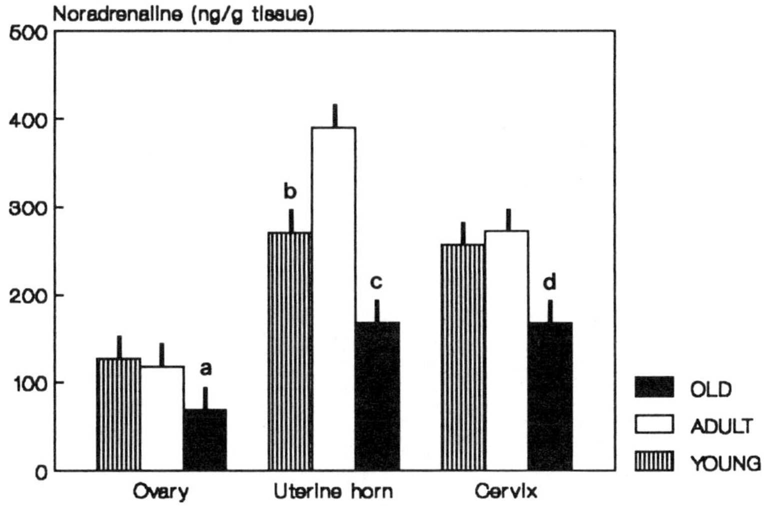

FIGURE 1. Noradrenaline concentrations in the ovary, uterus (uterine horn), and cervix of young (3 months), adult (12 months), and old (24 months) rats. Catecholamine concentrations were measured by high-pressure liquid chromatography with electrochemical detection. Values are means ± SEM. a, $p < 0.01$ vs. young or adult; b, $p < 0.001$ vs. adult; c, $p < 0.001$ vs. adult or young; d, $p < 0.001$ vs. adult or young.

aged rats.[32] With the study of Bronzetti et al.,[30] it now appears that both a central and a peripheral noradrenergic system decline occurs as the rat ages.

III. THE MALE REPRODUCTIVE SYSTEM

Unlike females, males do not experience an abrupt decline in their fertility or other sexual functions. This means that there is no complete exhaustion of germ cells, the total loss of fertility and precipitous decrease in circulating gonadal hormones that occurs in women. Men do, however, experience myriad reproductive difficulties starting at midlife.

It is important to know whether age induces changes in male reproduction. These alterations may be due to changes which are primarily gonadal, due to alterations in the hypothalamic-pituitary axis or to degeneration in the central nervous system.

Masculine sexuality declines beginning in the third decade of life.[33-35] Fewer than 35% of men 80 years of age or older are sexually active. In many cases, the decline in sexual activity is due to disease.[36] For example, 30 to 60% of diabetic men experience impotence.[36] Impotence is an occasional side effect of drug therapy[37] and is very frequent after prostatectomy, especially radical prostatectomy.[38]

There is no direct relationship between blood testosterone levels and male sexuality.[35] There is no significant decrease in the levels of circulating tes-

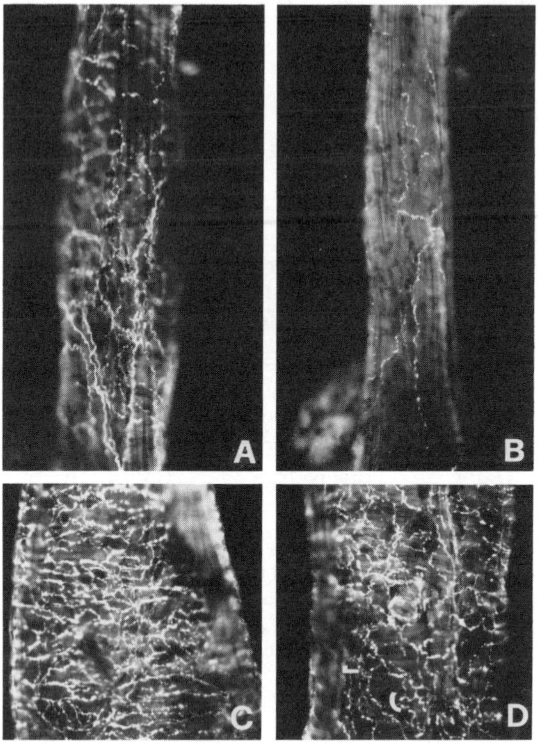

FIGURE 2. Glyoxylic acid histofluorescence for the demonstration of tissue stores of catecholamines. Noradrenergic perivascular plexus in whole mounts of the ovaric (A, B) and uterine (C, D) arteries. (A, C) 3-Month-old rat; (B, D) 24-month-old-rat. (Original magnification × 105.)

tosterone as men enter midlife. Thus, some men experience both decreased blood plasma levels of testosterone and a decreased sexual function, while other men do not.[35] Most investigators now feel that decreases in circulating testosterone levels do not cause impotence in males.

Likewise, there are few changes in the blood plasma levels of other sex steroids and the pituitary hormones. While LH and FSH may be elevated in elderly men, the levels of increase are usually much less than those reported in women of the same age.[39]

A. TESTES

Most researchers agree that the interstitial cells of Leydig, which produce the male androgens, decline in number as the male ages.[40] Sperm continues to be produced in the seminiferous tubules virtually to the end of the male's life.[41] However, as the male ages, various age-dependent structural testicular alterations occur which lead to decreased fertility. Some of these changes include a thickening of the basement membrane of the seminiferous tubules, frank sclerotization of the tubules, and degeneration secondary to atherosclerosis of the blood vessels supplying the testes. Sperm motility also appears to decrease with age, but there is no loss in the ability of motile sperm to penetrate and fertilize the ovum.[39] While the oldest documented case of paternity is 95 years,[42] there is a gradual, age-dependent decrease in the number of sperm produced[40] and normal sperm in the semen.[43] On the other hand, the concentration of sperm in the ejaculate and the volume of the ejaculate appear to increase with aging,[44] possibly due to a reduced frequency of sexual intercourse.

That the age of the male parent is a contributing factor to birth defects is shown by a higher fetal death rate and a higher incidence of Down's syndrome in offspring produced by older men.[45,46] Some decrease in fertility may be due to an age-induced decrease in adrenergic and muscarinic influences in the regulation of testicular function.[47]

B. VAS DEFERENS

The vas deferens is supplied with an adrenergic innervation. It appears that the distribution of the adrenergic fibers differs, depending on whether the proximal or the distal vas deferens is being examined, with a larger number of fibers being present in the distal portion of the duct. A recent investigation showed an increased number of fluorescent bundles in the vas deferens of old (20 months of age) rats compared to younger controls. On the other hand, the younger animals had a higher responsiveness to α-agonists, leading the investigators to hypothesize that in the vas deferens of the rat, there is an inverse correlation between the number of adrenergic nerve fibers and the responsiveness of the receptors.[48]

C. PENIS

Vasoactive intestinal polypeptide (VIP), a 28-amino acid peptide first isolated from pig intestine,[49] is now known to function as a neurotransmitter in penile erection in several different mammalian species, including man.[50] Polak and co-workers[51] observed that VIP is present in nerve fibers and/or nerve endings in penile erectile tissue and penile vasculature. This peptide is released when erection is induced and, when applied exogenously, causes tumescence.[52,53] In fact, it is the impairment of penile VIPergic innervation which is the probable cause of impotency in diabetic men.[54] Amenta and co-workers studied VIP levels and distribution in the penis of old Wistar rats in 1987 using radioimmunoassay and immunofluorescence.[55] They concluded that no significant changes in tissue VIP concentrations or distribution occurred in the penis of 30-month-old (considered to be old) rats compared to 3-month-old (young adult) rats. They concluded that the age-dependent impairment of male sexual function is not dependent on modifications of the VIPergic innervation of penile tissue.

IV. THE ENDOCRINE SYSTEM

A. HYPOTHALAMUS-PITUITARY

For some time, the effects of aging on the hypothalamic-pituitary axis have been under investigation. While alterations in this portion of the central nervous system can have profound reproductive ramifications (see above), the pituitary is discussed in Section 4 of this chapter because most physiologists consider the pituitary to be the "master gland" of the body.

At this time, the complicated age-dependent alterations of the hypothalamic-pituitary axis are still not completely understood. Using the female rat as the experimental animal, numerous conflicting reports on LH releasing hormone (RH) and gonadotropin levels have been reported; thus, age-dependent increases, decreases, or no changes have been reported in hypothalamic LHRH levels.[56] A decreased *in vitro* LHRH release from neurosecretory granules has been reported in old rats.[57]

A decreased level of dopaminergic function has likewise been reported in old female rats.[58] This reduction leads directly to decreased dopamine synthesis,[59] causing a reduced input of dopamine into the pituitary portal circulation.[60] Dopamine release stimulates the release of LHRH; consequently, an age-dependent reduction in the synthesis of dopamine is one of the reasons that a decreased LHRH level has been reported in the plasma levels of old female rats.[61] Age-dependent changes in other neurotransmitter systems such as noradrenaline and serotonin have also been reported.[62]

A recent work by Bestetti and colleagues[6] investigated functional and morphological changes in the hypothalamic-pituitary-gonadal axis in aged female rats. They reported that the number, total area, and immunoreactivity of LHRH-labeled axon cross sections in the lateral median eminence of the hypothalamus were less in aged rats than in control animals. Also of interest was their report that in old rats, the number of nucleated FSH-labeled cells and the total FSH area and immunoreactivity were almost twice that of young animals. They concluded that hypothalamic and pituitary disturbances may be a significant contributor to the alteration of hormones leading to reproductive decline in old female rats.

Vasopressin (VP) is a neuropeptide which is synthesized in the hypothalamus and released from the neurohypophysis into the general circulation, where it causes an antidiuretic effect by acting directly on the renal tubules. VP also acts as a neurotransmitter or neuromodulator in the rodent and the primate brain.[64] As aging occurs in the human brain, there is a remarkable decline in the number of VP-producing cells in the suprachiasmic nucleus of the hypothalamus. It is precisely this nucleus which regulates the circadian rhythm in man. In patients with Alzheimer's disease (AD), this degeneration was several orders of magnitude greater, and could be the factor causing the disturbances in the sleep-wake cycles and the endocrine rhythms associated with AD. On the other hand, Goudsmit et al.[64] observed no age-dependent or AD-related alterations in the hypothalamic-pituitary axis and found that the total number of cells in the axis remained constant. They also noted that the VP cells in this portion of the central nervous system increase their activity during senescence, probably as a compensatory mechanism for the decreased sensitivity of the kidney to the neuropeptide. Goudsmit's work was confirmed in a later study.[65]

B. THYROID

Few studies of age-dependent changes in the thyroid have been published in the literature. We do know that, like other glands in the body, the thyroid gland's metabolism and secretion decrease as the individual ages.

Frolkis et al.[47] recently reported a decreased adrenergic control of thyroidogenesis in rats who had entered old age. This is not surprising, since thyroidogenesis is influenced by adrenergic innervation. In sympathectomized rats, thyroid activity was greatly reduced,[66] and there was also a decrease in the transport-organic phase of iodine metabolism.[67] These alterations appear to be transitory, as aging animals were able to increase the amount of calcitonin and iodinated hormones produced.[66]

C. ADRENAL GLANDS

Using anesthetized Wistar rats, Ito et al.[68] and Sato[69] measured the secretion of adrenaline and noradrenaline from the adrenal gland under resting conditions. Both studies reported that the rate of adrenaline and noradrenaline release gradually increased in animals which were 10 months old when the levels of these neurotransmitters were compared to baseline values measured at 3 months of age. Maximum values two to four times the baseline data were measured at 2.5 years of age. It appears that in old age, the density of innervation to the adrenal remains the same as that measured in younger adult rats,[70] and this is the probable explanation for the continued high values reported in old age. Frolkis et al.,[47] however, reported a weakening of α-adrenoceptor and muscarinic cholinergic control of the glucocorticoid function of the adrenal.

In 1988, Souto et al.[71] studied the response of rats to insulin. In 1- and 3-month-old animals which had undergone insulin challenge, noradrenaline content did not differ between experimentals and controls. Adrenaline decrease, on the other hand, was significant in adults and in neonatal animals. The adrenaline decrease in the neonatal animals is most likely due to the gradual innervation and maturation of the adrenal chromaffin cells; these cells most likely experience an age-dependent decline, which would explain the marked decrease observed in the old animals.

Tomlinson and Coupland[70] investigated the adrenal medulla in Wistar rats from birth to old age using ultrastructural and biochemical methods. They also concluded that there is a gradual innervation and maturation of the adrenal chromaffin cells. Further data on the topic are reported in Chapter 4 of this volume.

REFERENCES

1. **Sherman, B. M. and Korenman, S. G.,** Hormonal characteristics of the human menstrual cycle throughout reproductive life, *J. Clin. Invest.,* 55, 699, 1975.
2. **Sherman, B. M., West, J. H., and Korenman, S. G.,** The menopausal transition: analysis of LH, FSH, estradiol and progesterone concentrations during menstrual cycles of older women, *J. Clin. Endocrinol. Metab.,* 42, 629, 1976.
3. **Metcalf, M. G., Donald, R. A., and Livesey, J. H.,** Classification of menstrual cycles in pre- and postmenopausal women, *J. Endocrinol.,* 91, 1, 1981.
4. **McClontock, M. K.,** Menstrual synchrony and suppression, *Nature (London),* 229, 224, 1971.
5. **Cutler, W. B., Garcia, C. R., and Krieger, A. M.,** Sexual behavior, frequency and menstrual cycle length in mature premenopausal women, *Psychoneuroendocrinology,* 4, 279, 1979.
6. **Cutler, W. B., Garcia, C. R., and Krieger, A. M.,** Sporadic sexual behavior and menstrual cycle length in women, *Horm. Behav.,* 14, 163, 1980.
7. **Graham, C. A. and McGrew, W. C.,** Menstrual synchrony in female undergraduates living on a coeducational campus, *Psychoendocrinology,* 5, 245, 1980.
8. **Quadagno, D. M., Shubeita, H. E., Deck, J., and Francoeur, D.,** Influence of male social contacts, exercise and all-female living conditions on the menstrual cycle, *Psychoendocrinology,* 6, 239, 1981.
9. **Gray, R. H.,** Biological and social interactions in the determination of late fertility, *J. Biosocial Sci.,* 6 (Suppl.), 97, 1979.
10. **Fergusson, I. L. C., Taylor, R. W., and Wastson, J. M.,** *Records and Curiosities in Obstetrics and Gynecology,* Balliere Tindall, London, 1982.
11. **Eaton, J. W. and Mayer, A. J.,** The social biology of very high fertility among the Hutterites. The demography of a unique population, *Hum. Biol.,* 25, 206, 1953.
12. **Kinsey, A., Pomeroy, W., Martin, C., and Gebhard, P.,** *Sexual Behavior in the Human Female,* W. B. Saunders, Philadelphia, 1953.
13. **Morrell, M. J., Dixen, J. M., Carte, C. S., and Davidson, J. M.,** The influence of age and cycling status on sexual arousability in women, *Am. J. Obstet. Gynecol.,* 148, 66, 1984.
14. **Block, E.,** Quantitative morphological investigations of the follicular system in women, variations at different ages, *Acta Anat.,* 14, 108, 1952.
15. **Richardson, S., Senikas, V., and Nelson, J. F.,** Follicular depletion during the menopausal transition; evidence for accelerated loss and ultimate exhaustion, *J. Clin. Endocrinol. Metab.,* 65, 1231, 1987.
16. **Finch, C. E., Felicio, L. S., Mobbs, C. V., and Nelson, J. F.,** Ovarian and steroidal influences on neuroendocrine aging processes in female rodents, *Endocr. Rev.,* 5, 467, 1984.
17. **Nelson, J. F. and Felicio, L. S.,** Reproductive aging in the female: an etiological perspective, *Rev. Biol. Res. Aging,* 2, 251, 1985.
18. **Metcalf, M. G., Donald, R. A., and Livesey, J. H.,** Pituitary-ovarian function in normal women during the menopausal transition, *Clin. Endocrinol.,* 14, 245, 1981.
19. **Grady, R. R., Charlesworth, M. C., and Schwartz, N. B.,** Characterization of the FSH-suppressing activity in follicular fluid, *Recent Prog. Horm. Res.,* 38, 409, 1982.
20. **Conn, P. M., Cooper, R., McNamara, C., Rogers, D. C., and Shoenhardt, L.,** Qualitative change in gonadotropin during normal aging in the male rat, *Endocrinology,* 106, 1549, 1980.
21. **German, J.,** Mongolism, delayed fertilization, and human sexual behavior, *Nature (London),* 217, 516, 1968.
22. **Leridon, H.,** *Human Fertility: The Basic Components,* University of Chicago Press, Chicago, 1977, 62.
23. **Albright, F., Smith, P. H., and Richardson, A. M.,** Postmenopausal osteoporosis, its clinical features, *JAMA,* 116, 2465, 1941.

24. **Ferrante, F., Bronzetti, E., Cavallotti, C., Ricci, A., and Amenta, F.,** The noradrenergic innervation of the ovary in old rats, *Mech. Ageing Dev.*, 54, 55, 1990.
25. **Mohsin, S. and Pennefather, J. N.,** The sympathetic innervation of the mammalian ovary, *Clin. Exp. Pharmacol. Physiol.*, 6, 335, 1979.
26. **Unsicker, K.,** Innervation of endocrine tissues, in *Ultrastructure of Endocrine Cells and Tissues,* Motta, P. M., Ed., Martinus Nijhoff, Boston, 1984, 321.
27. **Clegg, P. C.,** The effect of adrenergic blocking agents on the guinea-pig uterus in vitro and a study of the histology of the intrinsic myometrial nerves, *J. Physiol. London,* 169, 73, 1962.
28. **Sjoberg, N. O.,** The adrenergic transmitter of the female reproductive tract: distribution and functional changes, *Acta Physiol. Scand. Suppl.*, 305, 1, 1967.
29. **Krantz, K. E.,** Innervation of the human uterus, *Ann. N.Y. Acad. Sci.*, 75, 770, 1959.
30. **Bronzetti, E., Collier, W. L., Ferrante, F., Felici, L., and Amenta, F.,** Age-related changes of the noradrenergic innervation of the uterine vasculature: a fluorescence histochemical study, *Arch. Gerontol. Geriatr.*, 10, 303, 1990.
31. **Lu, K. H., Hopper, B. R., Vargo, T. M., and Yen, S. S.,** Chronological changes in sex steroid gonadotropin and prolactin secretion in aging female rats displaying different reproductive states, *Biol. Reprod.*, 21, 193, 1979.
32. **Wise, P. M.,** Aging of the female reproductive system, *Rev. Biol. Res. Aging,* 1, 195, 1983.
33. **Martin, C. E.,** Sexual activity in the aging male, in *Handbook of Sexology,* Money, J. and Musaph, H., Eds., ASP Bio-Medical Press, Amsterdam, 1977.
34. **Verwoerdt, A., Pfeiffer, E., and Wang, H. S.,** Sexual behavior in senescence. II. Patterns of sexual activity and interest, *Geriatrics,* 24, 137, 1969.
35. **Tsitouras, P. D., Martin, C. E., and Harman, S. M.,** Relationship of serum testosterone to sexual activity in healthy elderly men, *J. Gerontol.*, 37, 288, 1982.
36. **Comfort, A.,** Sexuality in later life, in *Handbook of Mental Health and Aging,* Birren, J. E. and Sloane, R. B., Eds., Prentice-Hall, Englewood Cliffs, NJ, 1980, 885.
37. **Lipson, L. G.,** Special problems in treatment of hypertension in the patient with diabetes mellitus, *Arch. Intern. Med.*, 144, 1829, 1984.
38. **Walsh, P. C. and Donker, P. J.,** Impotence following radical prostatectomy: insight into etiology and prevention, *J. Urol.*, 128, 492, 1982.
39. **Nieschlag, E., Lammers, U., Freischem, C. W., and Wickings, E. J.,** Reproductive functions in young fathers and grandfathers, *J. Clin. Endocrinol. Metab.*, 55, 676, 1982.
40. **Neaves, W. B., Johnson, L., Porter, J. C., Parker, C. R., and Petty, C. S.,** Leydig cell numbers, daily sperm production, and serum gonadotropin levels in aging men, *J. Clin. Endocrinol. Metab.*, 59, 756, 1984.
41. **Harman, S. M. and Talbert, G. B.,** Reproductive aging, in *Handbook of the Biology of Aging,* Finch, C. E. and Schneider, E. L., Eds., Van Nostrand, New York, 1984, 457.
42. **Seymour, F. I., Duffy, C., and Koernev, A.,** A case of authenticated fertility in a man of 94, *JAMA,* 105, 1423, 1935.
43. **Blum, V.,** Das problem des mannlichen Klimakteriums, *Wien. Klin. Wochenschr.*, 2, 1133, 1936.
44. **MacLeod, J. and Gold, R. F.,** The male factor in fertility and infertility. VII. Semen quality in relation to age and sexual activity, *Fertil. Steril.*, 4, 194, 1953.
45. **Parsons, P. A.,** Parental age and the offspring, *Q. Rev. Biol.*, 39, 258, 1964.
46. **Sonneborn, T. M.,** The human early fetal death rate in relation to age of father, in *The Biology of Aging,* Strehler, B. L., Ed., American Institute of Biological Sciences, Washington, D.C., 1960, 288.
47. **Frolkis, V. V., Gorban, E. N., and Moroz, E. V.,** Peculiarities of neural regulation of the thyroid, adrenocortical and testicular functions in old age, *Mech. Ageing Dev.*, 51, 89, 1990.
48. **Pellegrini, A., Soldani, P., Breschi, M. C., Martinotti, E., and Paparelli, A.,** Adrenergic innervation of the ductus deferens in young and aging rats: morpho-functional investigation, *Acta Histochem.*, 89, 67, 1990.

49. **Said, S. I. and Mutt, V.,** Polypeptide with broad biological activity: isolation from small intestine, *Science,* 169, 1217, 1970.
50. **Ottesen, B., Wagner, G., Virag, R., and Fahrenkrug, J.,** Penile erection: possible role for vasoactive intestinal polypeptide as a neurotransmitter, *Br. Med. J.,* 288, 9, 1984.
51. **Polak, J. M., Gu, J., Mina, S., and Bloom, S. R.,** VIPergic nerve in the penis, *Lancet,* 2, 217, 1981.
52. **Willis, T., Ottesen, B., Wagner, G., Sundler, F., and Fahrenkrug, J.,** Vasoactive intestinal polypeptide (VIP) as a possible neurotransmitter involved in penile erection, *Acta Physiol. Scand.,* 113, 545, 1981.
53. **Willis, E., Ottesen, B., Wagner, G., Sundler, F., and Fahrenkrug, J.,** Vasoactive intestinal polypeptide (VIP) as a putative neurotransmitter in penile erection, *Life Sci.,* 33, 383, 1983.
54. **Crowe, R., Lincoln, J., Blacklay, P. F., Pryor, J. P., Lumley, J. S. P., and Burnstock, G.,** Vasoactive intestinal polypeptide-like immunoreactive nerves in diabetic penis: a comparison between streptozotocin-treated rat and man, *Diabetes,* 32, 1075, 1983.
55. **Amenta, F., Cavallotti, C., De Rossi, M., Ferrante, F., and Geppetti, P.,** Vasoactive intestinal polypeptide levels and distribution in the penis of old rats, *J. Neural Transm.,* 70, 137, 1987.
56. **Steger, R. W. and Peluso, J. J.,** Hypothalamic-pituitary function in the old irregular cycling rat, *Exp. Aging Res.,* 5, 303, 1979.
57. **Rice, G. E., Cho, R. F., and Barnea, A.,** Aging-related reduced release of LH-releasing hormone from hypothalamic granules, *Neurobiol. Aging,* 4, 217, 1983.
58. **Demarest, K. T., Riegle, G. D., and Moore, K. E.,** Characteristics of dopaminergic neurons in the aged male rat, *Neuroendocrinology,* 31, 222, 1980.
59. **Demarest, K. T., Moore, K. E., and Riegle, G. D.,** Dopaminergic neuronal function, anterior pituitary dopamine content, and serum concentrations of prolactin, luteinizing hormone and progesterone in the aged female rat, *Brain Res.,* 247, 347, 1982.
60. **Reymond, M. J. and Porter, J. C.,** Secretion of hypothalamic dopamine into pituitary stalk blood of aged female rats, *Brain Res. Bull.,* 7, 69, 1981.
61. **Meites, J., Steger, R. W., and Haung, H. H.,** Relation of neuroendocrine system to the reproductive decline in aging rats and human subjects, *Fed. Proc.,* 39, 3168, 1980.
62. **Wise, P. M. and Ratner, A.,** Effect of ovariectomy on plasma LH, FSH, estradiol, and progesterone and medial axial hypothalamic LHRH concentrations in old and young rats, *Neuroendocrinology,* 30, 15, 1980.
63. **Bessetti, G. E., Reimond, M. J., Blanc, F., Boujon, C. E., Furrer, B., and Rossi, G. L.,** Functional and morphological changes in the hypothalamo-pituitary-gonadal axis of aged female rats, *Biol. Reprod.,* 45, 221, 1991.
64. **Goudsmit, E., Filers, E., and Swaab, D. F.,** Changes in vasopressin neurons and fibers in aging and Alzheimer's disease reversibility in the rat, *Prog. Clin. Biol. Res.,* 317, 1193, 1989.
65. **Vogels, O. J., Broere, C. A., and Nieuwenhuys, R.,** Neuronal hypertrophy in the human supraoptic and paraventricular nucleus in aging and Alzheimer's disease, *Neurosci. Lett.,* 109, 62, 1990.
66. **Riaschikov, S. N., Glumova, V. A., Markov, V. N., and Bazhenov, A. N.,** Dynamics of postnatal histogenesis of the thyroid in normal rats and after sympathectomy, *Ontogenetics.,* 19, 258, 1988.
67. **Riashchikov, S. N., Glumova, V. A., Markov, V. N., and Zelenin, A. A.,** Morphofunctional characteristics of the thyroid gland in intact and partially sympathectomized rats of different age groups, *Biull. Eksp. Biol. Med.,* 104, 351, 1987.
68. **Ito, K., Sato, A., Sato, Y., and Suzuki, H.,** Increase in adrenal catecholamine secretion and adrenal sympathetic nerve unitary activities with aging in rats, *Neurosci. Lett.,* 69, 263, 1986.

69. **Sato, A.,** Sympathetic regulation of adrenal medullary function during aging, *Fiziol. Zh. Kiev,* 36, 94, 1990.
70. **Tomlinson, A. and Coupland, R. E.,** The innervation of the adrenal gland. IV. Innervation of the medulla from birth to old age. A descriptive and quantitative morphometric and biochemical study of the innervation of chromaffin cells and adrenal neurons in Wistar rats, *J. Anat.,* 169, 209, 1990.
71. **Souto, M., Piezzi, R. S., and Bianchi, R.,** Catecholaminergic responses of neonatal adrenal gland to insulin, *J. Neural Transm.,* 73, 115, 1988.

Clinical and Therapeutic Perspectives

Chapter 14

AGE-RELATED DISORDERS OF THE AUTONOMIC NERVOUS SYSTEM

Paolo Calabresi, Letizia M. Cupini, Nicola B. Mercuri, and Giorgio Bernardi

TABLE OF CONTENTS

0-8493-6981-9/93/$0.00 + $.50

I. INTRODUCTION

The use of new methods in the study of autonomic nervous system functions has recently provided information concerning the pathophysiology of the neuronal, cardiovascular, and secretory systems. In addition, in recent years, a large number of studies have been performed to clarify the role of different functional systems in the pathogenesis of age-related diseases. However, the link between biochemical, pharmacological, genetic, and morphological mechanisms underlying the age-related dysfunctions of the autonomic nervous system is not yet fully understood. In the present review, we survey the diseases of the autonomic nervous system in which age-related processes have been implicated.

II. CLINICAL TESTS

The clinical tests to evaluate patients with autonomic nervous system disorders include the thermoregulatory sweat test (TST), quantitative sudomotor axon reflex test (Q-SART), blood pressure and heart rate recordings, heart rate responses to deep breathing and the Valsalva maneuver, plasma catecholamine measurements, detection of pupillary size, and urodynamic testing techniques.

The integrity of sympathetic efferent pathways can be assessed by tests of sudomotor activity. There are several of these. Sweat can be weighed after it is absorbed by small squares of filter paper. Powdered charcoal dusted on the skin will cling to moist areas and not to dry ones. In the galvanic skin-resistance test, a galvanometer indicates the resistance offered by the skin to the passage of weak galvanic current through the skin. An increase in sweating lowers the resistance; anhydrosis raises it. A color indicator such as quinizarin may be used.

Other procedures that may be applied to test circulatory control include the cold pressor test, in which the pressor response to immersion of a hand for 1 min in ice water is tested.

Digital blood flow is reflexively reduced by an inspiratory gasp and reflexively increased by radiant heat applied to the trunk. Cutaneous vasoconstriction or vasodilation may be measured using finger or hand plethysmography, or may be inferred from heat elimination from the fingertips.

Bladder function is best assessed by the cystometrogram, i.e., by measuring intravesicular pressure as a function of the volume of sterile saline solution permitted to flow by gravity into the bladder. A relatively simple apparatus is available for this purpose. The rise of pressure as 500 ml of fluid is allowed to flow gradually into the bladder and the emptying contractions of the detrusor can be recorded by a manometer. A quick and simple way of determining bladder atony is to measure the residual urine (by catheterization of the bladder) immediately after voluntary voiding or by estimating its volume by intravenous pyelography.

III. CLINICAL ENTITIES

A. ORTHOSTATIC HYPOTENSION IN THE ELDERLY

This clinical condition is caused by a dysfunction of the circulation upon assuming an upright posture. The diagnosis of orthostatic hypotension (OH) requires an orthostatic reduction in systolic blood pressure of >30 mmHg or mean arterial pressure of >20 mmHg. Several geriatric patients have significant orthostatic falls of blood pressure which may be coupled with dizziness and transient loss of consciousness.[1,2] A dysfunction of the baroreflex arc causing a failure of noradrenaline release has been indicated as one of the possible causes of OH in the elderly.[2,3] However, the relative influence of altered baroreflex sensitivity or of a disturbed sympathetic outflow leading to OH in the elderly remains to be determined, and in addition, other factors have been indicated such as organic brain diseases, heart diseases, varicose veins, and drugs.

It has also been postulated that multifactorial causes underlie OH in the elderly. Although *in vitro* studies of contractile responses of human arterial strips showed no age-related change in responsiveness,[4,5] changes in β-receptor function have been clinically implicated in OH.[6,7] Alteration in the structure of the blood vessels and a consequent loss of elasticity of the arterial walls with age could lead to loss of baroreflex sensitivity and to OH in the elderly.[8,9] In addition, hyponatremia and/or hypokalemia may occur in OH, suggesting a possible role of electrolytic dysfunction; in this condition, electrolytic replacement causes clinical improvement.[10]

Recently, Mathias et al.[11] reported that postprandial hypotension (PPH), which frequently occurs in autonomic failure, may aggravate OH. The authors studied the effect of a meal on supine and postural blood pressure (BP) responses in patients with chronic autonomic failure, and found that food lowers BP to a variable extent in different groups. A possible pathophysiological mechanism to explain PPH has been postulated: food ingestion increases splanchnic blood flow and gastrointestinal fluid secretion. This alteration of fluid volume distribution compromises cardiac filling, decreases cardiac output, and leads to systemic hypotension.[12,13]

Severe OH can be distinguished in idiopathic OH (or primary autonomic failure, PAF) and multiple system atrophy with autonomic failure (MSA, or Shy-Drager syndrome). The condition is designated MSA when there are associated clinical features of striatonigral or olivopontocerebellar involvement, and PAF when the central nervous system is uninvolved.[14,15]

B. PRIMARY AUTONOMIC FAILURE

At present, the possible link between PAF, MSA, Parkinson's disease, and the age-related degenerative processes of the autonomic nervous system is not fully understood. However, we believe that characterization of the pathophysiological and clinical alterations observed in these diseases may provide some information about degenerative processes involving the autonomic system in "normal" aging.

On the basis of biochemical, pharmacological, and pathological studies, PAF is considered a primary disorder of postganglionic sympathetic efferent neurons, whereas MSA is a primary disorder of the central preganglionic sympathetic nervous system.[16-22] Differentiation of these two types of primary OH is based on pharmacological evidence. In PAF, plasma levels of noradrenaline are subnormal while the patient is supine because of the failure of the damaged nerve terminals to synthesize or release catecholamines. When the patient stands up, the noradrenaline levels do not rise as they do in the unaffected person. There is also hypersensitivity to injected noradrenaline. On the other hand, in central preganglionic OH (MSA or Shy-Drager), the resting noradrenaline levels in the plasma are normal, but on standing, there again is no rise; moreover, in MSA, the sensitivity to exogenously administered noradrenaline is normal. In both conditions, the plasma levels of dopamine-β-hydroxylase are subnormal. However, there is evidence for preganglionic sympathetic alterations in PAF as well as in MSA; for example, (1) morphological evidence of preganglionic failure has been found in PAF,[20,23] (2) preganglionic sudomotor failure has been reported in PAF,[15] and (3) there is a significant subset of patients with OH associated with normal supine noradrenaline that fails to rise on standing, a pattern suggestive of failure of preganglionic sympathetic activation.[17]

PAF may affect most autonomic systems with increasing age. Subjects may have problems with thermoregulatory responsiveness (anhydrosis and sudomotor failure), impotence, and impairment of gastrointestinal, urinary, lacrimal, and pupillary functions.[24,25] It must be stressed that older subjects have autonomic failure involving cardiac autonomic control, sudomotor control, and noradrenaline levels as well as symptomatic OH. Heart rate responses to deep breathing and the Valsalva maneuver were reduced in PAF as well as MSA patients.[15]

Clinical criteria for the diagnosis of PAF are imperfect, and much of the controversy about the site and selectivity of the lesion possibly relates to the heterogeneity of what is designated PAF. The entity probably comprises subsets of patients who have OH in common. A small subset begins with PAF and progresses to MSA,[14] some patients develop diffuse autonomic failure with increasing age,[24] and still others develop OH without obvious autonomic failure.[26]

C. MULTIPLE SYSTEM ATROPHY OR SHY-DRAGER DISEASE

In MSA or Shy-Drager, the dysautonomia coexists with cerebellar and extrapyramidal signs. Shy-Drager is a disorder characterized by degeneration of the preganglionic lateral horn neurons of the thoracic spinal segments; this pathological condition is often associated with striatonigral degeneration and/or olivopontocerebellar atrophy.

Striatonigral degeneration is a syndrome that closely resembles the clinical features of Parkinson's disease, but with a different pathological basis: instead of nigrostriatal pathway degeneration (as in Parkinson's disease), morphological changes of striatal (mainly putaminal) neurons projecting to the substantia nigra are observed. Olivopontocerebellar degeneration is a sporadic (Dejerine-Thomas type) or hereditary (Menzel type) disorder characterized by neuronal degeneration in the cerebellar cortex, pons, and inferior olives in which the main feature is ataxia — first in the legs, then the arms and hands and bulbar musculature — a symptomatology common to all the cerebellar atrophies. A recent study[27] utilizing quantitative evaluation of pontine volume by computer tomography indicated that patients with pontine atrophy show significantly more autonomic dysfunction and extrapyramidal signs; in addition, this group of patients had a later onset and shorter duration of the disease, and impairment of higher brain functions. In contrast, the group without these manifestations had relatively normal pontine volume, a greater familial tendency for cerebellar degeneration, and longer duration of the disease, suggesting a close similarity with cerebello-olivary atrophy of the Holmes type.

The autonomic dysfunctions observed in MSA are caused by a preganglionic disorder; this idea is supported by experimental and clinical evidence. (1) In contrast to the marked reduction of resting noradrenaline observed in acute panautonomic neuropathy[28] and diabetic autonomic neuropathy,[29] MSA is characterized by a normal resting noradrenaline level, suggesting a preganglionic site for MSA. (2) The decreased response of noradrenaline levels to standing in MSA suggests the failure to activate central sympathetic pathways.[17] (3) Morphometric studies report a marked reduction in preganglionic sympathetic neurons in MSA patients.[16,30,31] Cohen et al.[15] reported that although the postganglionic adrenergic function is spared in MSA patients, postganglionic sudomotor failure and anhydrosis can be observed in several MSA patients, as assessed by the TST and Q-SART. The authors hypothesized that the postganglionic sympathetic sudomotor failure in MSA is due to a transynaptic effect.

D. AUTONOMIC NERVOUS SYSTEM DYSFUNCTION IN PARKINSON'S DISEASE

Autonomic dysfunction in Parkinson's disease (PD) was first reported by James Parkinson in 1817.[32] Abnormalities of salivation, sweating, bladder and bowel functions, OH, sialorrhea, seborrhea, dysphagia, heat and cold intolerance, and impotence are common features of this disease.[33,34] Some of these disorders are common in old age and have characteristics similar to normal aging and PD, yet data on their respective frequencies are not available. For this reason, it is difficult to know whether in a particular patient these disorders are a manifestation of age, disease, therapy, or a combination of these factors. Another problem that may occur, particularly regarding patients in whom the autonomic problems predominate, is whether the disturbances are part of MSA.

The pathological basis of OH in PD patients is not clear. In cases where parkinsonian features occur as a part of MSA, OH is believed to result from cell loss in the intermediolateral column of the spinal cord.[23] However, the site of the lesion responsible for OH in idiopathic PD has not yet been established with certainty. Appenzeller and Goss[35] postulated that hypothalamic lesions were the basis of dysautonomia of PD, while Vanderhaeghen et al.[36] suggested that lesions in the sympathetic ganglia may play an important role in the pathophysiology of OH in PD. Rajput and Rozdilski[37] found Lewy bodies and cell loss in the sympathetic ganglia of most cases of PD. They observed that the severity of the alterations of these structures related to the degree of OH, and concluded that the lesions of the sympathetic ganglia may play a major role in the production of OH in PD. On the other hand, Lewy bodies have also been found in the hypothalamic nuclei of PD patients,[38,39] suggesting a possible role of some hypothalamic nuclei in the genesis of autonomic disorders in PD.

Certain cardiovascular reflexes widely considered to be indices of autonomic function have been examined by a number of authors in order to study vegetative dysfunction in PD. The results have often been controversial, owing to differences in the selection of patients and controls, assessment methods, age of the studied subjects, and the effect of drugs.

The hypotension in PD is particularly common following meals.[40] Bannister et al.[41] studied the fall of blood pressure during insulin hypoglycemia in subjects with sympathetic dysfunction due to "idiopathic autonomic failure" and found a dose-related fall in BP during insulin infusion with euglycemia. They suggested that a hypotensive effect of insulin occurring independently of a blood glucose fall results in PPH in patients with sympathetic failure.

Hakusui et al.[42] demonstrated a remarkable effect of vasopressin in preventing PPH: a vasopressin infusion of 0.3 U/min prior to 75 g glucose loading completely inhibited a BP fall.

An important factor in the evaluation of OH in patients with PD is the possible contribution of drugs. In addition to OH, Barbeau et al.[43] found that PD patients had a lower mean blood pressure (BP) than age-matched controls. Several antiparkinsonian agents, including L-DOPA, may act through a central mechanism similar to that of the antihypertensive drug α-methyl-DOPA, although the peripheral conversion to dopamine may also contribute to the hypotension. This component will be eliminated by the use of decarboxylase inhibitors. Direct-acting dopamine agonists such as bromocriptine or lisuride are particularly potent in precipitating hypotension by causing peripheral vasodilation. Activation of dopamine receptors located on sympathetic terminals powerfully inhibits stimulation-induced noradrenaline release. This action is abolished by treatment with the peripheral D_2 receptor antagonist domperidone. Calne et al.[44] reported a significant increase in the postural fall of BP in L-DOPA-treated patients. Goetz et al.[45] studied autonomic functions before and after antiparkinsonian drugs (L-DOPA, amantadine, or dopamine receptor agonists). When motor disability was maximal ("off"), patients had a higher resting pulse rate, greater orthostatic fall of BP, and decreased responses to

Valsalva and cold pressure stimuli than their spouse-controls. After treatment, when function was optimal (''on''), the cardiovascular reflex abnormalities remained, but were no worse, suggesting that these drugs do not exacerbate a preexisting OH. Although it was initially suggested that anticholinergic drug therapy was a possible cause of hypotension in PD,[46] later studies[47] failed to demonstrate any effect of these drugs on BP.

An abnormal R-R interval in electrocardiography has been observed in PD. Kuroiwa et al.[48] studied systolic blood pressure, R-R interval, and R-R interval variance in 16 patients with PD. The authors detected a significant correlation between postural hypotension and anhydrosis that was related to a vasomotor function of sympathetic activity. Low R-R interval variability in the supine position correlated with bladder-bowel disturbance, indicating impaired parasympathetic activity.

Abnormalities in sweating and temperature regulation are common among PD patients.[33,49] In 1888, Gowers[50] described a hemiparkinsonian patient with excessive head sweating involving only the parkinsonian side. Impairment of sweating has also been reported by several authors.[35,37,51] Goetz et al.[45] demonstrated increased sweating in the head and neck in response to heat stimulation in 31 chronically treated PD patients. Sweating abnormalities resolved with anti-PD medication, suggesting that dopamine may play a role in the mechanisms underlying impaired thermoregulation in PD patients.

Abnormal temperature control regulation has been documented in PD patients. Appenzeller and Gross[35] have demonstrated impaired peripheral vasodilation and deficient sudomotor functions in response to heat in these patients. Accidental hypothermia and absence of shivering following experimental lowering of the body temperature have also been reported in PD patients.[52] Elliott et al.[53] studied vascular responses in the hands of five patients with bilateral PD who were not taking medication and in six patients with bilateral PD who were receiving L-DOPA therapy alone. Heat elimination from one hand produced by placing the contralateral arm in a water bath at 44°C was measured. The patients, whether or not receiving medication, had very low heat elimination compared with controls. In two patients with unilateral disease, the vascular responses in the affected hand were reduced compared with those in the unaffected side. The authors suggested the existence of abnormal vascular control in PD caused by autonomic dysfunction, in either the central nervous system or afferent sympathetic pathways. Komatsubara[54] found no significant differences in skin temperature between 35 PD patients ranging in age from 43 to 78 years of age and 35 healthy controls before placing their hands in ice water but the recovery rate at 10 min after cooling was lower than that of the controls. They suggested that the patients with a shorter duration of illness had blood vessels supersensitive to cold stimulation due to peripheral sympathetic dysfunction and that those with long duration had vascular paralysis.

Abnormalities of vasomotor and sudomotor functions in PD patients can be caused by deranged posterior hypothalamic functions. This hypothesis is suggested by two findings: (1) the posterior hypothalamus regulates sudomotor and vasomotor systems[55] and (2) in pathological studies on PD patients, the lateral posterior and tuberomammillary nuclei show the highest Lewy body hypothalamic degeneration.[38] However, it should be stressed that neurons involved in the control of sweating are also located in extrahypothalamic areas such as the mesencephalon, pons, medulla oblongata, spinal cord, limbic system, basal ganglia, and cerebral cortex.[56]

Among the autonomic dysfunctions, pupil abnormalities are frequently observed in PD patients, either due to the disease itself or to the antiparkinsonian drugs. Investigation of pupillary changes may be important, since pupillary size and its responses can be measured accurately and repeatedly using noninvasive techniques. The size of the pupil is commonly believed to be reduced in PD, but, again, reference should be made to the miosis common in old age.[57] In fact, if the correction is made for this factor, pupil size appears to be normal in PD.[58] Orskov et al.[59] performed continuous measurements of the area of the left pupil with an infrared-sensitive television camera. The cholinergic functions of latency, maximal contraction velocity, and amplitude of the light reflex were unchanged in the patients. Redilation of the pupil after elicitation of the light and dark reflexes, which are considered to be a sympathetic function, also remained unchanged.

Kirby and Fowler[60] summarized bladder dysfunction in PD as follows: patients with PD commonly suffer from frequency of micturition and urge continence, and an urodynamic study revealed a high incidence of inhibition from the basal ganglia, especially from the globus pallidus and substantia nigra.

The occurrence of sexual impotence is widely quoted as one of the autonomic dysfunctions in PD. Ward and Wilkinson[61] studied 19 patients (13 men and 6 women) ranging from 45 to 78 years of age. Sexual self-images were said to be affected by PD in ten men and four women. However, most patients continued to experience sexual arousal, and the majority currently attempt sexual intercourse. Nine men had erectile problems and seven had ejaculatory failure. The authors suggested that most patients in this population retain sexual interest and remain sexually active, and that therefore erectile and ejaculatory dysfunction in PD may result from neurological and pharmacological rather than psychological factors.

A recent study reviewed the clinical and autonomic features of all patients with extrapyramidal and cerebellar disorders studied in the Mayo Autonomic Reflex Laboratory from 1983 to 1989.[62] The patients were grouped into the following categories: PD (Parkinson-plus), MSA, hereditary multisystem degeneration, progressive supranuclear palsy, nonfamilial cerebellar degeneration, and nonspecific sporadic multisystem degeneration. Severe autonomic failure occurred in 97% of patients with MSA and 53% with nonspecific sporadic multisystem degeneration. Autonomic involvement was mild or absent in PD, while parkinsonism-plus and nonfamilial cerebellar degeneration patients had moderate autonomic failure. Orthostatic BP reduction, percentage of anhydrosis on the TST, Q-SART, forearm response, and heart rate response to deep breathing strongly regressed with severity. A response to L-DOPA treatment in patients with PD was more likely if cerebellar signs and cognitive deficits were absent. The presence of L-DOPA-induced dyskinesia was also a marker for a clinically favorable L-DOPA response.

REFERENCES

1. **Caird, F. I., Andrews, G. R., and Kennedy, R. D.,** Effect of posture on blood pressure in the elderly, *Br. Heart J.*, 35, 527, 1973.
2. **Johnson, R. H. and Spalding, J. M. K.,** *Disorders of the Autonomic Nervous System,* Blackwell Scientific, Oxford, 1974.
3. **White, N. J.,** Heart rate changes on standing in elderly patients with orthostatic hypotension, *Clin. Sci.*, 58, 411, 1980.
4. **Scott, P. J. W. and Reid, J. L.,** The effect of age on the responses of human isolated arteries to noradrenaline, *Br. J. Clin. Pharmacol.*, 13, 237, 1982.
5. **Stevens, M. J., Lipe, S., and Moulds, R.,** The effect of age on the response of human isolated arteries and veins to noradrenaline, *Br. J. Clin. Pharmacol.*, 14, 750, 1982.
6. **Elliott, H. E., Summer, D. J., McLean, K., et al.,** Effect of age on the responsiveness of vascular adrenoreceptors in man, *J. Cardiovasc. Pharmacol.*, 4, 388, 1982.
7. **Kelly, J. and O'Malley, K.,** Adrenoreceptor function and ageing, *Clin. Sci.*, 66, 509, 1984.
8. **Collins, K. J., Exton-Smith, A. N., and James, M. H.,** Functional changes in autonomic nervous responses with ageing, *Age Ageing,* 9, 17, 1980.
9. **MacLennan, W. J., Hall, M. R. P., and Timothy, J. I.,** Postural hypotension in old age: is it a disorder of the nervous system or blood vessels?, *Age Ageing,* 9, 25, 1980.
10. **Cox, J. R., Admani, A. K., Agarwal, M. L., et al.,** Postural hypotension-body fluid compartments and electrolytes, *Age Ageing,* 2, 112, 1973.
11. **Mathias, C. J., Holly, E., Armstrong, E., Shareef, M., and Bannister, R.,** The influence of food on postural hypotension in three groups with chronic autonomic failure — clinical and therapeutic implications, *J. Neurol. Neurosurg. Psychiatry,* 54, 726, 1991.
12. **Hakusui, S., Sugiyama, Y., Iwase, S., Hasegawa, Y., Koike, Y., Mano, T., and Takahashi, A.,** Postprandial hypotension: microneurographic analysis and treatment with vasopressin, *Neurology,* 41, 712, 1991.
13. **Lipsitz, L. A. and Fullerton, K. J.,** Postprandial blood pressure reduction in healthy elderly, *J. Am. Geriatr. Soc.*, 34, 267, 1986.
14. **Thomas, J. E. and Schirger, A.,** Idiopathic orthostatic hypotension. A study of its natural history in 57 neurologically affected persons, *Arch. Neurol.*, 22, 289, 1970.
15. **Cohen, J., Low, P., Fealey, R., Sheps, S., and Jiang, N. S.,** Somatic and autonomic function in progressive autonomic failure and multiple system atrophy, *Ann. Neurol.*, 22, 692, 1987.
16. **Bannister, R. and Oppenheimer, D. R.,** Degenerative diseases of the nervous system associated with autonomic failure, *Brain,* 95, 457, 1972.
17. **Ziegler, M. G., Lake, C. R., and Kopin, I. J.,** The sympathetic-nervous-system defect in primary orthostatic hypotension, *N. Engl. J. Med.*, 296, 293, 1977.
18. **Polinsky, R. J., Kopin, I. J., Ebert, M. H., and Weise, V.,** Pharmacologic distinction of different orthostatic hypotension syndromes, *Neurology,* 31, 1, 1981.

19. **Polinsky, R. J., Goldstein, D. S., Brown, R. T., Keiser, H. R., and Kopin, I. J.,** Decreased sympathetic neuronal uptake in idiopathic orthostatic hypotension, *Ann. Neurol.*, 18, 48, 1985.
20. **Low, P. A., Thomas, J. E., and Dyck, P. J.,** The splanchnic autonomic outflow in Shy-Drager syndrome and idiopathic orthostatic hypotension, *Ann. Neurol.*, 4, 511, 1978.
21. **Matthews, M.,** Assessing the peripheral ganglia in autonomic failure, in *Autonomic Failure*, 2nd ed., Bannister, R., Ed., Oxford University Press, New York, 1988, 521.
22. **Dotson, R., Ochoa, J., Marchettini, P., and Cline, M.,** Sympathetic neural outflow directly recorded in patients with primary autonomic failure, *Neurology*, 40, 1079, 1990.
23. **Johnson, R. H., Lee, G. de J., Oppenheimer, D. R., and Spalding, J. M. K.,** Autonomic failure with orthostatic hypotension due to intermediolateral column degeneration. A report of two cases with autopsies, *Q. J. Med.*, 35, 276, 1966.
24. **Collins, K. J.,** Autonomic failure in the elderly, in *Autonomic Failure*, Bannister, R., Ed., Oxford University Press, New York, 1983, 489.
25. **Foster, K. G., Ellis, F. P., Dore, C., et al.,** Sweat responses in the aged, *Age Ageing*, 5, 91, 1976.
26. **Robinson, B. J., Johnson, R. H., Lambbie, D. G., and Palmer, K. T.,** Do elderly patients with an excessive fall in blood pressure on standing have evidence of autonomic failure?, *Clin. Sci.*, 64, 587, 1983.
27. **Chida, K., Tamura, M., Kamikura, I., Takasu, T., and Goto, N.,** A quantitative evaluation of pontine volume by computed tomography in patients with cerebellar degeneration, *Neurology*, 40, 1241, 1990.
28. **Low, P. A., Dyck, P. J., Lambert, E. H., et al.,** Acute panautonomic neuropathy, *Ann. Neurol.*, 13, 412, 1983.
29. **Christensen, N. J.,** Plasma catecholamines in long-term diabetics with and without neuropathy and in hypophysectomized subjects, *J. Clin. Invest.*, 51, 779, 1972.
30. **Kennedy, P. G. E. and Duchen, L. W.,** A quantitative study of intermediolateral column cells in motor neuron disease and the Shy-Drager syndrome, *J. Neurol. Neurosurg. Psychiatry*, 48, 1103, 1985.
31. **Oppenheimer, D. R.,** Lateral horn cells in progressive autonomic failure, *J. Neurol. Sci.*, 46, 393, 1980.
32. **Parkinson, J.,** *An Essay on the Shaking Palsy*, Sherwood, Neely and Jones, London, 1817.
33. **Perkin, G. D.,** Autonomic function, in *Research Progress in Parkinson's Disease*, Pitman Medical, Kent, 1981, 111.
34. **Thorner, M. O.,** Dopamine is an important neurotransmitter in the autonomic nervous system, *Lancet*, 1, 662, 1975.
35. **Appenzeller, O. and Goss, J. E.,** Autonomic deficits in Parkinson's syndrome, *Arch. Neurol.*, 24, 50, 1971.
36. **Vanderhaeghen, J. J., Perrier, O., and Sternon, J. E.,** Pathological findings in idiopathic orthostatic hypotension, *Arch. Neurol.*, 22, 207, 1970.
37. **Rajput, A. H. and Rozdilski, B.,** Dysautonomia in parkinsonism: a clinico-pathological study, *J. Neurol. Neurosurg. Psychiatry*, 39, 1092, 1976.
38. **Langston, J. W. and Forno, L. S.,** The hypothalamus in Parkinson's disease, *Ann. Neurol.*, 3, 129, 1978.
39. **Ohama, E. and Ikuta, F.,** Parkinson's disease: distribution of Lewy bodies and monoamine neuron system, *Acta Neuropathol.*, 34, 311, 1976.
40. **Micieli, G., Martignoni, E., Cavallini, A., Sandrini, G., and Nappi, G.,** Post-prandial and orthostatic hypotension in Parkinson's disease, *Neurology*, 37, 386, 1987.
41. **Bannister, R., Da Costa, D. F., Kooner, J. S., MacDonald, I. A., and Mathias, C. J.,** Insulin-induced hypotension in autonomic failure during euglycaemia in man, *J. Physiol.*, 382, 26P, 1987.
42. **Hakusui, S., Takahashi, A., Sugiyama, Y., Koike, Y., and Hasegawa, Y.,** Vasopressin effect and pathophysiological analysis of postprandial hypotension, *Clin. Neurol.*, 30, 672, 1990.

43. **Barbeau, A., Gillo-Joffroy, L., Boucher, R., Nowaczynski, W., and Genest, J.,** Renin-aldosterone system in Parkinson's disease, *Science,* 165, 291, 1969.
44. **Calne, D. B., Brennan, J., Spiers, A. S. D., and Stern, G. M.,** Hypotension caused by L-Dopa, *Br. Med. J.,* 1, 474, 1970.
45. **Goetz, C. G., Lutge, W., and Tanner, C. M.,** Autonomic dysfunction in Parkinson's disease, *Neurology,* 36, 73, 1986.
46. **Martilla, R. J. and Rinne, U. K.,** Level of blood pressure in patients with Parkinson's disease: a case control study, *Eur. Neurol.,* 16, 73, 1977.
47. **Aminoff, M. J., Gross, M., Laatz, B., Vakil, S. D., Petrie, A., and Calne, D. B.,** Arterial blood pressure in patients with Parkinson's disease, *J. Neurol. Neurosurg. Psychiatry,* 38, 73, 1975.
48. **Kuroiwa, Y., Shimada, Y., and Toyokura, Y.,** Postural hypotension and low R-R interval variability in parkinsonism, spino-cerebellar degeneration, and Shy-Drager syndrome, *Neurology,* 33, 463, 1983.
49. **Tanner, C. M., Goetz, C. G., and Klawans, H. L.,** Paroxismal drenching sweats in idiopathic Parkinson's disease response to propanolol, *Neurology,* 32, 162, 1982.
50. **Gowers, W. R.,** *A Manual of Diseases of the Nervous System,* Blakiston, Philadelphia, 1888.
51. **Aminoff, M. J., Wilcox, C. S., and Slater, J. D. H.,** Autonomic defects in paralysis agitans and in the Shy-Drager syndrome, *Acta Neurol. Scand.,* 51, 105, 1972.
52. **Gubbay, S. S. and Barwick, D. D.,** Two cases of accidental hypothermia in Parkinson's disease with unusual EEG findings, *J. Neurol. Neurosurg. Psychiatry,* 29, 459, 1966.
53. **Elliott, K., Cote, L. J., Frewin, D. B., and Downey, Y. A.,** Vascular responses in hands of Parkinson's disease patients, *Neurology,* 24, 857, 1974.
54. **Komatsubara, C.,** Studies on autonomic dysfunction in Parkinson's disease. I. Evaluation of peripheral vasomotor function using the ice water tolerance test in Parkinson's disease, *Auton. Nerv. Syst.,* 25, 319, 1988.
55. **Myers, R. W.,** The hypothalamus, Charles C Thomas, Springfield, IL, 1969.
56. **Figa-Talamanca, L., Gualandi, C., Di Meo, L., Di Battista, G., Neri, G., and Lo Russo, F.,** Hyperthermia after discontinuance of L-Dopa and bromocriptine therapy: impaired dopamine receptors, a possible cause, *Neurology,* 35, 258, 1985.
57. **Korczyn, A. D., Laor, N., and Nemet, P.,** Sympathetic pupillary tone in old age, *Arch. Ophthalmol.,* 94, 1905, 1976.
58. **Korczyn, A. D., Rubenstein, A. E., and Yahr, M. D.,** The pupil in Parkinson's disease, in Proc. 8th Int. Symp. Parkinson's Disease, New York, 1985, P43.
59. **Orskov, L., Jakobsen, J., Dupont, E., de Fine Olivarius, B., and Christensen, N. J.,** Autonomic function in parkinsonian patients relates to duration of disease, *Neurology,* 37, 1173, 1987.
60. **Kirby, R. S. and Fowler, C. J.,** Bladder and sexual dysfunction in diseases affecting the autonomic nervous system, in *Autonomic Failure,* Bannister, R., Ed., Oxford University Press, Oxford, 1988, 419.
61. **Ward, C. D. and Wilkinson, S.,** Sexual dysfunction in Parkinson's disease, in 9th Int. Symp. Parkinson's Disease, Jerusalem, 1988, 60.
62. **Sandroni, P., Ahlskog, J. E., Fealey, R. D., and Low, P. A.,** Autonomic involvement in extrapyramidal and cerebellar disorders, *Clin. Auton. Res.,* 1, 147, 1991.

Chapter 15

THERAPEUTIC IMPLICATIONS OF AGE-RELATED CHANGES IN THE AUTONOMIC NERVOUS SYSTEM

Gary A. Ford and Brian B. Hoffman

TABLE OF CONTENTS

0-8493-6981-9/93/$0.00 + $.50

I. INTRODUCTION

Changes in the autonomic nervous system of the aging human result in alterations in the physiological state of many organs and would be expected to influence the response of elderly patients to disease and drugs. Although the effects of many changes in the autonomic nervous system have been defined in experimental settings, the physiological and therapeutic implications of many of these findings remain unclear. In part, this reflects the failure of many early studies to differentiate changes in the autonomic system due to acquired disease as distinct from aging, and the methodological difficulties in studying the autonomic nervous system in aging humans. In this chapter, we review the changes in the aged peripheral autonomic nervous system that may be relevant to clinical care of the elderly. Aging changes in brain adrenergic and cholinergic function are not discussed. We have emphasized the results from human studies, but have discussed the results of animal studies where this is relevant or data on humans are lacking. Finally, we discuss avenues of future research that are likely to lead to a better understanding of the clinical importance of aging changes in autonomic function.

II. BAROREFLEX FUNCTION

Heart rate is modulated predominantly through alterations in sympathetic and parasympathetic tone. The changes occurring with aging are discussed in detail in Chapter 7 of this volume. Carotid sinus and aortic arch baroreceptors influence autonomic nervous system activity through afferent input to the vasomotor center in the brainstem. The impairment in baroreflex responses is a well-characterized finding in the elderly which is manifested in a number of ways.[1,2] Pressor responses to phenylephrine infusion have been used to study reflex-induced bradycardia in humans.[3] In the elderly, a smaller response is seen in the reflex bradycardia that occurs in response to increases in blood pressure. The cause of this impairment in baroreflex responsiveness could be due to impaired function of the carotid body receptors, an alteration in central processing of the baroreceptor response, or a diminished cardiac or vascular response to sympathetic nervous system activity. The reduction in baroreceptor responsiveness is more marked in elderly patients with hypertension, suggesting that reduced elasticity of the carotid artery secondary to aging or atherosclerosis may contribute to diminished baroreceptor function.[4]

Similarly, there is a decreased tachycardiac response to hypotension produced by drugs or postural change in the elderly.[5-8] Minaker et al.[9] studied the cardiovascular and neurohumoral responses of healthy normotensive young and elderly subjects to nitroprusside infusions (nitroprusside is a vasodilator which acutely lowers blood pressure and induces a variety of neurohumoral responses). Similar increases in plasma concentrations of noradrenaline and renin were observed following a 20% reduction in mean arterial blood pres-

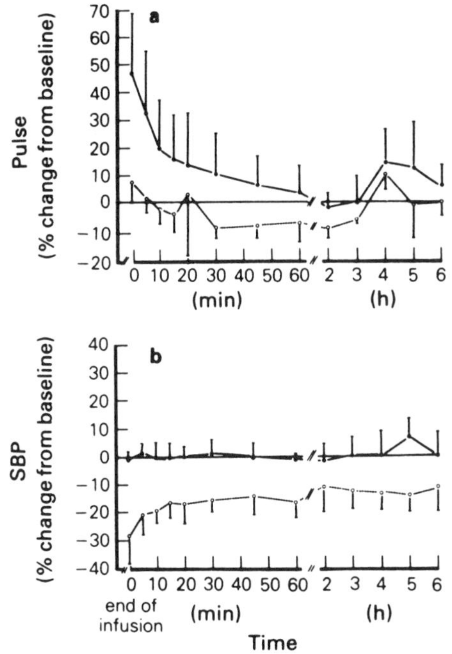

FIGURE 1. Changes in pulse and systolic blood pressure in young (●) and elderly (○) volunteers following i.v. administration of 2.5 mg of nifedipine. (From Robertson, D. R. C., Waller, D. G., Renwick, A. G., and George, C. F., *Br. J. Clin. Pharmacol.*, 25, 297, 1988. With permission.)

sure; however, increases in heart rate for any given reduction in blood pressure were much greater in the young. These data suggest that the sensing function of carotid baroreceptors is relatively preserved in the elderly since the neurohumoral responses are similar to those found in the young; this suggests that the problem lies at the level of the response of the heart and vasculature to sympathetic nervous system activation. However, changes in plasma renin and noradrenaline are crude measures of sympathetic nervous system activity, and a reduced baroreceptor response or alteration in central processing could contribute to impaired cardiovascular responses in the elderly.

This reduced baroreceptor function results in alterations in the response to some cardiovascular drugs. Robertson et al. studied the response of young and elderly normotensive subjects to oral and intravenous nifedipine.[10] Following intravenous dosing which eliminated the influence of aging alterations in first-pass hepatic metabolism, the young subjects maintained systolic blood pressure with a compensatory tachycardia (Figure 1). In contrast, the elderly experienced a pronounced fall in supine blood pressure, but did not demonstrate a tachycardia in response to the fall in blood pressure. Following oral dosing, no young subjects developed orthostatic hypotension, whereas three of the six elderly subjects without preexisting orthostatic hypotension demonstrated a greater than 20 mmHg fall in systolic blood pressure on standing. In a further study involving two elderly subjects with preexisting asymptomatic orthostatic hypotension, systolic blood pressure fell below 90 mmHg following oral administration of 10 mg of nifedipine, with resulting syncope in one patient. These studies demonstrate that the reduction in baroreceptor function is a double-edged sword with respect to the responsiveness of the elderly to calcium channel blockers. Elderly subjects demonstrate a greater hypotensive response, but this is at the price of impairment of postural blood pressure control. The response to α-adrenergic antagonists is similar, with elderly normotensive subjects showing a greater hypotensive response and reduced increase in heart rate following doxazosin.[11] Further combined pharmacokinetic and pharmacodynamic studies of these and other hypotensive agents are required in elderly patients with cardiovascular disease to determine whether this response to calcium channel blockers is seen with other agents such as angiotensin-converting enzyme inhibitors and β-adrenergic receptor antagonists. The technique of concentration-effect analysis has been used in an attempt to obtain a clearer assessment of drug responsiveness, but the confounding influence of aging changes in baroreceptor responsiveness may make it difficult to produce pharmacodynamic models to relate drug concentration at an "effect compartment" with hypotensive effect.[12]

Clinical observation suggests that the elderly tolerate treatment for hypertension and heart failure less well than younger patients because of an increased incidence of orthostatic hypotension, although detailed studies comparing different patient groups are lacking. This has led geriatricians to be cautious in initiating such treatment in the elderly because of the perception that such adverse effects outweigh the benefits of treatment for older patients. The risk/benefit ratio of antihypertensive therapy in the elderly has been very difficult to objectively evaluate because of the lack of large, well-controlled studies. The recent publication of three large-scale studies of antihypertensive therapy in the elderly (SHEP, STOP, and MRC trials) suggests that the risk/benefit ratio for treatment, at least in relatively healthy elderly hypertensives, may be more advantageous than previously generally thought.

The SHEP study (Systolic Hypertension in the Elderly Program) is the only large prospective randomized study to examine whether treating systolic hypertension in the elderly reduces the risk of stroke.[13] Hypertensives (N = 4736; BP: 160 to 219, <90 mmHg) aged 60 years and above, including patients already receiving treatment, were treated with a stepped care regime of chlorthalidone and then atenolol. They constituted 1% of a screened population, and were therefore a selected, possibly atypical group, since the prevalence of systolic hypertension is probably greater than 10% in this age range. Patients on entry to the study had a mean systolic blood pressure of 170/77 mmHg with a 1- and 3-min orthostatic fall of 5 and 6 mmHg, respectively. Mean systolic blood pressure was lowered to 144 mmHg by treatment (vs. 155 mmHg in the placebo group), with an impressive 36% reduction in stroke incidence and 27% reduction in major cardiovascular events; 13% of treated patients withdrew because of adverse effects. Reviewing those adverse effects that could be releated to orthostatic hypotension, the prevalence of faintness on standing (12.8 vs. 10.6%), but not feelings of unsteadiness or imbalance (33.7 vs. 32.9%), were increased with treatment, and there was a nonsignificant trend toward an increased incidence of falls during active treatment (12.8 vs. 10.4%).

The STOP study (Swedish Trial in Old Patients with Hypertension) was a prospective randomized study in 1627 hypertensives (BP: 180 to 230, 90 to 120 mmHg), aged 70 to 84 years, comparing three β-adrenergic antagonists (atenolol, metoprolol, and pindolol) and one combination diuretic (hydrochlorothiazide and amiloride) against placebo, which demonstrated a significant reduction in stroke morbidity and mortality.[14] Exclusion criteria included orthostatic hypotension with more than a 30 mmHg fall in systolic blood pressure and "other severe or incapacitating illnesses". Active and placebo patient groups had an 8 mmHg orthostatic fall in systolic BP at baseline. Active treatment lowered blood pressure by 29/17 mmHg compared to 2/7 mmHg in the placebo group. Only 7% of patients withdrew from active treatment (compared to 6% of patients receiving placebo) for subjective side effects. No patient reportedly withdrew because of symptomatic orthostatic hypotension.

The Medical Research Council (MRC) trial of treatment of hypertension in older adults studied 4396 hypertensives (BP: 160 to 209, <115 mmHg) aged 65 to 74, in a primary care setting, comparing atenolol and a combination diuretic (hydrochlorothiazide and amiloride) to placebo, with addition of nifedipine in the treatment groups if further therapy was required.[77] Patients in this study were recruited from a screening program, and the results of this study are arguably of more relevance to clinical practice, as the subjects are probably representative of hypertensive patients in primary care subject to the exclusion criteria for the study (treated heart failure or angina, diabetes, asthma, renal impairment, and serum potassium abnormalities). Patients were treated to achieve a target systolic blood pressure of 150 to 160 mmHg. Therapy with atenolol was associated with significantly more withdrawals (30%) for suspected major side effects than diuretic (15%) or placebo (3.7%). Withdrawals for symptoms probably due to orthostatic hypotension (dizziness) constituted a relatively small proportion of withdrawals (which included gout, muscle cramps, dyspnoea, lethargy, headaches, and bradycardia), but were higher in atenolol-treated patients (atenolol 10.1, diuretic 7.4, and placebo 1.2 withdrawals per 1000 patient years).

These results demonstrate that despite the liability of the elderly to develop orthostatic hypotension, this does not appear to be a significant clinical problem in the majority of elderly hypertensives. Initiation of treatment for diastolic and systolic hypertension in the elderly should not be withheld because of excessive concerns about adverse drug effects, but the results of the MRC trial would suggest it is advisable to closely monitor patients for orthostatic hypotension and other adverse effects, although orthostatic hypotension is not a frequent cause for discontinuation of therapy.

III. ORTHOSTATIC HYPOTENSION

The increased prevalence of orthostatic hypotension in the elderly, up to 30% in populations aged over 65 years of age, is largely due to the presence of concomitant disease states such as Parkinson's disease, diabetes, or drug therapy.[15] The prevalence in healthy elderly in the community, not taking drugs, is extremely low.[16] Where rigorous selection of healthy elderly subjects has been undertaken, and investigation of elderly normotensive subjects with normal lung and maximal exercise stress tests have been studied, no difference in blood pressure response to standing or tilting has been found compared to young subjects.[17] The prevalence of orthostatic hypotension increases with higher values of supine blood pressure. Logistic regression analysis of data from the second National Health and Nutrition Examination Survey demonstrated that the level of supine blood pressure was strongly related to orthostatic falls in blood pressure, whereas age was not an independent risk factor for orthostatic hypotension,[18] suggesting that the age-related increase in the prevalence of orthostatic hypotension may be explained in large part by age-related increases in systolic blood pressure.

As discussed in the previous section, the presence of orthostatic hypotension and increased risk to develop falls in blood pressure with circulatory stress may limit the use of certain therapies for hypertension and heart failure in susceptible elderly patients.

Consistent approaches to the measurement of blood pressure are very important in assessing possible orthostatic hypotension. The length of time lying flat before determination of supine blood pressure and the time and method by which standing blood pressure is determined exert a major influence on the recognition of orthostatic hypotension.[19] The maximal fall in standing blood pressure occurs at around 1 min after standing.[20] Using a single blood pressure measurement taken after 5 min of supine rest as baseline, followed by measurement of standing blood pressure in the arm lying relaxed at the side 1 min after the patient has achieved the upright position, is a practicable, sensible standardization to use in clinical practice. Although supine blood pressure readings decline with repeated measurement over 5 min, this does not significantly alter the orthostatic change in blood pressure, and pooling multiple baseline readings is probably not appropriate.[21] A decrease of greater than 20 mmHg systolic or 10 mmHg diastolic blood pressure is a reasonable criterion for postural hypotension. However, these criteria should be interpreted in the light of resting supine blood pressure because, as previously noted, subjects with higher supine blood pressure tend to have a larger postural fall in standing blood pressure. The interindividual variability in postural blood pressure measurements is extremely large, with many elderly subjects showing an increase in blood pressure on standing.[22] Large intraindividual variability in postural blood pressure change also exists, and should be taken into account during research and clinical practice. The day-to-day variability in postural systolic blood pressure change was studied in institutionalized elderly subjects and was found to be very large, primarily related to changes in basal supine blood pressure.

It has been suggested that elderly patients with symptoms of cerebral hypoperfusion without a significant change in blood pressure should be treated in the same way as patients with postural hypotension.[19] However, the attribution of cerebral hypoperfusion as a cause of postural symptoms is not straightforward, as other causes of standing dizziness such as vestibular or multiinfarct disease may produce similar subjective symptoms in patients. The use of cerebral blood flow imaging techniques may help define this entity more clearly. A number of techniques have been utilized to study cerebral blood flow, including nitrous oxide inhalation, xenon-133 inhalation, stable xenon contrast computerized tomography, and positron emission tomography. Such studies have demonstrated a decline in cerebral blood flow with increasing age, which may make the elderly more susceptible to decreases in cerebral perfusion.[23,24]

The mechanisms of orthostatic hypotension in the elderly remain uncertain, but are of critical importance if therapy is to be rationally designed to correct this problem. The complex cardiovascular adaptations that take place in response to standing make many mechanisms potentially possible. Continuing controversy exists about whether autonomic nervous system dysfunction or other factors such as reduced arterial wall compliance are responsible for age-related orthostatic hypotension.[25-29] Interpretation of published studies is difficult because it is very likely that differing populations of elderly patients with orthostatic hypotension were studied by various groups.

It is worth differentiating idiopathic (age-related) orthostatic hypotension from that associated with other forms of autonomic insufficiency. Elderly patients with idiopathic orthostatic hypotension can be differentiated from elderly subjects with a generalized autonomic neuropathy secondary to diseases such as diabetes or malignancy by the use of autonomic function tests. In idiopathic orthostatic hypotension, other autonomic function tests are normal for age, whereas this is not the case for orthostatic hypotension related to an underlying autonomic neuropathy where extensive sympathetic damage has occurred. Recent studies examining cardiovascular and neurohumoral responses to tilting in age-related orthostatic hypotension suggest that afferent baroreflex dysfunction, possibly due to the splinting of aortic and carotid baroreceptors by noncompliant arterial walls, may be the primary problem.[30]

It is important that age-related values for autonomic function tests are used. For example, healthy aging is associated with a marked alteration in heart rate variation at rest, and in response to standing, deep inspiration, and the Valsalva maneuver.[31,32] Autonomic function testing of heart rate variation is most easily performed using an instantaneous cardiac ratemeter, but can be accurately performed using a 50 mm/s single-lead electrocardiogram with

TABLE 1
Normal Heart Rate Response in Normal Subjects 20 to 75 Years of Age

	Age (years)											
Maneuver	**20**	**25**	**30**	**35**	**40**	**45**	**50**	**55**	**60**	**65**	**70**	**75**
Resting	7	7	6	6	5	5	4	4	4	4	3	3
Inspiration	11	10	9	8	7	6	5	5	4	3	3	2
Valsalva	22	20	19	17	16	15	13	12	11	9	9	8
Standing	15	13	12	11	9	8	7	6	5	5	4	3

Note: Heart rate response reported in beats per minute; lower 95% confidence limits for difference between maximum and minimum heart rate. Heart rate was recorded for 60 s at rest and following standing, for 10 s during a single deep inspiration, and for 45 s following a 15-s Valsalva maneuver.

From O'Brien, I. A. D., O'Hare, P., and Corral, R. J. M., *Br. Heart J.*, 55, 348, 1986. With permission.

R-R interval measurement. Heart rate variability following these maneuvers can be expressed as the heart rate difference (maximum − minimum) or a heart rate ratio (maximum/minimum). The results for heart rate difference (lower 95% confidence interval) in a wide age range (18 to 85 years) of 310 healthy volunteers from the studies of O'Brien et al.[31] are shown in Table 1. Heart rate was measured continuously during four procedures, as follows: (1) quiet supine rest for 60 s, (2) during a single deep breath (the subject is asked to inhale deeply over 5 s and then to exhale over 5 s), (3) following a Valsalva maneuver (15 s of forced expiration into a mouthpiece connected to a sphygmomanometer with a maintained pressure of 40 mmHg; heart rate is recorded from the beginning of the maneuver and continued for a further 45 s following the maneuver), and (4) standing for 60 s with heart rate recording beginning immediately after the subject is standing fully erect.

Cardiovascular responses to isometric exercise are particularly useful as a test of efferent sympathetic function, as this autonomic response is unaltered with healthy aging.[33-35] An increase in diastolic blood pressure of less than 10 mmHg during 5 min of 30% of maximum voluntary contraction of the dominant hand using a strain gauge dynamometer is indicative of efferent sympathetic damage. In clinical practice, a reasonable approach to measuring autonomic function would be to utilize measurements of heart rate variability using the methods and lower normal values established by O'Brien et al.,[31] with measurement of supine and 1-min standing blood pressure following 10 min of rest, and blood pressure response to 5 min of isometric hand grip. If a strain gauge dynamometer is unavailable, a rolled up sphygmomanometer cuff held in the hand can be used as a less accurate alternative.

Appropriate therapy for age-related orthostatic hypotension remains unclear. In general, this problem appears to be relatively mild compared to the problems experienced by elderly patients wih autonomic neuropathy. Asymptomatic elderly patients with orthostatic hypotension who require treatment with agents such as tricyclic antidepressants, calcium channel blockers, and nitrates should be monitored carefully during treatment. This is particularly relevant in hypertensives, who have a higher incidence of orthostatic hypotension. Whether certain therapies such as calcium channel blockers or angiotensin-converting enzyme inhibitors are better tolerated by elderly hypertensives and produce fewer problems than thiazide diuretics and β-adrenergic antagonists remains unclear.

For treatment of symptomatic orthostatic hypotension, the withdrawal of any drugs that may worsen orthostatic hypotension is important. Correction of hypovolemia, anemia, and hypokalemia is useful.[36] Advice on avoiding sudden changes in posture is also likely to be helpful. Elastic stockings are frequently prescribed, although they are difficult to use and there are no controlled studies demonstrating benefit in elderly patients with idiopathic orthostatic hypotension. Head-up tilt of the patient's bed at night improves orthostatic hypotension by stimulation of renin release, leading to increased sodium retention at night.

Most authors recommend fludrocortisone, commencing with a dose of 0.1 mg daily and increasing to up to 0.3 mg daily. Supine hypertension may be a problem in these patients; head-up tilt at night is very important in patients receiving mineralocorticoids. Further benefit may be obtained with the addition of indomethacin, 25 mg three times daily. This is probably the most useful therapeutic approach in the majority of elderly subjects after the previous nondrug approaches have been used. However, these recommendations are based on studies in younger patients with autonomic failure, who may well have a different underlying pathophysiology. Fludrocortisone increases plasma volume by sodium retention; it also increases vascular α-adrenergic receptor sensitivity to noradrenaline.[37] Controlled studies in well-defined groups of symptomatic elderly patients are required, with particular attention to measurements of changes in cardiac output and peripheral resistance induced by treatment. Inhaled ergotamine overcomes the problem of poor bioavailability of oral preparations and the practical difficulties of subcutaneous and intramuscular administration. Inhaled ergotamine has been shown to be effective in patients with severe autonomic failure, and merits further study.[38] However, the use of ergotamine and other pressor agents, such as α-adrenergic receptor agonists, in the elderly may be contraindicated because of coexisting ischemic heart disease which is commonly found in the elderly.

IV. POSTPRANDIAL HYPOTENSION

In healthy young individuals, a large increase in splanchnic blood flow occurs after food ingestion.[39] Blood pressure is maintained by an increase in heart rate and stroke volume combined with a reduction in blood flow to skeletal muscle.[40] Plasma noradrenaline concentrations increase and microneurography studies of skeletal muscle confirm an increase in sympathetic nervous system activity.[41]

Patients with autonomic failure have been found to have substantial falls (around 50 mmHg) in supine and standing blood pressure after meals.[42] Postprandial hypotension was subsequently recognized as a common hemodynamic finding in the elderly, and also in hypertensive subjects.[43] It was initially recognized in frail institutionalized elderly patients, but further work has established its occurrence in healthy elderly subjects.[44-46] Whether postprandial hypotension is a common cause of falls or syncope in the elderly is unclear. In the study by Lipsitz et al.,[44] institutionalized elderly, both fallers and nonfallers, had mean falls in blood pressure postprandially of about 25 mgHg. The maximal fall in blood pressure occurs about 1 h after food ingestion. One study found that postprandial hypotension was greater in fallers compared to nonfallers.[47] Hypertensive elderly subjects demonstrate larger postprandial falls in blood pressure than normotensive subjects.[48] Surprisingly, postural stress and antihypertensive treatment do not seem to have an additive effect on postprandial hypotension.[48] High carbohydrate meals produce a larger postprandial fall in blood pressure compared to high protein or high fat meals.[49] The mechanisms responsible for postprandial hypotension have not been fully delineated, but inability to compensate for increased splanchnic blood flow mediated by vasodilatory effects of gut hormones and possibly insulin probably play a role.[50]

Indomethacin and octreotide attenuate the fall in postprandial blood pressure when administered to subjects with autonomic failure.[42,51] Octreotide also prevents the postprandial fall in healthy elderly subjects.[52] Octreotide abolishes food-induced hypotension in autonomic failure patients without changing cardiac output or peripheral blood flow and abolishes the food-induced increase in superior mesenteric blood flow in healthy subjects.[39,53] This suggests that octreotide acts on the splanchnic circulation. This effect is probably mediated through inhibition of the release of insulin, neurotensin, and other gastrointestinal hormones. Octreotide is an expensive and relatively impractical form of therapy because of the necessity for subcutaneous administration. Caffeine has also been investigated as a therapy, with variable results. Administration at the end of, but not before, the meal appears to be effective.[54,55] In patients with falls or dizziness who have identified postprandial hypotension, caffeine taken at the end of the meal is a simple, palatable therapy.

The phenomenon of postprandial hypotension influences decisions about routine treatment of hypertension in the elderly. Unless 24-h ambulatory blood pressure recordings are utilized, meal-induced changes in blood pressure may modify blood pressure readings obtained in the clinic. Consequently, it may be important that recordings be standardized to be taken either before or at least 3 h after meals in elderly patients. However, most of the large multicenter studies which have demonstrated benefits of therapy for hypertension in the elderly, did not standardize or control for the taking of blood pressure readings in relation to meals. Certainly it would seem sensible that postprandial hypotension be taken into account in the design of hemodynamic studies in the elderly. As for orthostatic hypotension, the clinical significance of postprandial hypotension remains uncertain. It is tempting to speculate that it could be a precipitating factor in ischemic stroke and myocardial infarction. Anecdotal case reports of stroke and myocardial infarction following precipitous lowering of blood pressure in elderly hypertensives, and the more recent J-shaped curve association of survival with treatment of diastolic blood pressure readings in hypertension studies, have lent support to this possibility.[56,57] The J-shaped curve describes the relationship of mortality to blood pressure following hypotensive therapy in hypertension, where an increase in mortality has been observed in hypertensives with the lowest treatment blood pressures (<90 mmHg) compared to those with slightly higher treatment blood pressures. Interpretation of this finding is controversial, since the lower blood pressure may be a manifestation of other concurrent disease. Alterations in the threshold at which cerebral blood flow is maintained by cerebral autoregulation occur in hypertension.[58] Such changes, in combination with cerebrovascular disease, could render the elderly more susceptible to cerebral hypoperfusion when systemic blood pressure is decreased excessively.[59]

V. β-ADRENERGIC RESPONSIVENESS

Reduced β-adrenergic responsiveness has been found with increasing age in a number of organs, most notably the heart and vasculature.[60-63] The age-related reduction in cardiac chronotropic β-adrenergic responsiveness is likely to account for the reduction in maximal heart rate achieved during exercise in elderly subjects.[64-66] The average maximal heart rate (beats per minute) achieved by healthy elderly individuals is approximately 220 − age (in years). Increased physical fitness is associated with a further reduction in maximal heart rate in elderly subjects, which may reflect desensitization of cardiac β-adrenergic responses, from regular activation of the sympathetic nervous system.[67] It has been demonstrated that, at least in rigorously screened healthy elderly individuals, this age-related reduction in maximal heart rate does not impair maximal cardiac output during exercise, as the elderly compensate by generating a greater increase in stroke volume at lower heart rates.[68]

The reduced response of elderly tissues to β-adrenergic agonists has led to the expectation that the elderly would respond less well to β-adrenergic receptor antagonists for the treatment of hypertension and angina. Application of receptor theory would not necessarily predict this to be the case, since the response of a tissue to an antagonist is dependent on the degree of agonist stimulation, and receptor binding affinity to an antagonist need not be altered even if the efficacy of agonists in activating β-adrenergic responses were decreased in an elderly subject. In the preliminary studies by Vestal et al.,[63] a reduced effect of propranolol on isoprenaline cardiac chronotropic respon-

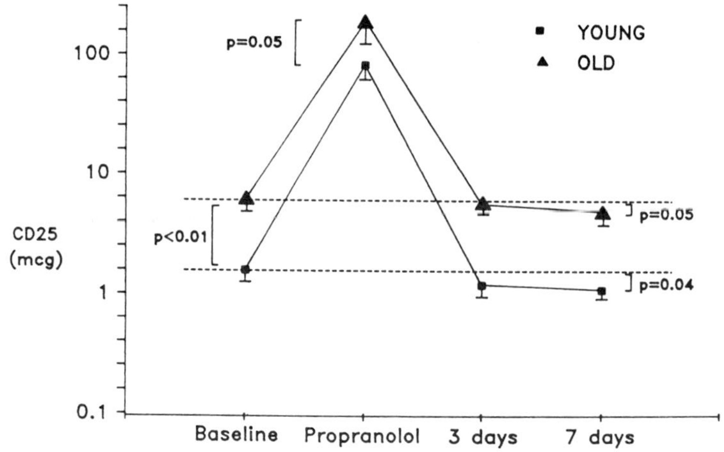

FIGURE 2. Cardiac β-adrenergic chronotropic sensitivity in eight young and nine elderly healthy subjects at baseline after 2 weeks of therapy with propranolol and at 3 and 7 d following withdrawal of propranolol.[71]

siveness was observed. This difference does not appear to be due to an age-related alteration in the pharmacokinetic handling of the two isomers of propranolol.[69] In contrast, similar studies using timolol found no change in the calculated binding affinity of cardiac β-adrenergic receptors for timolol.[70] It is possible that a change in the relative ratio of β_1- and β_2-cardiac receptors could account for the change in the calculated Kd seen with propranolol.

The withdrawal hypersensitivity that occurs following β-blocker withdrawal appears to be similar in young and elderly individuals.[71] The increased sensitivity of the heart to β-adrenergic agonists following β-adrenergic antagonist withdrawal is thought to be due, at least in part, to the increase in β-adrenergic receptor density that occurs during β-blocker therapy, although additional postreceptor changes could also contribute to this phenomenon. This increased sensitivity can last for 10 to 14 d, during which time receptor density returns to pretreatment values. Anecdotal cases of worsening angina and myocardial infarction following rapid β-receptor antagonist withdrawal have been reported.[72-74] Studies of changes in lymphocyte adrenergic receptor density show a similar increase in density in young and elderly subjects when treated with an antagonist.[75] These changes in lymphocyte receptor density mirror the response of cardiac tissue. Studies of cardiac β-adrenergic sensitivity, by measurement of cardiac chronotropic sensitivity to isoprenaline following propranolol withdrawal, demonstrate a similar increase in sensitivity in young and elderly subjects (Figure 2).[71] These data also argue against the age-related reduction in cardiac β-adrenergic responsiveness being due to desensitization from increased sympathetic nervous system activity, as this hypothesis would predict that a greater increase in cardiac β-adrenergic sensitivity would occur following β-receptor antagonist withdrawal. It would be prudent to reduce β-receptor antagonist dosage gradually in elderly patients with cardiac disease, although in practice, complications of rapid withdrawal of β-adrenergic antagonists are probably uncommon. Slow withdrawal of β-receptor antagonists with long plasma half-lives or intrinsic sympathomimetic activity is probably not necessary.

There have been suggestions that hypertensive elderly subjects are less responsive than their younger counterparts to β-adrenergic receptor antagonists in hypertension.[76] Because there are inherent difficulties in standardizing two different age groups of hypertensive patients and the mechanism(s) by which β-receptor antagonists lower blood pressure is uncertain, this is a difficult hypothesis to test. β-Adrenergic antagonists have been used as initial or secondary treatment in the larger controlled studies of treatment of both systolic and diastolic hypertension in the elderly, with good efficacy and impressive reductions in rates of stroke.[13,14] However, the MRC study of the treatment of hypertension in older adults failed to show a reduction in stroke and coronary events in the atenolol-treated group compared to the placebo group.[77] This was despite achieving a reduction in blood pressure similar to that of the diuretic-treated group who did demonstrate a benefit. The β-receptor antagonist-treated group required a greater use of nifedipine as an additional hypotensive agent, and there was a higher withdrawal rate because of adverse effects, compared to the diuretic group. Taking the results of these large trials together, it would seem that although β-adrenergic antagonists are reasonably effective at lowering blood pressure to target treatment levels in elderly hypertensive patients, there is probably an increased incidence of adverse effects compared to diuretic therapy, and some evidence of less efficacy in terms of stroke prevention. There may be some reduction in the sensitivity of the elderly to β-receptor antagonists, but more rigorous studies in selected groups of hypertensive patients with determination of plasma drug concentrations are required to address this question. The findings of the MRC study suggest that surrogate endpoints (simply measuring blood pressure) may be an insufficient basis on which to choose optimal therapies for hypertension in the elderly.

The use of β-adrenergic antagonists in the treatment of angina and post-myocardial infarction secondary prevention has not been specifically studied in the elderly. This issue is complicated because the perception of angina may be diminished with increasing age.[78] This may be related to an age-related reduction in the sensory innervation of the heart or could reflect a more generalized, possibly central, reduction in sensory function.[79] Such changes could account for the increased frequency of asymptomatic or so-called "silent" ischemia in the elderly.[80] Whether using β-receptor antagonists with intrinsic sympathomimetic activity, such as pindolol, oxprenolol, and acebutalol, is advantageous in the elderly is unclear. Theoretically, the lower reduction in resting heart rate and cardiac output that occurs with intrinsic sympathomimetic activity may be preferable in elderly patients at risk of fatigue from a reduced cardiac output. These theoretical advantages must be set against the observation that β-receptor antagonists with intrinsic sympathomimetic activity are ineffective in the secondary prevention of death from postmyocardial infarction.[81] Comparisons of the therapeutic effects of differing β-receptor antagonists in elderly patients with angina have not been performed.

Although alprenolol was reported to be less effective in older patients in reducing reinfarction following myocardial infarction, the larger Norwegian timolol study demonstrated that the benefits of β-receptor antagonist therapy in reducing reinfarction and mortality following myocardial infarction were not reduced in older patients (65 to 75 years of age); consequently, consideration should be given to commencing β-receptor antagonist therapy in elderly patients following myocardial infarction.[82-84] However, many elderly patients fall into the high-risk group of patients with reduced left ventricular function, who would otherwise benefit the most from β-antagonist therapy following myocardial infarction, but are unable to tolerate these drugs because of moderate or severe heart failure, bradycardia, hypotension, airway obstruction, or greater than first-degree heart block. The benefits of early therapy with intravenous atenolol in acute myocardial infarction were not reduced with increasing age.[85]

Whether pulmonary β-adrenergic responses are reduced with aging in humans has not been conclusively established, partly because of the methodological difficulties in standardizing the study of pulmonary bronchodilator responses between different age groups. Because of the widespread use of β-adrenergic receptor agonists in elderly patients with obstructive pulmonary disease, this is an important issue that merits further attention since inhaled β-receptor agonists have become a mainstay of treatment for asthma and chronic obstructive airway disease. Airway reactivity to methacholine increases with age[86,87] and has been shown in some studies to be negatively associated with β-adrenergic sensitivity.[88] However, when the relationship between aging, lymphocyte responsiveness to isoprenaline, and airway reactivity was specifically studied in healthy adults, no reduction in airway reactivity or lymphocyte responses to isoprenaline was found, and no correlation existed between airway reactivity and β-adrenergic lymphocyte responsiveness.[89] More recent work has demonstrated a negative correlation between airway reactivity and monocyte high-affinity β-adrenergic receptor density, but no relationship with total monocyte β-adrenergic receptor density.[90] Although a change in high-affinity β-receptor density need not be associated with a reduced cAMP or tissue response, these findings are consistent with the hypothesis that a reduction in the proportion of high-affinity pulmonary β-receptors would predispose the elderly to asthma. Further studies of pulmonary responses to β-receptor agonists and nonadrenergic bronchodilators in older asthmatics are required to determine the most effective approach to therapy in these patients.

It might be expected that the elderly would show less systemic adverse effects with inhaled β-receptors agonists, such as tremor and tachycardia, as chronotropic and finger tremor responses to infused β-receptor agonists are reduced with aging. However, one study found that hypokalemic and chronotropic responses to inhaled salbutamol were similar in young and elderly healthy volunteers.[91] One uncontrolled study of cardiac responses to inhaled β-receptor agonist therapy in elderly patients with severe chronic airway obstruction found cardiac arrhythmias in 5 of 19 patients during a 24-h period.[92] The safety of currently prescribed β-receptor agonist therapies in elderly patients with respiratory disease is unclear at this time. In clinical practice, attention to inhaler technique and use of appropriate delivery systems is the most important aspect of bronchodilator use. Problems of limited lung function, inability to hold breath, arthritis, poor hand-breath coordination, and cognitive impairment make efficient inhaler use by certain elderly patients very difficult.[93] Studies suggest that only half of the elderly patients attending an outpatient clinic use meter-dose pressurized inhalers correctly.[94,95] After teaching sessions, one third of patients remained unable to use pressurized inhalers correctly.[95] The development of alternative dry powder, automated inhalers, and spacer devices allow an effective method for delivering inhaled bronchodilators to be found for most elderly patients.[94] Although inhaler therapy achieves higher drug concentrations at the desired site of action, the systemic effects on the heart and other organs that clearly occur with currently prescribed β-agonists should be taken into account. Many elderly patients with obstructive airway disease have coexisting cardiac disease, and the resulting β-adrenergic-mediated increases in myocardial oxygen consumption may be deleterious.

Aging alterations in gut sympathetic nervous system activity and β-adrenergic responsiveness have not been studied in detail, but could influence gut motility in the elderly.

It has been suggested that aging selectively affects β_1-adrenergic responses, leaving β_2-adrenergic responsiveness relatively unchanged.[96,97] This may occur within the aging heart, and it is possible that certain tissues do not demonstrate a reduced β-adrenergic responsiveness with increasing age. However, these results cannot be simply explained by aging selectively affecting β_1-adrenergic responses, since an age-related reduction in β-adrenergic responsiveness has clearly been demonstrated in some vascular beds and lymphocytes.[60,62,98]

VI. α-ADRENERGIC RESPONSIVENESS

As discussed in Chapter 6 of this volume, the majority of evidence suggests that there is little change in cardiac and vascular α-adrenergic responsiveness with aging. Studies have not established whether there is any aging alteration in the effectiveness of inotropic support, such as dopamine or dobutamine, through cardiac α-adrenergic receptor stimulation. Dobutamine has complex cardiovascular effects due to the effects of the racemic mixture on α- and β-adrenergic receptors. Because of the reduced β-adrenergic responsiveness of the aged heart, it could be expected that higher plasma concentrations of dobutamine might be required in the elderly or that more marked effects on the peripheral vasculature via α-adrenergic receptors might be apparent. The finding that the positive inotropic effect of dobutamine is reduced in severe heart failure where cardiac β-adrenergic responsiveness is diminished would support this concept, but specific studies have not been undertaken in the elderly.[99]

α-Adrenergic receptor antagonists may induce orthostatic hypotension, which may be a limiting factor in their use for hypertension and prostatic outflow obstruction. In view of the epidemiological association between falls and orthostatic reductions of greater than 20 mmHg in systolic blood pressure, it would seem advisable to discontinue or reduce any therapy that produces orthostatic changes greater than this.[100] It is not known whether the elderly are more susceptible to the first-dose hypotensive effect of α-adrenergic antagonists or whether there is any aging alteration in the response of the elderly hypertrophied prostate to α-adrenergic antagonists. If prostatic α-receptors are a different subclass than those on vascular smooth muscle, it is possible that selective α-adrenergic receptor antagonists could be developed which would have less adverse effects on blood pressure.

VII. CHOLINERGIC RESPONSIVENESS

Aging changes in the cholinergic system have been less well studied compared to the adrenergic system. Gastrointestinal and cardiovascular function may be affected by aging changes in cholinergic function. Aging changes in gut cholinergic function are reviewed fully in Chapter 10 of this volume. Gastric emptying is dependent on vagal stimulation; the rate of gastric emptying may modify drug absorption. Results of the effect of aging on gastric emptying have been variable. Some studies have found slower emptying of solids and liquids in elderly subjects when assessed by isotopic or ultrasound techniques.[101,102] Other studies have found that the initial rate of gastric emptying, in the first 5 min, is greater in healthy elderly subjects, but that subsequent gastric emptying rates are similar.[103,104] The inhibitory effects of atropine on the early adaptive phase of gastric emptying were found to be greater in a small study of young and elderly subjects, suggesting that resting gastric vagal tone may be increased in the elderly.[104] This same study also found that atropine had a greater effect on reducing salivary flow in elderly subjects. Although gastric emptying influences the absorption of some orally administered drugs, there do not appear to be any clinically significant aging changes in the absorption of passively absorbed drugs, although factors such as small bowel mucosal surface area and splanchnic blood flow also influence drug absorption.

Achalasia is an uncommon disorder of the esophagus characterized by failure of the lower esophageal sphincter to relax after swallowing, which may be secondary to abnormal vagal function at the lower esophageal sphincter as many patients have abnormal gastric acid responses to insulin stress testing.[105] The prevalence of achalasia increases with age, and age-related changes in gastrointestinal vagal function may be a factor in the pathogenesis of this disorder.[106] Studies of other vagal responses in elderly patients with achalasia would be of interest.

Pseudoobstruction (large-bowel dilation in the absence of obstruction) is a complication of a number of medical disorders such as hypothyroidism and uremia; it may also occur following tricyclic antidepressant overdosage.[107] Pseudoobstruction occurs almost exclusively in the elderly, except as a rare occurrence following tricyclic antidepressant overdosage in younger individuals. The high prevalence of constipation in elderly subjects also suggests that there may be clinically important alterations in human gut motility in old age in addition to the effects of poor diet and immobility.[108] In animal models, an age-related decrease in the number of neurones in the myenteric plexus has been found, which could contribute to these problems.[109,110] Data on humans are lacking, and studies of the effect of aging on small bowel and colonic cholinergic function would be of interest. Cisapride might be a potentially useful therapy for pseudoobstruction or refractory constipation, as it increases acetylcholine release at the myenteric plexus, although this has not been specifically investigated in the elderly.

The vagus has a critical role in modulating cardiac function. Resting heart rate is strongly influenced by the slowing effect of vagal tone. Tachycardia responses induced by the cholinergic antagonist atropine are diminished with increasing age. However, this could be secondary to the diminished β-adrenergic responsiveness of the aging heart rather than simply reflecting differences in vagal tone.[111,112] Heart rate variability is dependent on beat-to-beat changes in sympathetic and parasympathetic activity. Heart rate variability during the respiratory cycle decreases with increasing age, suggesting that vagal influences on the heart demonstrate less beat-to-beat variability.[31,32] Power spectral analysis of heart rate variation, which allows separate analysis of the parasympathetic and sympathetic influences, demonstrates that both parasympathetic and sympathetic activity show less variability with increasing age.[113] As previously discussed, heart rate responses to a Valsalva maneuver are similarly decreased in elderly subjects. Heart rate responses to methacholine, a muscarinic agonist, are diminished in aging rats, suggesting that reduced cholinergic responsiveness of cardiac tissue may be responsible for the diminution in vagal modulation observed with aging.[114] The clinical significance of these reduced vagal influences is unclear.

VIII. SYNCOPE

Syncope is a common problem in the elderly, and may be an unrecognized cause of falls due to the associated amnesia for the period of loss of consciousness that occurs.[115,116] Syncope can be classified as due to cardiovascular and noncardiovascular causes. Cardiovascular syncope can be further divided into reflex and cardiac causes.[117]

A retrospective analysis of syncope in a large group of very elderly institutionalized patients revealed a 10-year prevalence of 23% and 1-year incidence of 7%.[118] Prospective follow-up of the same population found an annual incidence of 6% and recurrence rate of 30%. Investigation of these subjects found a noncardiovascular cause in half the subjects and a cardiovascular cause in 20% of subjects, with the remaining 30% of cases unexplained. Although patients who developed syncope were more functionally disabled than those without syncope, no difference in mortality or hospitalization rates were found between the two groups, suggesting that in this group of severely functionally impaired patients, syncope is a manifestation of comorbid disease, and does not result in increased mortality.

Aging changes in autonomic function are relevant to reflex cardiovascular syncope in the elderly, which is discussed here. Taken in isolation, decreased β-adrenergic, but maintained α-adrenergic vascular responsiveness in the elderly should tend to protect against orthostatic hypotension and vasovagal syncope, as a state of peripheral vasoconstriction might be expected to predominate. In addition, a reduced cardiac β-adrenergic responsiveness should result in less marked cardiac contractility without activation of the Bezhold-Jarisch reflex. In practice, reflex cardiovascular syncope is actually commonly seen in geriatric practice.

Studies indicate that true vasovagal syncope is uncommon in the healthy elderly, suggesting that age-related changes such as these do protect against syncope.[118,119] Because of the rarity of vasovagal episodes in healthy elderly subjects compared to its occurrence in significant numbers of otherwise healthy young adults, the specificity of bradycardic or hypotensive responses to tilt testing may make tilt testing a useful investigation in elderly patients with syncope.[5,120,121] In tilt testing, blood pressure and heart rate are monitored in patients during passive tilting to 70° upright, for up to 45 min. A symptomatic fall in blood pressure is considered a positive response, representing vasovagal syncope which may be the cause of syncopal episodes. It should be noted that instrumentation such as arterial or venous cannulation substantially increases the likelihood of fainting in subjects. Isoprenaline infusions have been added to tilt testing as a further provocative test.[122] Given the known changes in vascular and cardiac responsiveness to β-adrenergic stimuli, it is somewhat surprising that investigators using this isoprenaline provocation have not studied the effect of age on the response to isoprenaline in tilt testing.[123] The provocative effects of isoprenaline could be due to peripheral vasodilation leading to reductions in preload and afterload or to an increase in cardiac contractility with activation of ventricular mechanoreceptors. If the former mechanism is responsible, the use of a nonadrenergic vasodilator such as nitroglycerin might be more appropriate. The latter mechanism results in activation of the Bezold-Jarisch reflex, where vigorous contraction of a small ventricle stimulates intracardiac vagal mechanoreceptors which input into the brainstem, leading to activation of vagal tone with bradycardia and withdrawal of sympathetic tone, leading to peripheral vasodilation.[115,124] If the latter mechanism is operating, infusing isoprenaline to achieve an age-defined tachycardia might make the test more specific and sensitive. One controlled study of the use of isoprenaline during upright tilt testing for the investigation of unexplained syncope in a group of young patients found a positive response to tilting with isoprenaline in 75% of patients compared to 45 or 65% (depending on the protocol used) of nonsyncopal control subjects.[125] Such a low specificity suggests that isoprenaline provocation during tilt testing is unlikely to be clinically useful in the investigation of syncopal younger subjects. Further studies are needed to clarify the utility of such testing in the investigation of elderly patients with syncope.

In contrast to the problem of reduced baroreflex responsiveness found in healthy elderly subjects, there has been an increased recognition of the problem of carotid sinus hypersensitivity in some elderly patients.[126,127] This increased sensitivity can lead to hypotension through either a cardioinhibitory response, leading to vagally mediated bradycardia, or a vasodepressor response, leading to hypotension through peripheral vasodilation secondary to withdrawal of sympathetic activity. Symptoms due to cardioinhibitory carotid sinus syndrome respond well to pacemaker implantation. Because of the increased dependence of cardiac output on atrial contraction in the elderly, due to loss in ventricular compliance, and the occurrence of retrograde atrial activation with ventricular pacing, a dual-chamber device should be used to avoid symptoms from the so-called "pacemaker syndrome".[128-130] Because of the loss of atrioventricular synchrony, symptoms of dyspnoea, dizziness, palpitations, and neck pulsations may result in this syndrome.[131]

The pathophysiology of carotid sinus syndrome has not been established. Alterations in central modulation of the baroreflex, possibly related to cerebral ischemia, seem most likely, as these patients have a high incidence of hypertension, atrial fibrillation, and ischemic heart disease. Arginine vasopressin responses during tilting are similar in carotid sinus syndrome patients compared to age-matched controls. This suggests that changes in the carotid sinus itself are not responsible, as an increased sensitivity of the carotid sinus itself to blood pressure changes might be expected to result in a greater release of arginine vasopressin, which is controlled by afferent glossopharyngeal and vagal inputs into the brainstem.[132] However, alterations in cholinergic or adrenergic responsiveness of the heart and vasculature might also play a role in the pathogenesis. Like orthostatic hypotension, the contribution of this phenomenon to falls, syncope, and symptoms of dizziness in the elderly is unclear. In one series, one fourth of elderly patients with dizziness, syncope, or falls had carotid sinus syndrome.[126] Symptoms may be precipitated by head movement or vagal stimuli. The attribution of symptoms to carotid sinus hypersensitivity is not straightforward, as up to one third of asymptomatic elderly men with coronary artery disease or hypertension demonstrate carotid sinus hypersensitivity.[133,134] Drug therapy with cardioinhibitory effects, such as digoxin, β-receptor antagonists, and L-type calcium channel antagonists, can exacerbate or produce carotid sinus syndrome.

IX. INTERACTION OF OTHER DRUGS WITH AUTONOMIC NERVOUS SYSTEM

A. TRICYCLIC ANTIDEPRESSANTS

Many studies have demonstrated the considerable efficacy of tricyclic antidepressant drugs in treating depression in the elderly. Unfortunately, traditional antidepressant drugs, such as amitryptyline, are poorly tolerated by many elderly individuals. The anticholinergic effects due to antagonism of peripheral muscarinic cholinergic receptors lead to blurred vision, dry mouth, constipation, and urinary retention at usual therapeutic doses in most individuals. α-Adrenergic antagonist effects and noradrenaline reuptake inhibition effects contribute to postural hypotension and sinus tachycardia, respectively, in some patients. Elderly individuals with coexisting cardiac disease are more likely to develop symptomatic orthostatic hypotension and cardiac conduction problems. A prospective nonrandomized study of imipramine (mean dose, 90 mg daily) and nortryptyline (mean dose, 250 mg daily) in depressed patients found the incidence of severe postural hypotension (>50 mmHg fall in systolic blood pressure or inability to stand) to be 7% with imipramine and 0% with nortryptyline in a group of patients (mean age, 56 years) with normal electrocardiograms.[135] In contrast, the incidence in an older group (mean age, 67 years) with cardiac failure or conduction disease was 32% with imipramine and 5% with nortryptyline. Tricyclic antidepressants with less marked anticholinergic and cardiovascular effects, such as desipramine and nortryptyline, are preferable. There is a concern that these drugs may promote ventricular arrhythmias in some patients. Elderly patients with preexisting heart block or bundle branch block should receive tricyclic antidepressants with particular caution. For these reasons, the newer antidepressants such as the serotonin reuptake inhibitors and atypical antidepressants such as trazadone and maprotiline have become increasingly popular agents to use for the treatment of depression. However, controlled studies of efficacy in elderly patients for these newer therapies is frequently lacking, and it would be unwise to presume that the risk/benefit ratio for these newer agents is better until more extensive controlled studies are published. A tricyclic antidepressant with fewer anticholinergic adverse effects and relative safety in overdose is probably the preferable first-line therapy to choose for elderly depressed patients without heart block or significant heart failure. Therapeutic drug monitoring may be helpful with imipramine and nortryptyline, where a therapeutic window has been described,[136] although it is likely that the newer agents with an improved margin between efficacy and side effects will make such monitoring unnecessary for most patients.

B. MONOAMINE OXIDASE INHIBITORS

These drugs, once used for the treatment of hypertension because of their effects on the autonomic nervous system, have become increasingly popular alternatives for the treatment of depression. Monoamine oxidase (MAO) is an enzyme existing in two main isoforms, termed A and B, present in brain, peripheral nervous system, and liver that plays an important role in the metabolic inactivation of catecholamines. An advantage of MAO inhibitors is the lack of anticholinergic effects compared to tricyclic antidepressants. The major adverse effect seen with MAO inhibitions is orthostatic hypotension, which is probably due to a central effect of these drugs on endogenous catecholamines in the vasomotor center. A further problem with their use is the interaction with ingested tyramine and other monoamines present in many foods such as mature cheese and red wine. As a result of inhibition of gut and hepatic MAOs, these amines are not degraded following absorption and may produce hypertensive crises by causing release of noradrenaline leading to stimulation of vascular α-adrenergic receptors. For similar reasons, L-DOPA should not be prescribed concurrently with MAOs. MAO inhibitors in combination with tricyclic antidepressants have been used, with many positive uncontrolled reports of efficacy, although controlled evidence of improved efficacy with this combination is lacking. Cases of severe toxicity with hyperpyrexia, convulsions, and coma have been reported. For these reasons, combined use in the elderly should probably be confined to patients with severe depression who can be closely monitored.

C. DOPAMINE

The effect of aging on vascular and cardiac responses to dopamine has not been studied in humans. One study found that elderly rats had reduced mesenteric vasculature vasodilation to dopamine.[137] Dopamine is primarily used at low-dose infusions to selectively maintain renal blood flow via its effects on renal artery dopamine receptors, but it also has important effects on β-adrenergic receptors. At higher concentrations, the effects on α- and β-receptors are more prominent, and a rise in peripheral resistance becomes apparent.

D. L-DOPA

The introduction of L-DOPA was a major therapeutic advance in the treatment of elderly parkinsonian patients. L-DOPA is converted by peripheral decarboxylase to dopamine, which acts on vascular dopamine receptors as well as α- and β-receptors. Although this has not been extensively studied, effects on vascular dopamine receptors appear to dominate, as orthostatic hypotension is the major adverse effect of L-DOPA therapy. Other mechanisms could be responsible for producing orthostatic hypotension during L-DOPA therapy. Dopamine is taken up into peripheral nerve endings and, by displacing noradrenaline, could reduce the effect of neuronally released noradrenaline by a "dilution" mechanism. Effects on central dopamine receptors in vasomotor or other areas could also be important. Cardiac arrhythmias may also occur and are thought to be secondary to stimulation of cardiac β-adrenergic receptors by dopamine. The toxicity of L-DOPA has been substantially diminished by the concurrent administration of a peripheral decarboxylase inhibitor, either carbidopa or benserazide. These inhibitors do not readily cross the blood-brain barrier and enable adequate CNS concentrations of dopamine to be achieved with lessened peripheral dopamine effects due to inhibited decarboxylation of L-DOPA in peripheral tissues.

X. CONCLUSIONS

Although many changes in autonomic function have been identified in aging humans, our present knowledge has not yet resulted in many therapeutic developments for the treatment of elderly patients, and the clinical significance of many findings remains unclear.

A more detailed understanding of basic physiological and pharmacological alterations in the aged autonomic nervous system in humans is required. This is particularly relevant to cholinergic influences of the heart. Indirect clinical data suggest that there are major changes in the aging cardiac cholinergic system which may contribute to many clinical syndromes seen in the elderly. Unlike the adrenergic system, where much has been learned because of the availability of useful pharmacological probes, very little is known about cholinergic innervation, receptor function, and interaction with the sympathetic innervation of the aged heart. The development of specific muscarinic receptor subtype agonists and antagonists may provide useful tools in future studies. Because of the inherent difficulties in studying the autonomic nervous system in elderly humans, animal models are likely to be of continuing importance in developing our understanding of the aged autonomic nervous system.

From a clinical perspective, further information is needed concerning aging alterations in the response of pulmonary and cardiac tissue and the clinical implications of these changes with respect to bronchodilator, hypotensive, and heart failure therapies. The high prevalence of hypertension in the elderly justifies establishing large multicenter studies comparing the efficacy of different hypotensive agents in preventing stroke and myocardial infarction, with careful assessment of quality-of-life measures and the incidence of adverse events. Until prospective studies comparing angiotensin-converting enzyme inhibitors, α-adrenergic antagonists, and L-type calcium channel blockers against thiazide diuretics are performed, the belief that these agents will produce an equivalent reduction in cardiovascular mortality remains speculative.

Further studies of the effect of hypotensive therapies on cerebral flow in the elderly are likely to prove useful in defining appropriate target ranges for blood pressure lowering in different patient groups. Falls and syncope in the elderly remain major problems in geriatric practice. Further work is required to establish the contribution of autonomic disorders to these problems and to develop appropriate therapeutic interventions.

REFERENCES

1. **Shimada, K., Tadao, K., Ogura, H., Sadakane, N., and Ozawa, T.,** Differences in age-dependent effects on blood pressure on baroreflex sensitivity between normal and hypertensive subjects, *Clin. Sci.,* 70, 489, 1986.
2. **Randall, O., Esler, M., Culp, B., Julius, S., and Zweifler, A.,** Determinants of baroreflex sensitivity in man, *J. Lab. Clin. Med.,* 91, 514, 1978.
3. **Gribben, B., Pickering, T. G., Sleight, P., and Peto, R.,** Effect of age and high blood pressure on baroreflex sensitivity in man, *Circ. Res.,* 29, 424, 1971.
4. **Baskett, J. J., Lewis, R. R., Beasley, M. G., and Gosling, R. G.,** Changes in carotid artery compliance with age, *Age Ageing,* 19, 241, 1990.
5. **Hainsworth, R. and Al-Shamma, Y. M. H.,** Cardiovascular responses to upright tilting in healthy subjects, *Clin. Sci.,* 74, 17, 1988.
6. **Johnson, R. H., Smith, A. C., Apalding, J. M. K., and Wollner, L.,** Effect of posture on blood pressure in elderly patients, *Lancet,* 1, 731, 1965.
7. **Schwartz, J. B. and Abernathy, D. R.,** Responses to intravenous and oral diltiazem in elderly and younger patients with systemic hypertension, *Am. J. Cardiol.,* 59, 1111, 1987.
8. **Smith, J. J., Hughes, C. V., Ptacin, M. J., Barney, J. A., Tristani, F. E., and Ebert, T. J.,** The effect of age on hemodynamic response to graded postural stress in normal men, *J. Gerontol.,* 42, 406, 1987.
9. **Minaker, K. L., Meneilly, G. S., Young, J. B., Lansberg, L., Stoff, J. S., Robertson, G. L., and Rowe, J. W.,** Blood pressure, pulse and neurohumoral responses to nitroprusside-induced hypotension in normotensive aging men, *J. Gerontol.,* 46, M151, 1991.
10. **Robertson, D. R. C., Waller, D. G., Renwick, A. G., and George, C. F.,** Age-related changes in the pharmacokinetics and pharmacodynamics of nifedipine, *Br. J. Clin. Pharmacol.,* 25, 297, 1988.
11. **Elliot, H. L., Meredith, P. A., Vincent, J., and Reid, J. L.,** Clinical pharmacological studies with doxazosin, *Br. J. Clin. Pharmacol.,* 21, 27s, 1986.
12. **Holford, N. H. G. and Sheiner, L. B.,** Understanding the dose-effect relationship: clinical application of pharmacokinetic-pharmacodynamic models, *Clin. Pharmacokinet.,* 6, 429, 1981.
13. **SHEP Cooperative Research Group,** Prevention of stroke by antihypertensive drug treatment in older persons with isolated systolic hypertension, *JAMA,* 265, 3255, 1991.
14. **Dahlof, B., Lindholm, L. H., Hansson, L., Scherstein, B., Ekbom, T., and Wester, P. O.,** Morbidity and mortality in the Swedish trial in old patients with hypertension. (STOP-Hypertension), *Lancet,* 338, 1281, 1991.
15. **Caird, F. J., Andrew, G. R., and Kennedy, R. D.,** Effect of posture on blood pressure in the elderly, *Br. Heart J.,* 35, 527, 1973.
16. **Mader, S. L., Josephson, K. R., and Rubenstein, L. Z.,** Low prevalence of postural hypotension among community-dwelling elderly, *JAMA,* 258, 1511, 1987.
17. **Kaijser, L. and Sachs, C.,** Autonomic cardiovascular responses in old age, *Clin. Physiol.,* 5, 347, 1985.
18. **Harris, T., Lipsitz, L. A., Kleinman, J. C., and Cornoni-Huntley, J.,** Postural change in blood pressure associated with age and systolic blood pressure. The National Health and Nutrition Examination Survey II, *J. Gerontol.,* 46, M159, 1991.
19. **Mader, S. L.,** Ageing and postural hypotension: an update, *J. Am. Geriatr. Soc.,* 37, 129, 1989.
20. **Macrae, A. D. and Bulpitt, C. J.,** Assessment of postural hypotension in elderly patients, *Age Ageing,* 18, 110, 1989.
21. **Mader, S. L., Palmer, R. M., and Rubenstein, L. Z.,** Effect of timing and number of baseline blood pressure determinations on postural blood pressure response, *J. Am. Geriatr. Soc.,* 37, 444, 1989.

22. **Lipsitz, L. A., Storch, H. A., Minaker, K. L., and Rowe, J. W.,** Intra-individual variability in postural blood pressure in the elderly, *Clin. Sci.*, 69, 337, 1985.
23. **Hagstadius, S. and Risberg, J.,** Regional cerebral blood flow characteristics and variations with age in resting normal subjects, *Brain Cognit.*, 10, 28, 1989.
24. **Leenders, K. L., Perani, D., Lammertsma, A. A., et al.,** Cerebral blood flow, blood volume and oxygen utilization. Normal values and effect on age, *Brain*, 113, 27, 1990.
25. **MacLennan, W. J., Hall, M. R. P., and Timothy, J. I.,** Postural hypotension in old age: is it a disorder of the nervous system or of blood vessels?, *Age Ageing*, 9, 25, 1980.
26. **Robinson, B. J., Johnson, R. H., Lambie, D. G., and Palmer, K. T.,** Do elderly patients with an excessive fall in blood pressure on standing have evidence of autonomic failure?, *Clin. Sci.*, 64, 587, 1983.
27. **Robinson, B. J., Stowell, L. I., Johnson, R. H., and Palmer, K. T.,** Is orthostatic hypotension in the elderly due to autonomic failure?, *Age Ageing*, 5, 288, 1990.
28. **Smith, S. A. and Fasler, J. J.,** Age-related changes in autonomic function: relationship with postural hypotension, *Age Ageing*, 12, 206, 1983.
29. **White, N. J.,** Heart rate changes on standing in elderly patients with orthostatic hypotension, *Clin. Sci.*, 58, 411, 1980.
30. **Tonkin, A. L., Wing, L. M. H., Morris, M. J., and Kapoor, V.,** Afferent baroreflex dysfunction and age-related orthostatic hypotension, *Clin. Sci.*, 81, 531, 1991.
31. **O'Brien, I. A. D., O'Hare, P., and Corral, R. J. M.,** Heart rate variability in healthy subjects: effect of age and the derivation of normal ranges for tests of autonomic function, *Br. Heart J.*, 55, 348, 1986.
32. **Waddington, J. L., McCulloch, M. J., and Sambrookes, J. E.,** Resting heart rate variability in man declines with age, *Experientia*, 35, 1197, 1979.
33. **Goldstraw, P. W. and Warren, D. J.,** The effect of age on cardiovascular responses to isometric exercise: a test of autonomic function, *Gerontology*, 31, 54, 1985.
34. **Johansson, S. R. and Hjalmarson, A.,** Age and sex differences in cardiovascular reactivity to adrenergic agonists, mental stress and isometric exercise in normal subjects, *Scand. J. Clin. Lab. Invest.*, 48, 183, 1988.
35. **Petrorfsky, J. S. and Lind, A. R.,** Aging, isometric strength and endurance and cardiovascular responses to static effort, *J. Appl. Physiol.*, 38, 91, 1975.
36. **Bradshaw, M. J. and Edwards, R. T. M.,** Postural hypotension — pathophysiology and management, *Q. J. Med.*, 231, 643, 1986.
37. **Davies, B., Bannister, R., Sever, P., and Wilcox, C. S.,** The pressor action of noradrenaline, angiotensin II and saralasin in chronic autonomic failure treated with fludrocortisone, *Br. J. Clin. Pharmacol.*, 8, 253, 1979.
38. **Biaggioni, I., Zygmunt, D., Haile, V., and Robertson, D.,** Pressor effect of inhaled ergotamine in orthostatic hypotension, *Am. J. Cardiol.*, 65, 89, 1990.
39. **Kooner, J. S., Peart, W. S., and Mathias, C. J.,** The peptide release inhibitor, octreotide (SMS 201-995) prevents the haemodynamic changes following food ingestion in normal human subjects, *Q. J. Exp. Physiol.*, 74, 569, 1989.
40. **Kelbaek, H., Munck, O., Christensen, N. J., and Godtfredsen, J.,** Autonomic nervous control of postprandial hemodynamic changes at rest and upright exercise, *J. Appl. Physiol.*, 63, 1862, 1987.
41. **Berne, C., Fagius, J., and Niklasson, F.,** Sympathetic response to oral carbohydrate administration: evidence from microelectrode recordings, *J. Clin. Invest.*, 84, 1403, 1989.
42. **Robertson, D., Wade, D., and Robertson, R. M.,** Postprandial alterations in cardiovascular haemodynamic dysfunctional states, *Am. J. Cardiol.*, 48, 1048, 1981.
43. **Heseltine, D. and Potter, J. F.,** Postprandial hypotension in elderly people, *Age Ageing*, 19, 233, 1990.
44. **Lipsitz, L. A., Nyquist, R. P., Wei, J. Y., and Rowe, J. W.,** Postprandial reduction in blood pressure in the elderly, *N. Engl. J. Med.*, 309, 81, 1983.
45. **Lipsitz, L. A. and Fullerton, H. J.,** Postprandial blood pressure reduction in healthy elderly, *J. Am. Geriatr. Soc.*, 34, 267, 1986.

46. **Peitzman, S. J. and Berger, S. R.,** Postprandial blood pressure decrease in well elderly persons, *Arch. Intern. Med.,* 149, 286, 1989.
47. **Johnsson, P. V., Lipsitz, L. A., Kelley, M., and Korstner, J.,** Hypotensive responses to common daily activities in institutionalized elderly. A potential risk for recurrent falls, *Arch. Intern. Med.,* 150, 1518, 1990.
48. **Jansen, R. W. M. M., Penteman, J. M., Van Lier, H. J. J., and Hoefnagels, W. H. L.,** Blood pressure reduction after oral glucose loading and its relation to age, blood pressure and insulin, *Am. J. Cardiol.,* 60, 1087, 1987.
49. **Potter, J. F., Heseltine, D., Hartley, G., Matthews, J., MacDonald, I. A., and James, O. F. W.,** Effects of meal consumption on the postprandial blood pressure, catecholamine and insulin changes in elderly subjects, *Clin. Sci.,* 77, 265, 1989.
50. **Jansen, R. W. M. M. and Hoefnagels, W. H. L.,** Hormonal mechanisms of postprandial hypotension, *J. Am. Geriatr. Soc.,* 39, 1155, 1992.
51. **Hoeldtke, R. D., O'Dorisio, T. M., and Boden, G.,** Treatment of autonomic neuropathy with a somatostatin analogue SMS 201-995, *Lancet,* 2, 602, 1986.
52. **Jansen, R. W. M. M., Peeters, T. L., Lenders, J. W. M., van Lier, H. J. J., van Laaar, A., and Hoefnagels, W. H. L.,** Somatostatin analogue (SMS 201-995) prevents the decrease in blood pressure after oral glucose loading in the elderly, *J. Clin. Endocrinol. Metab.,* 68, 752, 1989.
53. **Raimbach, S. J., Cortelli, P., Kooner, J. S., Bannister, R., Bloom, S. R., and Mathias, C. J.,** Prevention of glucose-induced hypotension by the somatostatin analogue octreotide (SMS 201-995) in chronic autonomic failure: haemodynamic and hormonal changes, *Clin. Sci.,* 77, 623, 1989.
54. **Lenders, J. W. M., Morre, H. L. C., and Smits, P., et al.,** The effects of caffeine on the post-prandial fall of blood pressure in the elderly, *Age Ageing,* 17, 236, 1988.
55. **Heseltine, D., Dakkak, M., and Woodhouse, K., et al.,** The effect of caffeine on postprandial hypotension in the elderly, *J. Am. Geriatr. Soc.,* 39, 160, 1991.
56. **Jansen, P. A. F., Schulte, B. P. M., Meyboom, R. H. B., and Gribnau, F. W. J.,** Antihypertensive treatment as a possible cause of stroke in the elderly, *Age Ageing,* 15, 129, 1986.
57. **Cruickshank, J. M.,** Coronary flow reserve and the J curve relation between diastolic blood pressure and myocardial infarction, *Br. Med. J.,* 297, 1227, 1988.
58. **Strangaard, S., Olesen, J., Skinhoj, E., and Lassen, N. A.,** Autoregulation of brain circulation in severe arterial hypertension, *Br. Med. J.,* 1, 507, 1973.
59. **Lipsitz, L. A.,** Altered blood pressure homeostasis in advanced age: clinical and research implications, *J. Gerontol.,* 44, M179, 1989.
60. **Pan, H. Y. M., Hoffman, B. B., Pershe, R. A., and Blaschke, T. F.,** Decline in β-adrenergic receptor mediated vascular relaxation with aging in man, *J. Pharmacol. Exp. Ther.,* 239, 802, 1986.
61. **Scarpace, P. J., Tumer, N., and Mader, S. L.,** β-Adrenergic function in aging: basic mechanisms and clinical implications, *Drugs Ageing,* 2, 116, 1991.
62. **Van Brummelen, P., Buhler, F. R., Kiowski, W., and Amann, F. W.,** Age-related decrease in cardiac and peripheral vascular responsiveness to isoprenaline: studies in normal subjects, *Clin. Sci.,* 60, 571, 1981.
63. **Vestal, R. E., Wood, A. J. J., and Shand, D. G.,** Reduced beta-adrenoreceptor sensitivity in the elderly, *Clin. Pharmacol. Ther.,* 26, 181, 1979.
64. **Astrand, L., Astrand, P., and Roodahl, K.,** Maximal heart rate during work in older men, *J. Appl. Physiol.,* 14, 562, 1959.
65. **Conway, J., Wheeler, R., and Sannerstadt, R.,** Sympathetic nervous activity during exercise in relation to age, *Cardiovasc. Res.,* 5, 577, 1971.
66. **Lakatta, E. G., Gerstenblith, G., Angell, C. S., Shock, N. W., and Weisfeldt, M. L.,** Diminished inotropic response of aged myocardium to catecholamines, *Circ. Res.,* 36, 262, 1975.

67. **Lester, M., Sheffield, L., Trammel, P., and Reeves, T.,** The effect of age and athletic training on the maximal heart rate during muscular exercise, *Am. Heart J.*, 76, 370, 1968.
68. **Rodeheffer, R. J., Gerstenblith, G., Becker, L. C., et al.,** Exercise cardiac output is maintained with advancing age in healthy human subjects: cardiac dilatation and increased stroke volume compensate for a diminished heart rate, *Circulation,* 69, 203, 1984.
69. **Zhou, H. H., Whelan, E., and Wood, A. J. J.,** Lack of effect of ageing on the stereochemical disposition of propranolol, *Br. J. Clin. Pharmacol.*, 33, 121, 1992.
70. **Klein, C., Gerber, J. G., Gal, J., and Nies, A. S.,** Beta-adrenergic receptors in the elderly are not less sensitive to timolol, *Clin. Pharmacol. Ther.*, 40, 161, 1986.
71. **Ford, G. A., Blaschke, T. F., and Hoffman, B. B.,** Beta adrenergically mediated cardiac chronotropic and vascular smooth muscle responses during propranolol therapy and withdrawal in young and elderly persons, *J. Gerontol.*, 47, M22, 1992.
72. **Diaz, R. G., Somberg, J. C., and Freeman, L. B.,** Withdrawal of propranolol and myocardial infarction, *Lancet,* 1, 1068, 1973.
73. **Miller, R. R., Olson, H. G., Amsterdam, E. A., and Mason, D. T.,** Propranolol withdrawal rebound phenomenon, *N. Engl. J. Med.*, 293, 416, 1975.
74. **Mizgala, H. F. and Counsell, J.,** Acute coronary syndromes following abrupt cessation of oral propranolol therapy, *Can. Med. Assoc. J.*, 114, 1123, 1976.
75. **Zahniser, N. R., Bier-Laning, C. M., Gerber, J. G., and Nies, A. S.,** Timolol induced upregulation of polymorphonuclear leukocyte $beta_1$-adrenergic receptors in the elderly, *Clin. Pharmacol. Ther.*, 45, 469, 1989.
76. **Buhler, F. R., Hulthen, U. L., Kiowski, W., and Bolli, P.,** Greater antihypertensive efficacy of the calcium channel inhibitor verapamil in older and lower renin patients, *Clin. Sci.*, 63, 439s, 1982.
77. **MRC Working Party,** Medical Research Council trial of treatment of hypertension in older adults: principal results, *Br. Med. J.*, 304, 405, 1992.
78. **Umachandran, V., Ranjadayalan, K., Ambepityia, G., Marchant, B., Kopelman, P. G., and Timmis, A. D.,** Aging, autonomic function, and the perception of angina, *Br. Heart J.*, 66, 15, 1991.
79. **Harkins, S. W., Price, D. P., and Martelli, M.,** Effects of age on pain perception: thermonociception, *J. Gerontol.*, 41, 58, 1986.
80. **Kannel, W. B. and Abbott, R. D.,** Incidence and prognosis of unrecognized myocardial infarction: an update on the Framingham study, *N. Engl. J. Med.*, 311, 1144, 1984.
81. **Yusuf, S., Peto, R., Lewis, J. A., et al.,** B-blockade during and after myocardial infarction. An overview of the randomised trials, *Prog. Cardiovasc. Dis.*, 27, 335, 1985.
82. **Andersen, M. P., Bechsgaard, P., Frederiksen, J., et al.,** Effect of alprenolol on mortality among patients with definite or suspected acute myocardial infarction: preliminary results, *Lancet,* 2, 865, 1979.
83. **The Norwegian Multicenter Study Group,** Timolol-induced reduction in mortality and reinfarction in patients surviving acute myocardial infarction, *N. Engl. J. Med.*, 304, 801, 1981.
84. **Gunderson, T., Abrahmsen, A. M., Kjeksus, J., et al.,** Timolol-related reduction in mortality and reinfarction in patients ages 65–75 years surviving acute myocardial infarction, *Circulation,* 66, 1179, 1982.
85. **ISIS-1 Group,** Randomized trial of intravenous atenolol among 16027 cases of suspected acute myocardial infarction, *Lancet,* 2, 57, 1986.
86. **Burney, P. G. J., Britton, J. R., Chinn, A. E., Tattersfield, A. E., Papacosta, A. O., Kelson, M. C., Anderson, F., and Corfield, D. R.,** Descriptive epidemiology of bronchial reactivity in an adult population study, *Thorax,* 42, 38, 1987.
87. **Hopp, R., Bewtra, A., Nair, N. M., and Townley, R. G.,** The effect of age on methacolione response, *J. Allergy Clin. Immunol.*, 76, 609, 1985.
88. **Kaliner, M., Shelhamer, J., Smith, L., Davis, P. B., and Venter, J. C.,** Autonomic nervous system abnormalities and allergy, *Ann. Intern. Med.*, 96, 349, 1982.

89. **Davis, P. B. and Byard, P. J.,** Beta-adrenergic responses and airway reactivity in healthy adults, *Mech. Ageing Dev.,* 54, 29, 1990.
90. **Connolly, M. and Vestal, R. E.,** personal communication.
91. **Lipworth, B. J., Tregaskis, B. F., and McDevitt, D. G.,** Comparison of hypokalemic, electro-cardiographic and haemodynamic responses to inhaled isoprenaline and salbutamol in young and elderly subjects, *Eur. J. Clin. Pharmacol.,* 40, 255, 1991.
92. **Higgins, R. M., Cookson, W. O. C. M., Lane, D. J., John, S. M., McCarthy, G. L., and McCarthy, S. T.,** Cardiac arrhythmias caused by nebulised beta-agonist therapy, *Lancet,* 2, 863, 1987.
93. **Allen, S. C. and Prior, A.,** What determines whether an elderly patient can use a metered dose inhaler correctly?, *Br. J. Dis. Chest.,* 80, 45, 1986.
94. **Diggory, P., Bailey, R., and Vallon, A.,** Effectiveness of inhaled bronchodilator delivery systems for elderly patients, *Age Ageing,* 20, 379, 1991.
95. **Armitage, J. M. and Williams, S. J.,** Inhaler technique in the elderly, *Age Ageing,* 17, 275, 1988.
96. **Kendall, M. J., Woods, K. L., Wiolkins, M. R., and Worthington, D. J.,** Responsiveness to β-adrenergic stimulation: the effects of age are cardioselective, *Br. J. Clin. Pharmacol.,* 14, 821, 1982.
97. **Klein, C., Hiatt, W. R., Gerber, J. G., and Nies, A. S.,** Age does not alter human vascular and non-vascular beta-2 adrenergic responses to isoproterenol, *Clin. Pharmacol. Ther.,* 44, 573, 1988.
98. **Dillon, N., Chung, S., Kelly, J., and O'Malley, K.,** Age and beta-adrenoceptor-mediated function, *Clin. Pharmacol. Ther.,* 27. 769, 1980.
99. **Fowler, M. B., Laser, J. A., Hopkins, G. L., Minobe, W., and Bristow, M. R.,** Assessment of the β-adrenergic receptor pathway in the intact failing human heart: progressive receptor down-regulation and subsensitivity to agonist response, *Circulation,* 74, 1290, 1986.
100. **Lipsitz, L. A.,** Orthostatic hypotension in the elderly, *N. Engl. J. Med.,* 321, 952, 1989.
101. **Horowitz, M., Madden, G. J., Chatterton, B. E., Collins, P. J., Harding, P. E., and Shearman, D. J. C.,** Changes in gastric emptying rates with age, *Clin. Sci.,* 67, 213, 1984.
102. **Moore, J. G., Tweedy, C., Christian, P. E., and Datz, F. L.,** Effect of age on gastric emptying of liquid-solid meals in man, *Dig. Dis. Sci.,* 28, 340, 1983.
103. **Kupfer, R. M., Heppell, M., Haggith, J. W., and Bateman, D. N.,** Gastric emptying and small bowel transit rate in the elderly, *J. Am. Geriatr. Soc.,* 33, 340, 1985.
104. **Rashid, M. U. and Bateman, D. N.,** Effect of intravenous atropine on gastric emptying, paracetamol absorption, salivary flow and heart rate in young and fit elderly volunteers, *Br. J. Clin. Pharmacol.,* 30, 25, 1990.
105. **Heaton, N. D., Garrett, J. R., and Howard, E. R.,** The enteric nervous system: structure and pathology, in *Autonomic Failure,* Bannister, R., Ed., Oxford University Press, Oxford, 1988, 238.
106. **Mayberry, J. F. and Atkinson, M.,** Studies of incidence and prevalence of achalasia in the Nottingham area, *Q. J. Med.,* 220, 451, 1985.
107. **Hyatt, R.,** Colonic pseudo-obstruction: an important complication in hospitalized elderly patients, *Age Ageing,* 16, 145, 1987.
108. **Brocklehurst, J. D. and Kahn, M. Y.,** A study of fecal stasis in old age and the use of 'Dorbanex' in its prevention, *Gerontol. Clin.,* 11, 293, 1969.
109. **Baker, D. M. and Santer, R. M.,** A quantitative study of the effects of age on the noradrenergic innervation of Auerbach's plexus in the rat, *Mech. Ageing Dev.,* 42, 147, 1988.
110. **Gabella, G.,** Fall in the number of myenteric neurons in aging guinea pigs, *Gastroenterology,* 96, 1487, 1989.
111. **Dauchot, P. and Gravenstein, J. S.,** Effects of atropine on the electrocardiogram in different age groups, *Clin. Pharmacol. Ther.,* 12, 274, 1971.

112. **Pfeifer, M. A., Weinberg, C. R., Cook, D., Best, J. D., Reenan, A., and Halter, J. B.,** Differential changes of autonomic nervous system function with age in man, *Am. J. Med.*, 75, 249, 1983.
113. **Schwartz, J. B., Gibb, W. J., and Tran, T.,** Aging effects on heart rate variation, *J. Gerontol.*, 46, M99, 1991.
114. **Kelliher, G. J. and Conahan, S. T.,** Changes in vagal activity and response to muscarinic receptor agonists with age, *J. Gerontol.*, 35, 842, 1980.
115. **Linzer, M.,** Syncope, *Am. J. Med.*, 90, 1, 1991.
116. **Lipsitz, L. A.,** The drop attack: a common geriatric symptom, *J. Am. Geriatr. Soc.*, 31, 617, 1983.
117. **Manolis, A. S., Linzer, M., Salem, D., and Estes, M.,** Syncope: current diagnostic evaluation and management, *Ann. Intern. Med.*, 112, 850, 1990.
118. **Lipsitz, L. A., Wei. J. Y., and Rowe, J. W.,** Syncope in an elderly, institutionalised population: prevalence, incidence, and associated risk, *Q. J. Med.*, 55, 45, 1985.
119. **Imholz, B. P. M., Dambrink, J. H. A., Karemaker, J. M., and Weiling, W.,** Orthostatic circulatory control in the elderly evaluated by non-invasive continuous blood pressure measurement, *Clin. Sci.*, 79, 73, 1990.
120. **Grubb, B. P., Temesy-Armos, P., Hahn, H., and Elliot, L.,** Utility of tilt-table testing in the evaluation and management of syncope of unknown origin, *Am. J. Med.*, 90, 6, 1991.
121. **Lipsitz, L. A., Marks, E. R., Koestner, J., Jonsson, P. V., and Wei, J. Y.,** Reduced susceptibility to syncope during postural tilt in old age, *Arch. Intern. Med.*, 149, 2709, 1989.
122. **Almquist, A., Goldenberg, I. F., Milstein, S., et al.,** Provocation of bradycardia and hypotension by isoproterenol and upright posture in patients with unexplained syncope, *N. Engl. J. Med.*, 320, 346, 1989.
123. **Waxman, M. B., Yao, L., Cameron, D. A., Wald, R. W., and Roseman, J.,** Isoproterenol induction of vasodepressor-type reaction in vasodepressor-prone persons, *Am. J. Cardiol.*, 63, 58, 1989.
124. **Mark, A. L.,** The Bezold-Jarisch reflex revisited: clinical implications of inhibitory reflexes originating in the heart, *J. Am. Coll. Cardiol.*, 1, 90, 1983.
125. **Kapoor, W. N. and Brant, N.,** Evaluation of syncope by upright tilt testing with isoproterenol, *Ann. Intern. Med.*, 116, 358, 1992.
126. **Kenny, R. A. and Traynor, G.,** Carotid sinus syndrome — clinical characteristics in elderly patients, *Age Ageing*, 20, 449, 1991.
127. **Murphy, A. I., Rowenbotham, B. J., Boyle, R. S., Thew, C. M., Fardoulys, J. A., and Wilson, K.,** Carotid sinus hypersensitivity in elderly nursing home patients, *Aust. N.Z. J. Med.*, 16, 24, 1986.
128. **Ausubel, K. and Furman, S.,** The pacemaker syndrome, *Ann. Intern. Med.*, 103, 420, 1985.
129. **Miyatake, K., Okamoto, M., Kinoshita, N., Owa, M., Nakasone, I., Sakakibara, H., and Nimura, Y.,** Augmentation of atrial contribution to left ventricular inflow with aging as assessed by intracardiac doppler flowmetry, *Am. J. Cardiol.*, 53, 586, 1984.
130. **Bryg, R. Y., Williams, G. A., and Labovitz, A. J.,** Effect of aging on left ventricular diastolic filling in normal subjects, *Am. J. Cardiol.*, 59, 971, 1987.
131. **Johnson, A. D., Laiken, S. L., and Engler, R. L.,** Hemodynamic compromise associated with ventriculoatrial conduction following transvenous pacemaker placement, *Am. J. Med.*, 65, 75, 1978.
132. **Kenny, R. A., Lyon, C. C., Ingram, A. M., Baylis, J., Lightman, S. L., and Sutton, R.,** Enhanced vagal activity and normal arginine vasodepressin response in carotid sinus syndrome: implications for a central abnormality in carotid sinus hypersensitivity, *Cardiovasc. Res.*, 21, 545, 1987.
133. **Sigler, L. H.,** The cardioinhibitory carotid sinus reflex, *Am. J. Cardiol.*, 12, 175, 1963.
134. **Smiddy, J., Lewis, H. D., and Dunn, M.,** The effect of carotid sinus massage in older men, *J. Gerontol.*, 27, 209, 1992.

135. **Roose, S. P., Glassman, A. H. Giardina, E. V. G., et al.,** Tricyclic antidepressants in depressed patients with cardiac conduction disease, *Arch. Gen. Psychiatry,* 44, 273, 1987.
136. Task Force on the use of laboratory tests in psychiatry, Tricyclic antidepressants — blood level measurements and clinical outcome: an APA Task Force report, *Am. J. Psychiatry,* 142, 155, 1985.
137. **Wanstall, J. C. and O'Donnell, S. R.,** Vasodilator responses to dopamine in rat perfused mesentery are age-dependent, *Br. J. Pharmacol.,* 98, 302, 1989.

Index

INDEX

A

B

C

D

I

J

K

L

W, X, Y